# THE
# Booles &
# THE Hintons

Two dynasties that helped
shape the modern world

George Boole (1815–64).

# THE Booles & THE Hintons

Two dynasties that helped
shape the modern world

## GERRY KENNEDY

ATRIUM

First published in 2016 by Atrium
Atrium is an imprint of
Cork University Press
Youngline Industrial Estate
Pouladuff Road, Togher
Cork T12 HT6V, Ireland

British Library Cataloguing in Publication Data
A CIP catalogue record for this book is available from the British
Library.
ISBN-978-1-78205185-5

Typeset by Burns Design
Printed in Poland by Hussar Books

www.corkuniversitypress.com

To a new brood of Booles:
Finn, Elliot, Mark, Scarlet,
Alice and Louie.

# CONTENTS

# ACKNOWLEDGEMENTS

One of the biggest problems with writing this book has been the dearth of archive material, especially the correspondence of some of the key figures. There is virtually none relating to Mary Everest Boole, who must have written copiously; similarly with Charles Howard Hinton and both Ethel and Wilfrid Voynich. I have no explanation for this.

I am very grateful, therefore, to a number of people who supplied me with invaluable material. Within the extended Boole/Hinton family my thanks go in particular to Marni Rosner, who provided generous continuing support, also Kevin Boole, Patricia Davison, Charlotte Goodhill and Natalie Protopapa. Margaret Till, of the Levetus family, close friends with Ethel and Wilfrid Voynich, provided a very helpful collection of letters that had required heroism of a high order in deciphering the latter's impossible handwriting. Other caches came from John Rollett, who patiently tried to talk me through the fourth dimension. Similarly Mark Blacklock did his best and loaned some useful material on Charles Howard Hinton. Richard Fremantle kindly sent me material that his mother, Anne Fremantle, had written about Ethel Voynich. Richard Herczynsky in Warsaw provided sources re Wilfrid Voynich's early life. I am grateful as well to Andrew Cook, an expert regarding Sidney Reilly, for advice. His researcher, Dimitry Belanovsky, very kindly offered me hospitality in Moscow and researched State Archives there. Richard Keeler was most obliging, showing me the ear and eye watercolours of Margaret Boole at Moorfields Hospital.

Perhaps the most indispensable resource came from Barbara Garlick in Australia. She kindly donated a copy of Evgenia Taratuta's, *The Fate of a Writer and the Fate of a Book*, translated from the Russian. Taratuta was besotted with the life of Ethel Voynich and had Soviet backing and resources to explore it. Dealing with Russian and Polish sources necessitated other translations. I am very grateful to Veronique Grattan, Elena Strygina and Jacek Klinowski for their help.

A number of kind people, especially David Thompson, Elizabeth Block and Paul Kennedy, ploughed through my tome, offering suggestions and guidance. Kind others include Pamela Blevins, David Downing, Donald Mitchell, Tom Raines, Donna Salisbury, Doug Sandle and Annie Winner. Xun Zhou was very helpful regarding the Chinese chapters. My thanks too to John Joubert for his assessment of Ethel Voynich's music.

Although I'm not able to list their names, I owe a debt of gratitude to librarians from all over the world, most especially those at the magnificent resource that is the British Library, but also University College and the Wellcome Institute, London; the Ashmolean, Oxford; Bristol University Library; Trinity College Library, Cambridge University; the Boole Library, University College Cork; and, in the US, the University of California, Los Angeles.

# FOREWORD

Nearly half a century of my life had gone by before I set out to explore the family connection between myself and the Boole and Hinton dynasties. A chance remark at a funeral provided the inspiration. In the same casual spirit I never formulated a coherent plan of research to write this book; it all simply unravelled as a sort of quest. Seeing that its course made me dizzy at times, I felt some guidelines for a reader might be something to hold onto.

I tackled the biographies of the Booles and Hintons fitfully or when travelling in their orbits allowed – or simply at my whim. Eventually the house-to-house search took in nearly twenty major characters and many more minor ones, spread over two centuries and three continents. It has been difficult to deal with this scale of personae and events chronologically; a certain shuffling of the pack has been inevitable. The inclusion of a joker in the shape of my diary excerpts from a round-the-world trip in 1983 complicated things further.

After an excursion to my own backyard and a glimpse of the mysterious *Voynich Manuscript* in chapters 1 and 2, we follow the lives of the mid-nineteenth-century dynastic founders: George Boole, his wife, Mary Everest Boole and James Hinton. The Booles' five daughters, as children, have a chapter devoted to them but it is Ethel, the youngest, and her partner, Wilfrid Voynich, who are a narrative mainstay. They dominate from the 1880s until her death in 1964. Their involvement in Russian anarchist politics, novels and the world of rare books takes us geographically from England to Russia and the US.

Interspersed is the key role of James Hinton's son, Charles Howard Hinton, and his esoteric dabblings with the fourth dimension. Married to the Booles' eldest daughter, Mary Ellen, his life leads us on to Japan in the 1890s and again to the US. The other three Boole daughters, Margaret, Alice and Lucy, grounded in England, make their adult appearances.

Now in the early twentieth century, we devote a chapter to the life of Cambridge scientist, Sir Geoffrey Taylor, Margaret Boole's famous son. Meanwhile the four sons of Charles Howard Hinton, George, Eric, Billy and 'Ted', propel us to the establishment, by Carmelita Hinton, of the progressive Putney School in Vermont and a chapter concerning Mexico. The final chapters are focused on Maoist China, from 1948 to the end of the Cultural Revolution, where Joan Hinton exiled herself after being involved with the making of the atomic bomb. Her brother, Bill, spent much time there too, leaving the legacy of the celebrated documentary, *Fanshen*.

The lives of the Booles and Hintons were unknown to me on my 1983 eastward global circumnavigation visiting Peace Movement activists. It was somewhat surprising therefore to find that the trajectory echoed the lives of some of the main players. My 1983 diary entries follow my own journey to Russia, China, Japan and across the US. There are also more recent excursions tracking the Boole/Hintons in Poland, Lithuania, post-Soviet Russia, China and America.

While I have engaged in a great deal of 'objective' research into the participants' lives, there is a strong tinge in the book of my own biography and a 'subjective' take on some of the historical issues that arise. This makes for a rather hybrid compilation: biography, travelogue and commentary. My 'excuse', if I need one, is that the personal 'discovery' element, plus my kinship with the characters, dealt me something of a free hand.

It may be possible to read some of the chapters and sections separately. The family trees and the *Voynich Manuscript* timeline should help navigation. It has been a long, unpredictable journey and often a hard one, taking me into subject areas that are not part of my academic background. It has been, however, a highly rewarding, once-in-a-lifetime experience.

**John Boole** (1777–1848): shoe-maker, George Boole's father. Married Mary Anne Boole (1770–1854).

**George Boole** (1815–64): mathematician, eldest son of John Boole.

**Mary Anne Boole** (1817–87): daughter of John Boole.

**William Boole** (1819–88): second son of John Boole.

**Charles Boole** (1821–1904): third son of John Boole.

**Thomas Roupell Everest** (1801–55): vicar, father of Mary Everest.

**George Everest** (1790–1866): surveyor of India. Brother of Thomas Everest.

**Mary Everest/Boole** (1832–1916): writer. Married George Boole, 1855. Mother of five daughters.

**Mary Ellen Boole/Hinton** (1856–1908): poet, eldest daughter of George Boole, wife of Charles Howard Hinton.

**Margaret Boole/Taylor** (1858–1935): known as Maggy, second daughter of George Boole.

**Sir Geoffrey Taylor** (1886–1975): known as G. I., scientist, first son of Margaret Taylor and Edward Ingram Taylor.

**Alicia Boole/Stott** (1860–1940): known as Alice, geometrician, third daughter of George Boole.

**Lucy Boole** (1862–1904): chemist, fourth daughter of George Boole.

**Ethel Boole** (1864–1960): known as ELV, novelist, fifth daughter of George Boole.

**Wilfrid Voynich** (1865–1930): revolutionary, antiquarian book dealer, husband of Ethel Boole.

**John Howard Hinton** (1791–1873): Baptist minister, married to Eliza Birt.

**James Hinton** (1822–75): surgeon and social philosopher, son of John Howard.

**Margaret Haddon/Hinton** (1826–1902): wife of James Hinton.

**Caroline Haddon** (1837–1905): writer, sister of Margaret.

**Charles Howard Hinton** (1853–1907): Known as Howard, writer on the fourth dimension, son of James. Married Mary Ellen Boole, 1880. Emigrated to Japan and US 1887. Had four sons.

**George Boole Hinton** (1882–1943): mineralogist/plant collector, eldest son of Charles Howard and Mary Ellen (known in text as 'Snr').

**Eric Boole Hinton** (1883–?): second son of Howard and Mary Ellen.

**William Howard Hinton** (1884–1909): third son of Howard and Mary Ellen. Known as Billy.

**Sebastian Hinton** (1887–1923): known as Ted, fourth son, lawyer, married Carmelita Chase.

**Howard Everest Hinton** (1912–77): entomologist, eldest son of George Boole Hinton. Born in Mexico, later lived in England.

**George Boole Hinton** (1913–97): mineralogist, second son of George Boole Hinton. Lived in Mexico.

**James Boole Hinton** (1915–2006): plant collector/writer, third son of George Boole Hinton. Lived in Mexico.

**Carmelita Chase/Hinton** (1890–1983): educator, founder of Putney School, Vermont, wife of Ted Hinton.

**Jean Rosner/Hinton** (1917–2002): social activist, eldest child of Ted and Carmelita Hinton.

**William Hinton** (1919–2004): known as Bill, second child, writer of *Fanshen*.

**Joan Hinton** (1921–2010): atomic scientist/agronomist, third child. Lived in China from 1948.

**Erwin Engst** (1922–2003): known as Sid, agronomist. Lived in China from 1946, married to Joan Hinton.

**Fred Engst** (b. 1952): teacher in China, eldest child of Joan and Sid.

**William Engst** (b. 1954): known as Billy, engineer in US, second child of Joan and Sid.

**Karen Engst** (b. 1956): third child of Joan and Sid.

**Marni Rosner** (b. 1952): daughter of Jean Rosner.

# Descendants of John Boole and Mary Ann Joyce

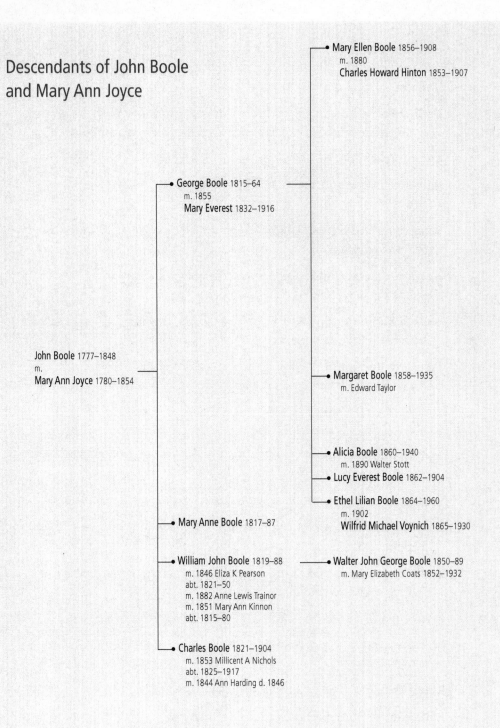

John Boole 1777–1848
m.
Mary Ann Joyce 1780–1854

George Boole 1815–64
m. 1855
Mary Everest 1832–1916

Mary Ellen Boole 1856–1908
m. 1880
Charles Howard Hinton 1853–1907

Margaret Boole 1858–1935
m. Edward Taylor

Alicia Boole 1860–1940
m. 1890 Walter Stott

Lucy Everest Boole 1862–1904

Ethel Lilian Boole 1864–1960
m. 1902
Wilfrid Michael Voynich 1865–1930

Walter John George Boole 1850–89
m. Mary Elizabeth Coats 1852–1932

Mary Anne Boole 1817–87

William John Boole 1819–88
m. 1846 Eliza K Pearson
abt. 1821–50
m. 1882 Anne Lewis Trainor
m. 1851 Mary Ann Kinnon
abt. 1815–80

Charles Boole 1821–1904
m. 1853 Millicent A Nichols
abt. 1825–1917
m. 1844 Ann Harding d. 1846

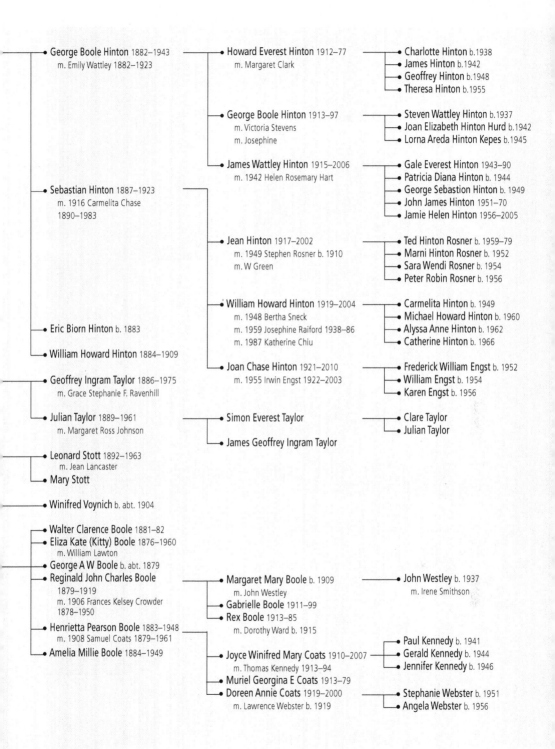

George Boole Hinton 1882–1943
m. Emily Wattley 1882–1923

Howard Everest Hinton 1912–77
m. Margaret Clark

Charlotte Hinton b.1938
James Hinton b.1942
Geoffrey Hinton b.1948
Theresa Hinton b.1955

George Boole Hinton 1913–97
m. Victoria Stevens
m. Josephine

Steven Wattley Hinton b.1937
Joan Elizabeth Hinton Hurd b.1942
Lorna Areda Hinton Kepes b.1945

James Wattley Hinton 1915–2006
m. 1942 Helen Rosemary Hart

Gale Everest Hinton 1943–90
Patricia Diana Hinton b. 1944
George Sebastion Hinton b. 1949
John James Hinton 1951–70
Jamie Helen Hinton 1956–2005

Sebastian Hinton 1887–1923
m. 1916 Carmelita Chase
1890–1983

Jean Hinton 1917–2002
m. 1949 Stephen Rosner b. 1910
m. W Green

Ted Hinton Rosner b. 1959–79
Marni Hinton Rosner b. 1952
Sara Wendi Rosner b. 1954
Peter Robin Rosner b. 1956

William Howard Hinton 1919–2004
m. 1948 Bertha Sneck
m. 1959 Josephine Raiford 1938–86
m. 1987 Katherine Chiu

Carmelita Hinton b. 1949
Michael Howard Hinton b. 1960
Alyssa Anne Hinton b. 1962
Catherine Hinton b. 1966

Eric Biorn Hinton b. 1883

William Howard Hinton 1884–1909

Joan Chase Hinton 1921–2010
m. 1955 Irwin Engst 1922–2003

Frederick William Engst b. 1952
William Engst b. 1954
Karen Engst b. 1956

Geoffrey Ingram Taylor 1886–1975
m. Grace Stephanie F. Ravenhill

Julian Taylor 1889–1961
m. Margaret Ross Johnson

Simon Everest Taylor

Clare Taylor
Julian Taylor

James Geoffrey Ingram Taylor

Leonard Stott 1892–1963
m. Jean Lancaster
Mary Stott

Winifred Voynich b. abt. 1904

Walter Clarence Boole 1881–82
Eliza Kate (Kitty) Boole 1876–1960
m. William Lawton
George A W Boole b. abt. 1879
Reginald John Charles Boole
1879–1919
m. 1906 Frances Kelsey Crowder
1878–1950
Henrietta Pearson Boole 1883–1948
m. 1908 Samuel Coats 1879–1961
Amelia Millie Boole 1884–1949

Margaret Mary Boole b. 1909
m. John Westley
Gabrielle Boole 1911–99
Rex Boole 1913–85
m. Dorothy Ward b. 1915

John Westley b. 1937
m. Irene Smithson

Joyce Winifred Mary Coats 1910–2007
m. Thomas Kennedy 1913–94
Muriel Georgina E Coats 1913–79
Doreen Annie Coats 1919–2000
m. Lawrence Webster b. 1919

Paul Kennedy b. 1941
Gerald Kennedy b. 1944
Jennifer Kennedy b. 1946

Stephanie Webster b. 1951
Angela Webster b. 1956

# Descendants of James Hinton and Margaret Haddon

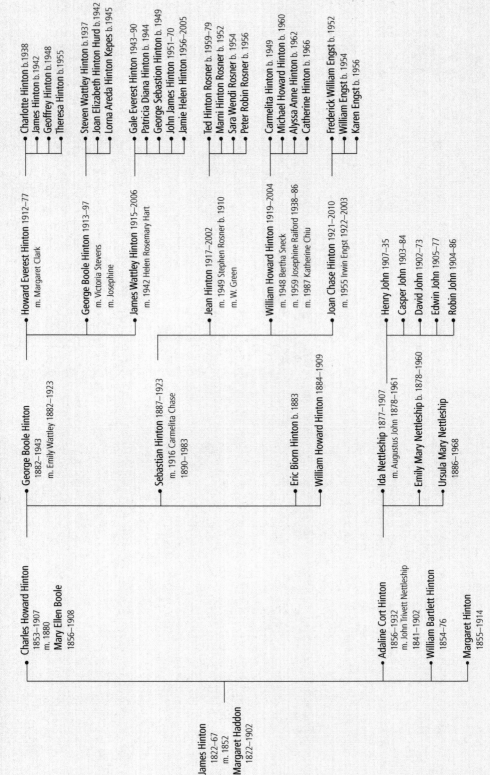

James Hinton
1822–67
m. 1852
**Margaret Haddon**
1822–1902

**Charles Howard Hinton**
1853–1907
m. 1880
**Mary Ellen Boole**
1856–1908

**George Boole Hinton**
1882–1943
m. Emily Wattley 1882–1923

**Howard Everest Hinton** 1912–77
m. Margaret Clark

Charlotte Hinton b.1938
James Hinton b.1942
Geoffrey Hinton b.1948
Theresa Hinton b.1955

**George Boole Hinton** 1913–97
m. Victoria Stevens
m. Josephine

Steven Wattley Hinton b.1937
Joan Elizabeth Hinton Hurd b.1942
Lorna Areda Hinton Kepes b.1945

**James Wattley Hinton** 1915–2006
m. 1942 Helen Rosemary Hart

Gale Everest Hinton 1943–90
Patricia Diana Hinton b. 1944
George Sebastion Hinton b.1949
John James Hinton 1951–70
Jamie Helen Hinton 1956–2005

**Sebastian Hinton** 1887–1923
m. 1916 Carmelita Chase
1890–1983

**Jean Hinton** 1917–2002
m. 1949 Stephen Rosner b. 1910
m. W. Green

Ted Hinton Rosner b. 1959–79
Marni Hinton Rosner b. 1952
Sara Wendi Rosner b. 1954
Peter Robin Rosner b. 1956

**William Howard Hinton** 1919–2004
m. 1948 Bertha Sneck
m. 1959 Josephine Raiford 1938–86
m. 1987 Katherine Chiu

Carmelita Hinton b. 1949
Michael Howard Hinton b. 1960
Alyssa Anne Hinton b. 1962
Catherine Hinton b. 1966

**Joan Chase Hinton** 1921–2010
m. 1955 Irwin Engst 1922–2003

Frederick William Engst b. 1952
William Engst b. 1954
Karen Engst b. 1956

**Eric Biörn Hinton** b. 1883

**William Howard Hinton** 1884–1909

**Adaline Cort Hinton**
1856–1932
m. John Trivett Nettleship
1841–1902

**Ida Nettleship** 1877–1907
m. Augustus John 1878–1961

**Emily Mary Nettleship** b. 1878–1960

**Ursula Mary Nettleship**
1886–1968

Henry John 1907–35
Casper John 1903–84
David John 1902–73
Edwin John 1905–77
Robin John 1904–86

**William Bartlett Hinton**
1854–76

**Margaret Hinton**
1855–1914

# GRAVY OVER A TABLECLOTH

Cliché number one: 'Everyone has a story to tell.' It's a cliché because it's self-evidently true; how would it be possible for anything animate not to have a story, a life that interacts with its environment? Silverfish under carpets probably have stories to tell, they're just not very vocal. The remarkable interwoven lives I have felt a need to take up and explore have been touched upon by others but the threads have never been pieced together.

Jim O'Brian told me part of his own story in a pub in Hayes, Middlesex in May 2006. Opposite me and my half pint he set down a full glass, plus the odd mix of a smart mobile and battered tobacco tin, from which he shaped a roll-up. Unshaved and rather hunched, he probably won't mind being called a bit shambling. The lunchtime atmosphere around us was unmistakably working class: thick cigarette smoke, unalleviated television, the clack of the pool table and stabs of cackling laughter.

I was wondering how to pluck up courage to talk to someone, so his presence was a gift. 'Do you know much about George Orwell?' I ventured, nodding at the picture of the author on the wall next to us. Not an unreasonable question in a pub calling itself 'The Famous George Orwell'. 'Course I do,' came the reply, 'his real name was Eric Blair and he lived in Hayes for a couple of years back in the thirties.

'The Famous George Orwell' pub, Hayes.

There's a plaque to him somewhere in the old village.' Did he know his books? 'I showed my kids the *Animal Farm* cartoon once, but it didn't mean anything political to them.' I decided not to ask his views on socialism, assuming he had any, and enquired instead how long he'd lived here. Sixty-two years, born and bred, left school early, worked on the fairgrounds for two years and settled down as a carpenter and landscape gardener; couldn't bear the 'plunk, plunk, plunk' of factory life, although there'd always been plenty of it round here. He wanted to work outside still but had given up because he needed a hip replacement. 'I can get down but I can't get up.'

Diplomatic niceties over, I felt I could mention that I was 'doing some writing' and wondered if he knew anything about the Woolpack Farm by Bournes Bridge over the railway. I'd sleuthed that it was thereabouts that Orwell had written his poem 'On a Ruined Farm near His Master's Voice Gramophone Factory', published in 1934. To his left Orwell saw 'The black and budless trees browned by acid smoke' and to the right, 'tumbling skeletons ... the factory towers, white and clear ... where the tapering cranes sweep round, and great wheels turn, and trains roar by like strong, low-headed brutes of steel'. 'All built over now,' said Jim, agreeing that my positioning was probably about right. I explained that it didn't really matter precisely, but that I enjoyed a historical sense of place and time. I wanted to add, but couldn't be so impolite as to condemn somewhere he'd lived so long, that for myself, as a teenager here in the 1950s, it had seemed bland and anonymous with little connection to the wider world. Hayes did not have a pedigree.

Cliché number two: 'Never go back.' So why had I? I was trying to place my background in relation to a wider dynasty of influential radical thinkers and doers that I'd recently discovered I was related to. Orwell was a kind of local but famous intermediary to get me used to the idea. It was gratifying that he too had arrived at his convictions not just by theorising but by acting, experiencing the lives of ordinary people in restaurants, hopfields, doss-houses, mines; even risking his life in a foreign civil war. Admittedly his time in Hayes had been far tamer, teaching at a small private school because he needed the work. Forever the social commentator, he couldn't resist expressing a distaste for Hayes and the suburbs in his poem and later more forcefully in one of his novels.

*Coming up for Air*,[1] published in 1939, was Orwell's attempt, distilled from his own restless stay in the town from 1930 to 1932, to characterise the unhappy, alienated life of the novel's hero, George Bowling. Hayes was then growing apace from a rural village, felicitously close to the Great

Western Railway, into a burgeoning satellite of London, the new industrial locus of England. The factories that Jim O'Brian hated typified the emerging mass consumer society, producing vacuum cleaners, radios, records, cars, razors and foods that one could live without: chocolates, ice-cream, biscuits and processed cheese. Orwell stayed distant from it all, projecting his loathing on to his literary alter ego, George, who even names the enemy locations: 'Do you know the look of these new towns that have suddenly swelled up like balloons in the last few years, Hayes, Slough, Dagenham and so forth?' He adds, 'I don't mind towns growing, so long as they do grow and don't merely spread like gravy over a tablecloth.' This condescension is mixed with compassion for 'the miles and miles of ugly houses, with people living dull, decent lives inside them'. People like my parents.

In the story, Bowling decides to go AWOL and nostalgically return to Little Binfield, a country town like Orwell's own Henley, in search of an old flame and a secret pond where he'd seen huge fish. Disillusion is of course inevitable: his childhood sweetheart has grown fat and the pool has been filled with rubbish from one of the nearby housing estates. Both Georges feel a pang for the loss of the simpler ways of the countryside – Bowling viewing his pond, Orwell viewing the HMV factory from the railway bridge in his poem.

Jim O'Brian's regrets are not so strong: Hayes is the fount of his identity. The high street has lost its family shops, given way to the usual corporate suspects: McDonalds, Iceland and Superdrug. Burtons, 'the fifty-shilling tailor', is now a travel agent's, a necessary facility for the urban jet-setters and the immigrant population that have brought spice to the town with its tandoori restaurants. Jim has something of the Georges about him, a lingering contact with the natural world. Apart from gardening, he tells me he used to watch birds from the church tower in Cranford Park. Mention of this immediately locates my one-time address. Our house lay a short walk from this huge open space that I'd played cricket in and cycled around, parts of which, by the River Crane, were wild enough to provide adventures with fallen trees and birds' nests. It also allowed us to behave in the same boyish, brutish way as in the novel, wilfully destroying the frogs that gathered in the swampy water.

I tell Jim I'm planning to revisit the spot later. Eyeing my backpack he warns, 'Don't go to the park with that.' 'Why?' I ask. 'Perverts,' he replies, 'they hang out there, it's well known, you'll get into trouble.' Astonished at the times-a-changing, we meander on to reminisce further. 'Do you remember a once-a-week local paper called the *Hayes News?*' I ask him. 'I used to deliver it as a kid.' He's amazed. 'So did I!' We recall the maverick owner who ran the Labour-leaning rag that we sold at doorsteps for tuppence-halfpenny on Friday nights with a halfpenny commission for our trouble. He smiles. 'Some days I'd do two rounds and keep the money from one of them; I was rich.' I laugh and congratulate him on his cheek and recount the reversed tale of an occasion when, worried about my poor sales, I'd hidden a quantity of papers rather than return them to the office. Naturally I paid the full cover price and earned almost nothing for my evening's effort. My family howled with mirth at my naïveté. Such innocence of the ways of the world, however, was not characteristic of my father.

Brave man for his time, he'd taken the plunge and run away from the deeply depressed Sunderland of his youth in the northeast of England where he'd worked in its shipyards. He was drawn by the allure of the cleaner and more secure employment in the London suburbs that Orwell so denigrated. As a sewing machine mechanic upholstering flash American cars he had met my secretarial mother and become, post-war, head of a household with three kids. He earned just enough with overtime to pay a mortgage, shell out for school uniforms and run a battered car. His income was supplemented by the part-time, pen-pushing of his wife at the concrete HMV factory so heavily castigated by Orwell.

There was a faint but discernible patina of the middle-class world in our family life. We may not have displayed shelves full of books, but in our 'best' front room stood a glass cabinet parading handed-down relics of more genteel forebears: a porcelain sugar bowl, vases with oriental designs, pewter ware and napkin rings; all redundant to our daily needs. Occasionally we'd receive a postcard from rarely met middle-class kin who went on holiday to the Alps or Venice – far from my mother's choice of faded Victorian coastal resorts. My father's real salt-of-the-earth working-class relatives were much nearer at hand as he'd been responsible for importing his four brothers from Sunderland to help make buses in Southall. Although he shared their liking for beer, football and minor gambling, we were kept away from their un-achieving influence. I liked their noisy, extended family ways that rarely extended to us.

I shook hands with Jim O'Brian and wished him well. I hope he got his operation before too long. I wanted to stand on the bridge where Orwell must have stood, to recapture some of his feelings. The still vast bulk of the HMV factory, despite demolition of some of the blocks, is impressive.

The old HMV building with 'low-headed brutes of steel'.

The farm he would have once seen in stark contrast to the giant factory complex lies now under a parcel depot. Recalling my days as an ardent trainspotter close to this site, I decided to take some photos of the 'low-headed brutes of steel': these days the elegant diesels that sweep by. I clicked away as three teenage girls, bare-midriffed, sashayed by me. Ten yards further on one of them turned and yelled, 'Pervert!' They presumably fancied that I was photographing them. I hadn't even reached the park yet.

The railway line had transported me to a grammar school in Ealing for six years. It was here that my parents' aspirations for me and my brother (not my sister) were moulded into middle-classdom. This they saw more in terms of cash than status; having experienced poverty first hand, assets came first. Scraping uncertainly by had shaped their firm belief in the welfare state and a radical outlook Orwell would have thought sound. My trade unionist father was articulate, with his factory-born, socialist

5

ideology. This found approval with his children despite the insistent Tory values imparted at our school. It did not find favour with my mother's two sisters, our aunts.

Aunt Muriel was mainly responsible for unwittingly providing my link to a radical political perspective she would probably have disapproved of. Looking back, her cropped hair, trouser suits, cigarette holder and 'spinster' status probably betokened a sapphic allegiance, unrecognised and unconsummated. She sold *chapeaux* for an upmarket department store and gob-smacked us all by recounting how she'd once personally counselled a television celebrity.

As children, more mundanely, we would look forward to her twice-yearly visits, mainly because she always left behind a two-shilling piece for each of us: a relative fortune. Her intellectual pretensions included a regular mentioning of a Boole family, from whom the contents of our front room display cabinet originated. Our grandmother, Henrietta, who had died when we were young, was a descendant of William Boole, the brother of the nineteenth-century mathematical genius, George Boole. He had married Mary Everest, niece of the man who'd calculated the height of the eponymous mountain. Muriel was very proud of having been given Everest as her middle name; highly suitable for a social climber.

Apparently then, my great-great-great uncle was a whizzkid at maths, but unlike Muriel it meant very little to us and our proletarian ways. I imagined Boole as a fusty boffin in a high-necked collar stroking a beard, ruminating on *pi*. Muriel produced a picture of him. He was remarkably handsome in fact with a determined mouth, high forehead and aquiline nose. In 1964, when my brother and I were indeed at university, Muriel attended the centenary of his death at Lincoln cathedral, where in honour of one of the city's most famous sons a stained glass window had been installed. She had come away from the jamboree with a family tree which not only had her own name proudly inscribed on it, but we three kids too. To my parents, also logged, it was of academic interest only, especially as none of us evinced any mathematical ability whatsoever. Sadly unfulfilled, Muriel died of her cigarettes in 1978.

Aunt Doreen, the youngest of the three sisters, was a different kettle of fish. Unassuming and motherly, she'd married an ex-military man of middle-class tastes. He worked in some capacity at the Meteorological Office, a fact that impressed me much more than having a mathematician as a distant relation. They lived in a tiny cottage in rural Bedfordshire where, aged six or seven, I would occasionally stay for a week, blissfully

Geoffrey Taylor in Lincoln at the 100th anniversary of George Boole's death. Gabrielle Boole is next to him, the author's aunt, Muriel, top row.

far from our crowded suburban home. These episodes imbued me with a nostalgic love of the countryside that resonates fully with Orwell. Riding on tractors, double helpings of treacle tart and the little library that first brought the enchantment of A. A. Milne left lasting impressions. Later in life I quit London for the arcadian hills of south Shropshire where Orwell himself had first fallen in love.

Doreen died more peacefully than her sister, in the year 2000. The funeral brought five cousins together. We convened afterwards at her little house nearby and enjoyed that strange mild euphoria that comes from having coped with one's emotions and being genuinely pleased to feel part of a wider, unaccustomed kinship. Chit-chat, catching up, tea and sandwiches followed. I was introduced for the first time to Laurie, my aunt's grandson. In among more chat he asked me whether I'd heard of a famous distant relative of ours. I expected to hear of George Boole, but he spoke of Wilfrid Voynich. News to me. He suggested that I should look him up on the internet; apparently Voynich had found some rare, peculiar book. I made a mental note of the name.

I walked on to our house on Coronation Road, its name neatly dating the estate to 1937. Cliché number three: 'Everything looks smaller than it did when you were young.' It failed to live up to its shallow promise. I had to count out front doors as they weren't numbered, to assign house number 35, the house I grew up in. It looked huge. I didn't remember the large mock-tudor, timbered gable at the front, half shared with next door.

Glancing up and down the road, contemporary manorial pretension could be spotted everywhere: false leaded windows, brass-effect lanterns, and even add-on columns, supporting nothing but some crazy baroque vision. How the 'Georges' would have snorted. Behind our house my economical father kept chickens and grew vegetables to help out our diet. It was still a struggle, however, keeping up with the bills and eventually acquiring momentous, now taken-for-granted items, such as a twin-tub washing machine, fridge and a telephone. These 'gadgets' were decried only by our winter visitor, Henrietta Boole's spouse, Sam, my maternal grandfather, who migrated to us from Devon every year.

The incomer's put-you-up bed shrunk the already chock-a-block bedroom I shared with my brother into a hostel. We also had to put up with his outspoken and cantankerous ways. His political views, bred in part by being a runt plumber in a clan of teachers, were far further to the left than my father's and probably the Politburo at the time. 'Capitalists', with the stress on the 'pit', were the source of all the world's woes alongside the foolish credit-borrowing (provider of the machine that washed his underwear) that was dragging us all back to the Great Depression. Fierce acrimonious rows regularly broke out. The cry, 'If you love communism so much, why don't you go and live in Russia,' issued regularly from my mother.

He was an all-round pain; getting to school on time was often frustrated by finding him locked in the toilet dipping his fried bread into a mug of cocoa while reading the *Daily Herald*. The mayhem inevitably helped form my own convictions, perhaps surprisingly not averse, leading to a bolshy stance at our stuffy grammar school and a desire to quit home and head north a.s.a.p. to industrial Leeds to mingle with the 'real' working class.

I did nothing about the Voynich reference for several months after my aunt's funeral. There was no particular reason for this except that I'd mentally placed it in a box marked, 'of possible interest sometime', the same enclosure that stored the dusty, minimal knowledge concerning

George Boole. Tapping in the keyword 'Voynich' on the computer made it clear that, as far as the world was concerned, the box label should read, 'urgent, to be followed up diligently'. There were hundreds of pages.

The *Voynich Manuscript* was discovered in 1912 near Rome by antiquarian bookseller Wilfrid Voynich, after whom it was named. The 200 vellum pages contain a neat but so far completely indecipherable handwriting, despite endless analysis. Accompanying the text were crude but compelling drawings of herb-like plants and naked nymphs bathing in pools. The illustrations, although not beyond the bounds of possibility, made no sense either. The ensemble had rightly been dubbed 'the most mysterious manuscript in the world'.

It was wonderful stuff that nobody much in the media world seemed to have come across. What was even more interesting was the dynastic link a further search revealed: Wilfrid was married to an Ethel Boole, the youngest of the five daughters of our pet mathematician, George Boole. It was immediately apparent that there was nothing remotely fusty about the family.

Wilfrid was a Polish nationalist and had escaped from a Russian prison camp in Siberia, finally meeting Ethel in London in 1890, herself immersed in a crusading campaign against the oppressive Russia of the Tsars. Although music was her lifelong love, she had written a novel of dark, nihilistic atheism that had later become an iconic best-seller in the communist world. In a film version it was set to music by Shostakovich. And that was just one of the daughters.

All five of them were brought up in poverty by their widowed, eccentric mother, Mary Everest Boole. They all developed independent and proto-feminist minds, excelling in medicine and (predictably) maths. Their father's talents may have been beyond my comprehension but his mark on the world cannot these days be avoided. The Boolean algebraic system underlies the whole digital revolution and the modern computer-driven world. His fame has been acknowledged by the naming of a crater on the moon after him.

Why had no one mentioned this brood to me in my wild, Shelleyish days? My life would have been changed. Was there a conspiracy on my Aunt Muriel's part to keep her curious nephews away from nasty lefties? Possibly. But maybe in a pre-internet world she just wasn't aware of the wider facts. Surfing the Booles much later, I discovered a further exciting link. The eldest of Boole's five daughters, Mary Ellen, had married one Charles Howard Hinton, who was crazily obsessed with the fourth dimension. He ended up living in the US, starting a Boole/Hinton line there. Mary Ellen's granddaughter, Joan Hinton, had, in 1945, worked on the first atomic bomb at Los Alamos before joining her brother, William, in China in 1948. They both dedicated their lives to the Maoist revolution. He fell foul of McCarthyism but eventually wrote a famous and influential radical book about China's agricultural upheaval.

The information that stuttered onto my computer screen in an almost random order was stunning. I dug out Muriel's battered copy of the Boole/Hinton family tree that had also been boxed and put away, looking for the first time at the links on the ancestral hanging mobile. It was surreal to see my family's names alongside those world-shatterers, both humbling and uplifting at the same time. They were internationalist and cosmopolitan but had a strong allegiance to radical egalitarian social change. Unusually, what came across was the enormous energy of the women of the clan.

Despite the class difference, our lives were interconnected not only by blood but a very similar worldview. Wasn't my father's move south from the depressed northeast of heroic proportions? Hadn't he fought daily as a shop steward for the rights of working men and state provision of education and health? He'd been there too by the roadside in Hayes to cheer on his sons walking with one of the first Aldermaston peace marches. I didn't really need Orwell to justify going back to Hayes; I felt as though we were possibly all in some kind of 'socialist' pageant together playing our parts.

For me, the role that I'd felt of value had been with the peace movement in the 1980s at the height of the Cold War. The sheer folly and madness of that era had taken me on travels to conferences and demos including a three-month global tour visiting other campaigners. The voyage had coincidentally taken in the major locations of some of my ancestors. In winter 1983 in 'Leningrad', 100 years after Ethel, I'd been, like her, smuggling suspect literature to dissidents. A week's journey away on the Trans-Siberian I could have had tea with Joan Hinton in Beijing. In Japan I found myself in places known by Mary Ellen Boole. I'd brushed right by the Boole and Hinton families unknowingly.

I felt slightly shifty hanging around outside the house lost in remembrances, and so walked on the quarter mile to Cranford Park. I knew very well that the pretty church of St Dunstan's (1086 AD), had now to be reached by a pedestrian subway under the M4. Driving by in the slow lane, one could almost touch its tower. The scene was otherwise unchanged except for the unnerving display of warning notices aimed at the predicted debauchees. *Anyone taking drugs in the park will be reported to the Police. Anyone loitering and causing annoyance will be asked to leave and may face prosecution. No metal detecting.* On a gibbet pole nearby was spiked a video camera. This was *1984* with a vengeance. The unpleasant sense of lost innocence was further compounded when I found that the overgrown *Swallows and Amazons* part of the Crane river I once birds-nested in had been sanitised and purged of its wilderness. I felt as let down as George Bowling.

On the train back to London I reflected on the day's outing. Cliché number four, 'You can't turn the clock back', is undeniably true, but I hadn't expected to. What I was looking for was a winding-up of a spring that would take me forward. Meeting Jim O'Brian and the ghost of Orwell had reminded me of the much larger story that I felt needed to be told.

# THE UGLY DUCKLING

Everyone likes a good mystery. The enigmatic unsolved *Voynich Manuscript* with its enticing look of solvability has drawn in many a detective. The naïve pictures and handsome script exude a siren-like, come-hither allure. It has indeed driven some to madness. I feel justified in promoting a reader's interest in it before knuckling down to a 'genetic' family biography. The eponymous volume had played a major part in not only Wilfrid Voynich's life, but having married Ethel Boole, hers as well. It deserves recognition after all for its key role in my narrative: without its stray mentioning I would never have begun this book at all. What follows in this chapter is an unavoidably dense précis of the manuscript's contents and history.

In April 2001, a year after first hearing of the *Voynich Manuscript* at my aunt's funeral, I contacted the Beinecke Rare Book and Manuscript Library of Yale University, which houses the volume, to make a BBC radio programme about it. All I received back was 'a copy of the catalogue records for your information' and a mere outline of the manuscript under the heading 'Cipher Manuscript MS 408, Central Europe (?), s. XV-ex XVI (?), plus items B–N'.

Eventually after much chivvying, an e-mail arrived stating that a room could be arranged to inspect their treasure. A request had also been made to record and ask the curator a few general questions. This was turned down flat. Nevertheless, in mid-July I drove from New York the 120 miles north to New Haven, Connecticut, the home of Yale. The place is a little like Oxford or Cambridge in so far as the university is a major industry and 'gown' merges physically with 'town'. Some of the buildings and their pseudo-gothic towers also looked Oxbridgean, although the occasional gilded dome reminded me of the cathedrals within the Kremlin.

The Beinecke Library itself, however, was ultra-modern, looking rather like a pile of black waffles stacked into a cube, the whole resting on four obelisk legs. I boldly lunged through the revolving doors, feeling somewhat nervous, as if on a blind date. I wasn't sure what to expect from a meeting in the flesh; an immediate falling-in-love or a limp disappointment? The Head of Public Services led me to a large classroom pushing a trolley load of Voynichobilia. She extracted the *Voynich Manuscript* from its protective outer and laid it on a slab of extruded foam, ready for inspection. With rubber-gloved hands she opened the volume and prepared to turn the vellum pages.

I'd like to say I gasped with a pleasure aroused from the realisation of a long-nurtured desire, but the artificiality of the situation and my stony-faced chaperone rather took the shine off things. In the unsettling hush, the yellowish folios at least made a satisfying creak and crackle. The *Voynich Manuscript* is pretty ineffable, and I sorely taxed my vocabulary attempting to record a sense of wonder at its strangeness. Eventually I ground to a halt and turned to ask some innocent questions. The radio silence was broken with, 'No comment.'

This taciturnity may have been due to the fact that the Beinecke's visage had already been egg-spattered by another manuscript it owned: the *Vinland Manuscript*, purporting to be a map made by the first Viking visitors to the New World. Why publicise another possible cuckoo-in-the-nest? All this has now changed: there are websites galore, numerous explanations have been offered and television programmes aired. (My own book, co-authored with Rob Churchill, *The Voynich Manuscript*,[1] appeared in 2004.) The tome does get to have a seductive hold on you. So what's all the fuss about?

It was first brought to the world's attention in 1912 by Wilfrid Michael Voynich. If the manuscript can be described as enigmatic, secretive, weird and wonderful, those same adjectives could equally be applied to the life of its discoverer. It reads not unlike an adventure novel. We know little about his early existence apart from his birth date: 31 November 1865 at Kovno, an important town in Lithuania, then under Russian rule. His Polish father was a 'petty official' or barrister. Wilfrid attended Moscow University from where he graduated in chemistry and became a licensed pharmacist. In 1885 he found himself engaged with a Polish revolutionary movement in Warsaw. Involved in a failed plot to free two of his comrades condemned to death, Voynich was imprisoned for two years in the Warsaw Citadel. Escaping from subsequent banishment to Siberia, he miraculously found his way to England in 1890 where he met and married Ethel Boole, youngest daughter of George Boole. Their partnership lasted until his death in 1930.

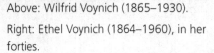

Above: Wilfrid Voynich (1865–1930).
Right: Ethel Voynich (1864–1960), in her forties.

What amazed me was their linking with the exciting world of revolutionary politics in late-nineteenth-century Russia. Ethel travelled there in 1887, spending two years making contact with opposition circles. In 1890 she joined forces in London with Wilfrid and a group of Russian political exiles to expose tsarist oppression. This involvement then dwindled, leaving Wilfrid to turn his resourcefulness and drive to the antiquarian book trade. Ethel turned to writing novels. H. P. Kraus, a US book dealer, was amazed by Voynich's success: 'He rose like a meteor in the antiquarian sky and his catalogues testify to his ability[2]'. His success allowed him rapidly to expand his business from London to offices in Europe and venture on safari-like forays hunting for rare and ancient books. It was this that led him to the famous manuscript that still bears his name. He wrote:

In 1912, during one of my periodic visits to the Continent of Europe, I came across a most remarkable collection of precious illuminated manuscripts. For many decades these volumes had laid buried in the chest in which I had found them in an ancient castle in Southern Europe. While examining the manuscripts, with a view to the acquisition of at least part of the collection, my attention was especially drawn by one volume. It was such an ugly duckling compared with the others that my interest was aroused at once. I found that it was entirely written in cipher. Even a necessarily brief examination of the vellum

14

upon which it was written, the calligraphy, the drawings and the pigments suggested to me as the date of its origin the latter part of the thirteenth century.[3]

So what was Wilfrid Voynich's strange find in that southern European castle? (The general public was not told of the true place, Villa Mondragone, Frascati, near Rome, for another fifty years.) For an analysis of the character and style of the volume, one can do no better than the thoughts of Mary D'Imperio in her definitive academic monograph *The Voynich Manuscript: An Elegant Enigma*: 'The impression made upon the modern viewer is one of extreme oddity, quaintness and foreignness – one might almost say unearthliness. The manuscript seems to stand totally apart from all other even remotely comparable documents.'[4]

On page after page one is confronted by pen and wash images of beguiling uniqueness. While the drawings may lack reality or artistic merit, it is the scale (over 200 images) and dynamism of the creativity that impresses as remarkable. The manuscript when viewed as a whole conveys a unity of style but can be broken into sections.

In the 'Botanical' sections, folios 1r–66v and 87r–96v (folios are *recto*, right-hand page, *verso* the other side), there is a passing, and sometimes quite close, resemblance to real plants from the natural world, but these are in the minority. The roots seem exaggerated, distorted or even expressionistic, but for what purpose remains unclear. Occasionally a plant seems to have mutated halfway through its growth, or undergone some form of grafting. One-third of the 126 specimens could be classed as 'strange', having at least one feature of root, stem or flower that does not look authentic. Another third have more than one feature that might be considered biologically unlikely or fantastic. A further third seem to exist in the realms of the possible. Often the images seem to be like a child's 'flip-book' where head, torso and legs can be variously combined to humorous effect. A sunflower was mooted for f93v and 33v dating the manuscript to post-Columbus, as the plant was introduced from the Americas. Folio 9v might be a viola, 56r a fern and 4v a myrtle. Ethel Voynich, herself a keen botanist, attempted her own identifications, such as scabious for 34r, castor oil plant for 6v and geranium for 36r.

In the section of folios 67r–73v, dubbed 'Astrological', are found circles, some with the moon or sun in the centre, from which radiate segments containing recognisable zodiac forms and naked women standing or sitting in tubs holding stars. Stranger still is the 'Biological' section, 75r–84v,

*The Voynich Manuscript*, folio 33v, 'sunflower'.

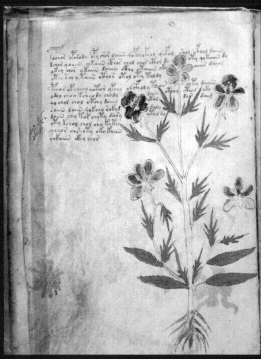

Folio 9v, 'viola'.

Left:
Folio 73v,
'astrological'
section.

Opposite:
Folio 78r,
'plumbing'
and 'nymphs'.

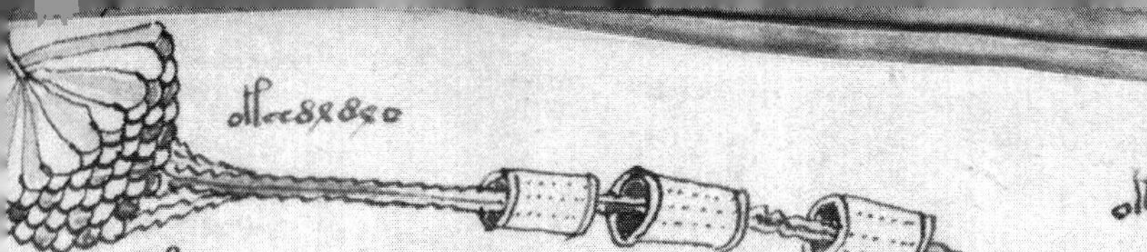

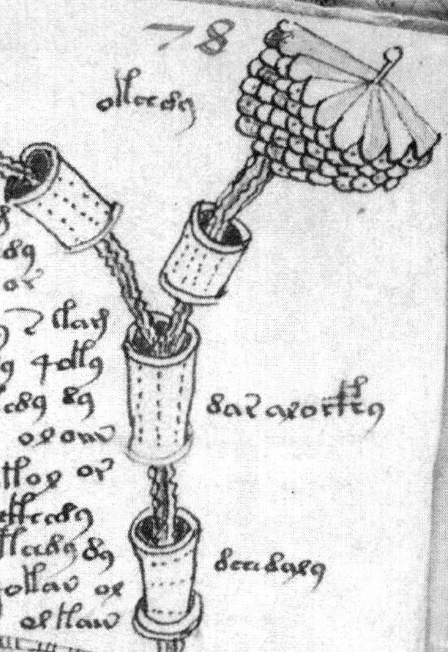

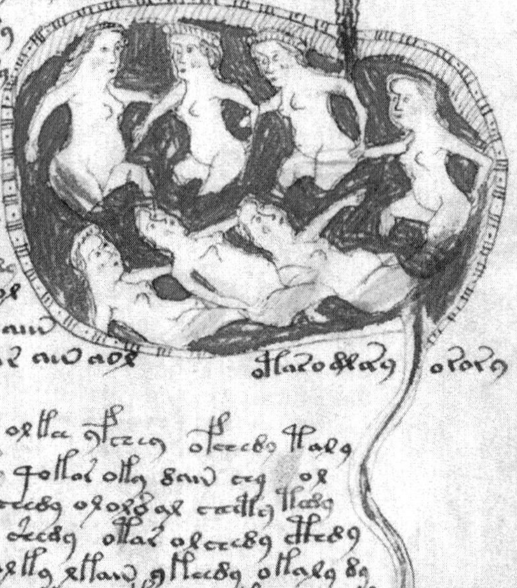

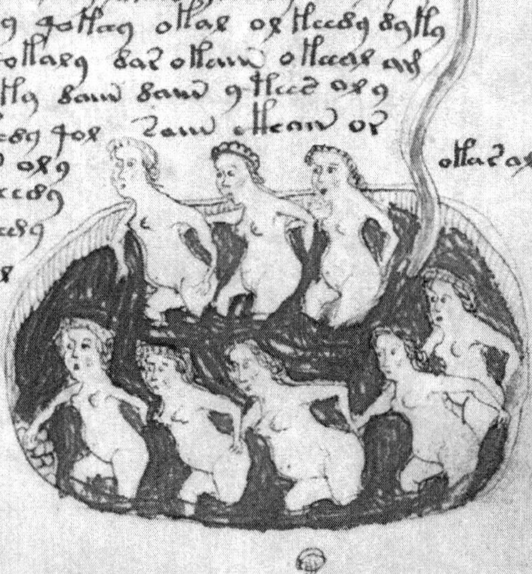

# The Trail of the Voynich Manuscript

**Roger Bacon** (1214–94)
Renowned medieval philosopher, believed by some to be the author of the Voynich Manuscript.

**Rudolf of Bohemia** (1552–1612)
Holy Roman Emperor, interested in occult science and collector of art.

**Jacobus de Tepenecz** (1575–1622)
Physician to Emperor Rudolf. Assumed owner of Voynich Manuscript 1608–22. His signature found on 1st folio.

**Georg Baresch** (1585–1662)
Alchemist, earliest confirmed owner of Voynich Manuscript. 1639 letter to Kircher in Rome was accompanied by a sample of the Voynich Manuscript's script.

**Johannes Marcus Marci** (1595–1667)
Physician and scientist, Rector of Prague University. Marci sends the Voynich Manuscript to Kircher accompanied by a letter dated 1665–66. It mentions Rudolf buying the Manuscript for 600 ducats.

**Athanasius Kircher** (1602–80)
Jesuit scholar and cryptographer in Rome. Voynich Manuscript not mentioned in his correspondence.

*Whereabouts of manuscript unkown between 1666 and 1912*

**Wilfrid Voynich** (1865–1930)
Polish nationalist, exiled to Siberia 1887. Arrived London 1890, became rare book dealer. Discovered Voynich Manuscript in Rome 1912. Tries to decrypt and sell for $160,000.

**Ethel Voynich** (nee Boole) (1864–1960)
Novelist and activist. Married Voynich 1892 (1902). Owned Voynich Manuscript after Voynich's death. Kept secret of where Voynich Manuscript discovered.

**William Romaine Newbold** (1865–1926)
Professor of Philosophy at University of Pennsylvania. His decryptions linked Voynich Manuscript to Roger Bacon. His theories were rejected.

**H P Kraus** (1907–88)
Specialist in rare books. Bought Voynich Manuscript 1961 for $24,500 from Ethel's companion, Anne Nill. Unable to find buyer, he donated it to Yale University.

**Beinecke Library, Yale**
Voynich Manuscript now held there under title 'Ms 408 Cipher manuscript. Central Europe [?]…'

featuring plump, naked nymphs dancing or reclining in baths or pools of limpid green liquid connected by streams, channels or 'plumbing' that appears more organic than architectural. Other individual figures stand alone in urns, arms upraised, either dispatching or receiving something via the interconnecting tubing (see also page 452). There is a possibly sexual air to these illustrations; not that they are designed to arouse, but that they give an underlying impression of fecundity, reproduction, birth and death. Streams of seeds or pollen issue from grape-like clusters of eggs producing flows that feed the pools in which the nymphs bathe.

A stand-alone fold-out 'Rosettes' section, 85r–86v, the Beinecke describes as containing 'an elaborate array of nine medallions, filled with stars and cell-like shapes, with fibrous structures linking the circles. Some medallions with petal-like arrangements of rays filled with stars, some with structures resembling bundles of pipes.' The last two sections, 99r–102v and 103r–117v, seem to be 'Pharmaceutical', containing species of what appear to be herbs and roots alongside what have been interpreted as medicine jars. 'Bullet-pointed' with little stars, the latter section is of continuous text.

Right: Folio 86v, the 'rosettes'.

Below: Folio 101v, from the 'pharmaceutical' section.

It is the enticing quality of the script and the intellectual challenge of 'breaking the code' that has led would-be solvers to concentrate their considerable and diverse talents on the manuscript. Mary D'Imperio describes a first impression: 'The script seems like a reasonable, workable, well-constructed system of writing, with a look of ease and natural flow. On closer inspection the surface of the simplicity vanishes, and a still more seductive and captivating character emerges.' In the 1970s a US Navy cryptographer, Prescott Currier, suggested that there are actually two 'languages' in the manuscript using the same script made by two separate hands. There are similarities between the Voynichese hand and other scripts but no exact match has ever been found. There are occasional Latin abbreviations in use during the Middle Ages and what appear to be early European versions of Arabic numerals.

It is not even clear exactly how many different characters the Voynich alphabet consists of. Estimates range between twenty-three and forty, but many of the basic shapes can be joined together to create compound symbols, while others re-occur with the addition of 'hooks', 'tails' or 'commas'. Still more are 'ligatured', joined together by a stroke bridging the top of two upright characters. The symbols do at least seem to form words, and the script runs from left to right. One oddity is the strange repetition of the same word consecutively.

Beyond these facts, there is little that can be ascertained. The drawings were probably produced first, as the lines of script end running up to the edges of the illustrations. Given that there are no corrections it is likely that the whole work is a copy of some original version. If the images refer to the text or are a decoy to lead the would-be reader on a false trail is unknown. The whole manuscript does seem to suggest an underlying meaning, not just a stream of nonsense.[5]

The last part of Wilfrid Voynich's life was to be dominated, in one way or another, by his 'ugly duckling', his hopes for intellectual recognition frustrated by its linguistic impenetrability, his dreams of financial security thwarted as he continued to search for a buyer. From the moment he acquired the codex, Voynich had been trying to trace its previous history and ultimately discover the identity of the author. There was at least one important clue. Attached to the cover of the Manuscript was a letter.

REVEREND AND DISTINGUISHED SIR; FATHER IN CHRIST:
This book bequeathed to me by an intimate friend, I destined for you, my very dear Athanasius as soon as it came into my possession, for

I was convinced it could be read by no one except yourself. The former owner of this book once asked your opinion by letter, copying and sending you a portion of the book from which he believed you would be able to read the remainder, but he at that time refused to send the book itself. To its deciphering he devoted unflagging toil, as is apparent from attempts of his which I send you herewith, and he relinquished hope only with his life. But his toil was in vain, for such Sphinxes as these obey no one but their master, Kircher. Accept now this token, such as it is and long overdue though it may be, of my affection for you, and burst through its bars, if there are any, with your wonted success. Dr Raphael, tutor in the Bohemian language to Ferdinand III, then King of Bohemia, told me the said book had belonged to the Emperor Rudolf and that he presented the bearer who brought him the book 600 ducats. He believed the author was Roger Bacon, the Englishman. On this point I suspend judgement; it is your place to define for us what view we should take thereon, to whose favour and kindness I unreservedly commit myself and remain at the command of your Reverence.

Johannes Marcus Marci of Cronland<br>Prague, 19 August, 1665? 6?

This single sheet, written in a scrawling, Latin hand, presumably by a scribe, not only introduced a cast of fascinating characters into the story, but also seemed to suggest a verifiable history for the manuscript. Rather surprisingly, Voynich admitted that, at first, he paid little attention to the letter, regarding it as of no consequence. He certainly changed his mind about its significance, however, and later unequivocally attributed it to 'Roger Bacon the Englishman'. Even without the Marci letter, Voynich stated that the moment he saw the manuscript he had made up his mind. The fact that it was written in cipher, Wilfrid thought, pointed to Bacon, the thirteenth-century friar and proto-scientist who might have had reasons to keep secrets from the Church and who had hinted at codes in his writing.

Who was the key mysterious 'bearer' to whom Emperor Rudolf had given 600 ducats and who believed Bacon to be the author? Voynich set to work diligently to find out and reported his findings in a lecture in April 1921 in Philadelphia, 'A Preliminary Sketch of the History of the Roger Bacon Cipher Manuscript'.[6] He came up with the 'nugget' prize: the Elizabethan magus, John Dee.

Dee must have indeed acquired manuscripts from the 1538 dissolution and pillaging of English monasteries. The *Voynich* could have been another to add to the large collection of works by Bacon that he owned. It was further true that a credulous Dee, under the spell of a con man, Edward Kelley, had decamped to Bohemia with him to make their fortunes at the court of Rudolf II. After many an adventure they fell out. Dee did get to meet Rudolf, but there is no compelling evidence that the former ever owned the volume. So what of the other characters from the letter to Marci, the most solid piece of evidence to date the manuscript? It is a convoluted tale.

'Dr' Raphael Mnishovsky (1580–1644) was in fact a lawyer and writer at the Bohemian court, but his appearance in the Marci letter to Kircher is really only relevant in so far as it alludes to Bacon. Of greater importance is the trail of connections from the sender of the letter, Johannes Marci (1595–1667), the personal physician of Rudolf. His connection with Athanasius Kircher (1602–80) dated from 1638 and was based on Kircher's ability as a cryptographer. A more mysterious figure in the chain is the, 'intimate friend' who bequeathed the manuscript to Marci. His identity was not ascertained until Rene Zandbergen, a foremost Voynichologist, located a letter to Kircher from the 'former owner', an alchemist, George Baresch, written in 1639 and bound in a collection at Villa Mondragone. It provides the path of transmission of the manuscript as it was Marci who had inherited his collection of material after his death.

Baresch, like Marci later, asks Kircher for help in understanding a manuscript in his possession which he conjectures is 'all of a medical nature … from the pictures of herbs, of various images, of stars and of other things which appear like chemical secrets …'. It should be noted, however, that his description of the manuscript, although close, is not entirely satisfactory. There is no mention of its most mysterious part – the naked nymphs and their 'plumbing'. There is no evidence of what Kircher thought of the manuscript, however, no mention of it in his letters, nor is it catalogued with any of his possessions.

One other item that prompted Wilfrid Voynich to ascribe the manuscript to the Prague period appears on its very first folio. In the 'Preliminary Sketch' … he reported that while preparing a photostat of it, one of the plates was accidentally under-exposed, revealing a faded signature. After being treated with chemicals the name Jacobus de Tepenecz emerged, a Bohemian pharmacist and member of Rudolph II's court. The granting of his title in 1608 and death in 1622 supports the time frame. Voynich assumed the manuscript, after Rudolph's purchase of it from Dee, had come into the possession of de Tepenecz, perhaps as a gift from Rudolph, and had then been passed on via Baresch to Marci and eventually Kircher.

Surprisingly, we have to wait until the nineteenth century for the 'ugly duckling' to turn up again. The story once more is convoluted. In 1870 after the Pope's dissolution of the Collegium Romanum, a Jesuit university, part of its library, including Kircher's letters, was given to the then General of the Society and ended up in another Jesuit college, Villa Mondragone, where Voynich found it. He acquired thirty manuscripts altogether, including the famous volume, and it seems was held to secrecy as part of the deal. The remaining history of the manuscript can be told quickly.

Voynich lived intermittently in New York from 1915; Ethel joined him in 1921. He set up business there and earnestly hoped to sell the manuscript. By the late 1920s he had almost given up hope. His letters reveal that he was permanently tottering on the edge of another financial crisis yet by all accounts he lived well, staying in expensive hotels. His health continued to deteriorate and in March 1930, Voynich died. In accordance with his will, the manuscript passed into the joint possession of his wife and his secretary, Anne Nill, with the proviso that it might be sold for $100,000 to any public institution, but not to a private collector. The volume spent the next thirty years in a safe deposit vault, although photocopies continued to circulate. In 1960 Ethel died and left the manuscript to Nill. She in turn died one year later having sold it, despite Wilfrid's wishes, to New York book dealer, Hans Kraus, for £24,500. He attached a price tag of $100,000, but no one was prepared to buy such a strange and uncertain object. He eventually donated it to Yale University where it still resides, an object of curiosity too tricky at the time of my visit for the Beinecke Library to subject it to any scientific tests.

A century after its discovery, the content of the strange artefact is not much nearer to an adequate explication. A great deal of attention has been paid to plant identification. Dr Edith Sherwood, a retired PhD in chemistry, is a leading practitioner, having attributed labels to all of the specimens bar two: 38r and 42r. She suggests that most are herbs or weeds typical of the Mediterranean and discounts the problematic roots as not being a key to recognition. Indeed they sometimes seem to be merely schematic: eyes appear in f17r, a snake wanders through the roots of f49r. Positive examples of hers, for example, are f23v, borage, primrose f87v and lunaria f34v. (Readers will have to judge for themselves by looking at images online at the Beinecke website.)

Apart from the astrological signs there are virtually no clearly recognisable cultural artefacts save a few scattered clues. A crossbow on folio 73v seems to point to German origin, possibly of the fourteenth century. Researchers have drawn attention to a small castle located in the fold-out 'Rosettes' folio 86v, that sports swallow-tail crenellations found in north Italy. Sergio Toresella, an expert on medieval herbals, also suggested a late-fifteenth-century north Italian look to the plant drawings which, when combined with the existence of medicinal baths in that part of the world, seem to provide some linking context.

The text has naturally been the other avenue of examination, but here again despite analysis by banks of computers and an array of eminent cryptographers nothing has emerged of real value. It continues to perplex, even to the verge of madness. One such victim, William Romaine Newbold, as we shall see in Chapter 15, convinced himself that the manuscript was indeed by Roger Bacon, who wrote about the microscope and telescope centuries before they were discovered. Such wishful thinking has pervaded most theories presented so far.

To name but a few: William Friedman, a US military cryptoanalyst, in the early 1940s suggested the manuscript to be a treatise in an artificial language. In 1945 Leonell Strong, a medieval researcher, posited authorship by Anthony Askham, a sixteenth-century English astrologer. Part of his translation reads, 'When skuge of tun'e –bag rip, seo uogon kum sli of se mosure-issued ped-stans skubent, stokked kimbo-elbow crawknot.' Robert Brumbaugh, a professor of philosophy at Yale, in 1978 advocated that the characters represent numerals capable of being interpreted into a Latin alphabet. In the same year, Harvard professor, John Stojko maintained that the manuscript was written in vowel-less Ukrainian. Part of his translation reads, 'Where after religion you believe in religion and wish that to Ora. Emptiness is that what baby God's Eye is fighting for.' Leo Levitov in 1987 hit on an old form of Flemish that referred to the suicide rites of Cathars. Others have suggested that the book was written in Chinese using a phonetic or semi-phonetic alphabet. Aliens have appeared on the stage too. Edith Sherwood has invoked a young Leonardo da Vinci using coded anagrams and Nick Pelling a first-half fifteenth-century architect, Antonio Avelino, escaping to Constantinople with his encoded secrets of military engineering. Rich Santa Coloma enlisted sixteenth-century Francis Bacon, whose book *The New Atlantis* might have been used as the basis for the manuscript as a *faux* ancient accompaniment to its fantasy island.

Perhaps in frustration the idea has surfaced that the manuscript is a hoax and contains nothing more than gibberish and fancy. The fact that it appears to contain sense yet evades intelligibility would seem appropriate for this scenario. Gordon Rugg of Keele University in 2004 seemed to have answered the question as to how anyone could manufacture such quantities of textual nonsense. He demonstrated the possibility with the use of a cardan grille, invented about 1550, consisting of a grid of cut-outs that could be endlessly moved over series of letters written below and built up into a coherent-looking text. Rugg felt it likely therefore that the manuscript is a well-fashioned hoax from around that time concocted to sell to the gullible. Others pointed out, however, that the writing conforms to a test known as Zipf's law, showing that it does contain an identifiable linguistic structure. In my volume about the manuscript I took it upon myself to sign up, with an uncertain hand, to the hoax theory designating Wilfrid Voynich as perpetrator. I was not the first to suggest this.

My suspicions were generated in part based on his character. Although Voynich was brave, ingenious and charming, he was also devious and dishonest. This was documented in his revolutionary days when he earned the reputation for being egoistic, through to his bookdealer days when he openly admitted to dodgy practices – perhaps standard for his trade. (We shall explore his personality more anon.) His skills as a forger were possibly an extension of his political experiences, when, as a chemist, he was busy forging passports and ID for his colleagues. His acquired knowledge about ancient manuscripts was prodigious. It was known too that he was able to acquire quantities of blank vellum.

Supposing Voynich, one mused, had come across the Baresch letter in the volumes of Kircher's correspondence at Mondragone sometime before 1912 and used it as a basic crib involving herbs, astrology and 'chemical secrets', adding in the intriguing naked nymphs, the plumbing and the baths. This he supplemented with filler text, perhaps not so differently as Rugg suggested. The whole confection is augmented by the Marci letter to Kircher full of indicative hearsay establishing the route to John Dee and back further to Bacon. If a sale had gone ahead and his authorship been refuted Voynich would no doubt have muttered *caveat emptor* all the way to the bank.

In the early 1920s Voynich did indeed ask for more research about Baresch from Bohemian sources, having stated emphatically in the 'Preliminary Lecture' … that the 'intimate friend' would be found. It is quite possible that he already knew of the existence of Baresch. Robert

Brumbaugh, while researching in the Beinecke Library files, found a translation of an 'unaddressed and undated carbon of a note from Prague to Voynich', among similar others from the Bohemian State archives. The note says that 'Marci probably inherited the manuscript from George Barschius, an alchemist ...',[6] i.e. Baresch. Brumbaugh did not follow up the clue; Voynich may well have done so before his 1921 lecture.

All this, however, was put in context in 2009 when an Austrian television company was allowed at last by the Beinecke Library at Yale to radio-carbon test samples of vellum from the manuscript. Their conclusion was that the date parameters of 1404 to 1438 were 95 per cent reliable. Furthermore, experts at the University of Chicago tested the inks and declared that they were based on iron gall, consistent with that era and applied fresh to the vellum. This information put the generally supposed date of the *Voynich Manuscript* back by a century. Out of the window go most of the elaborate theories. Might the manuscript be just one enormous 'doodle' from that time, the kind produced by a bored monk in some wintry monastery, perhaps with a leaning to 'outsider art'? Or more mundanely, might the obscure herbs be an ill-educated pharmacopoeia perhaps presenting heretical information about abortifacients? Or might it be, as Toresella suggested, a 'show-book' of a medieval snake-oil peddlar plying his wares? The problem with these less contentious assertions is the script.

In 2013, research by Marcello Montemurro and Damian Zanette from the University of Manchester hit the news. They re-endorsed Zipf's law stating that the manuscript strictly adheres to linguistic rules: the most frequently occurring word in natural languages will appear probably twice as often as the secondmost and three times as often as the third, etc. Such a pattern is not so apparent in randomised sequences. But how likely is it, one asks, that the script and illustrations evincing a west European origin hides an unidentified language? Hoax theory is again knocked – would some ancient prankster be able to concoct some apparently authentic but invented language that modern computer science cannot crack? On the other hand, in 2007, Andreas Schinner, an Austrian physicist showed that some statistical properties of the manuscript are similar to gibberish. There are too many regular repetitions of words, a fact not encountered in other languages, suggesting that some author is deliberately creating nonsense.

Toying with the *Voynich Manuscript* is like playing a pinball machine. Start the ball rolling and it will hit and light up images at every turn of its trajectory until it disappears down into a dark sump below. One recent player, Rich Santa Coloma, revived the hoax theory by claiming that large quantities of ancient blank vellum might have been found and used with traditional inks applied fresh to create a forgery. Voynich in 1908 had indeed acquired an Aladdin's cave of antiquarian material by buying the Libreria Franceschini, in Florence. This may be an assertion too far but at the risk of a certain obstinacy I still feel that Voynich may have played a dubious part in its history. It would be interesting to look more closely at the Marci letter itself to discover whether it might be a modern fabrication on re-used vellum to bolster the Bacon authorship. It is after all the main link specifically to the Prague period of the early seventeenth century.

A number of questions about the letter are raised. Why was it not incorporated into the other bound letters of Marci to Kircher? Why does it not allude to any details of the manuscript itself? Would that have been a clue too manifest? How was it in any case that Marci knew so much about the largely unknown Bacon to suggest him as author? Voynich himself decided at once that it was by Bacon but then dissembled that he only noticed it some time after the purchase. This is too much information, making him out to be a numbskull of a book dealer. But we know from his successes that this is not so. Helen Zimmern in an article in the *Pall Mall Gazette* in 1908 describes his prowess: 'Voynich never discarded a book until he'd stripped it of its cover.' His shrewdness produced results: in 1915 in Chicago he exhibited a map of Magellan he'd found in the binding of a Genoese book. Was his comment another disingenuous ploy to steer a path between public certainty and doubt?

The de Tepenecz signature on folio 1r also has a suspicious feel about it – was it not a lucky 'accidental' break enabling Voynich to tie in his story? In 2012 Rich Santa Coloma, while looking, like Brumbaugh, in the files at the Beinecke Library, came across a photostat of the faint de Tepenecz autograph before it was restored. If the signature had been there all along why hadn't Baresch, Marci or Kircher noticed it? The finger of suspicion points to the possibility that Voynich himself copied it from somewhere and added it. According to Rafal Prinke, the signature appears at least in another manuscript. To avoid the strained credibility of conveniently finding it, Voynich deliberately botched and masked the operation of 'recovering' it – again surely something no self-respecting book dealer would do to a priceless item. This may all seem a tangled tale, but wary forgers will go to great lengths to assert provenance.

My hints in the direction of a partial hoax involving the Marci letter a decade ago were not noticed by the valiant Voynich scholars. Another decade from now it may have been made irrelevant again by new research. The field is open to all comers. I never set myself up, however, to be an expert; I was merely one of the first twenty-first-century publicisers of its existence who, like Voynich himself, had stumbled upon something fascinating quite by chance – in his case in a chest of dead books, in mine at a funeral.

Since then the public interest in the *Voynich Manuscript* has proliferated madly. The riddle has infiltrated itself into general culture, with articles in magazines and scholarly places that make headlines every time a new player has 'cracking' news. Nick Pelling lists twenty-six novels based on the manuscript since 1969, and growing. The Beinecke Library itself commissioned a piece of music for voices by Stephen Gorbos performed in 2013, entitled, familiarly, *Such Sphinxes as These Obey No-one but Their Master*. It was set for 'Tuvan throat singing, belting and pop techniques, yodelling [and] Inuit throat singing'. Maybe the secrets of the indefatigably enigmatic *Voynich Manuscript*, like those throats, will have been cleared one fine day. Perhaps it should be just appreciated for its allure and sad beauty alone. Back in 2004 I was happy in part to do just that. The payoff for me was that a remark over a cup of tea had lifted a lid on the whole treasure trove history of the Boole family. It absorbed me eventually for a whole decade.

# BRINGING STARRY WISDOM DOWN

George Boole (1815–1864), father of the digital revolution.

In late November 2012 I found myself at King's Cross station, London, standing below the last destination board in my pursuit of the Boole and Hinton families. I had reserved a place for George Boole, my great uncle times three, at the end of the saga, because I felt that as I'd carelessly ignored his relevance to me for five decades, he could wait a while longer. Underlying this, if truth be told, was my apprehension at having to grapple with the world of mathematics that had made him famous. His genius regarding the subject has not been passed on it seems to my branch of the family.

This gestation of half a century in some ways parallels the period twice that length that Boole's ideas on the application of maths to logic had taken to form the vital basis of the digital revolution that now empowers the globe. It was his use of the binary system, as we shall see, that allied itself to computers. Those delving into that deeper world know of his Boolean 'operators', 'and', 'or' and 'not'.

The platform numbers at King's Cross are themselves somewhat bewildering. One of them is designated '0' and another 'nine and three quarters' – allocated to Harry Potter's Hogwart's Express. Its literary location is identified by half of a baggage trolley disappearing into a brick wall. Only slightly less fanciful is the fabled last resting place of the defeated Queen Boudica somewhere under platforms eight to ten. What did lure me onto the waiting train at platform four was the pleasure of fitting a person to their habitat. George Boole was born in 1815 in the medieval city of Lincoln, a place that I'd had no reason to visit before.

Some of my own personal mythologies were relevant too. The journey between King's Cross and Leeds had been a regular part of my ten-year stay in that city as a student and grown-up. En route northwards again I decided to stop off at a point on the east coast mainline that would offer a small epitaph to the history of a closer genetic tie.

In 1934 my father, Tom, aged twenty-one, had descended with a no doubt small suitcase onto whichever platform at King's Cross served his train arriving from Sunderland in the distant northeast. Three months after finishing an engineering apprenticeship at one of the Wearside shipyards, his job and several hundred others were sacrificed to the vagaries of the capitalist cycle. London as ever exercised its Whittington call. Tired after a long, smoky journey, he took himself to the station buffet for a sandwich and a mug of tea. 'Would you like a tray?' enquired the cockney lady behind the counter. The noun emerged squeezed into 'try'. London, thought my father, in his own accent, really is as fast as they said it would be.

No one would have ever accused him, of all people, of not having tried enough. In the relatively booming south of England he did find work. Two years later, however, given one hour's notice, he was told that his factory job in west London was terminated. The only vacancy the Labour Exchange could provide for him lay in Peterborough, halfway back from King's Cross on the track to home.

Here he alighted once more in a foreign town, all his belongings yet again in hand. He walked the half mile to the designated factory only to be told that owing to a clerical error the vacancy had been filled. 'I felt I'd been hit in the face with a brick,' he wrote in his sketchy account many years later. Shattered and suddenly rootless, he trudged back to the bus station, the cheaper travel option, to wherever next. Over a meal of pie and chips he decided to board whichever coach turned up first – south, back to possible unemployment, north, back to his family and an even more likely dole queue. 'I did not know it at the time but my whole future was decided for me,' he wrote. The Boolean operator 'or' had decisively come into play. As chance would have it, the southbound coach arrived first. My parents' futures, and hence mine, were decided by the road conditions of that day. I felt a small homage should be paid to his dice-playing.

The original bus station at Peterborough and the factory have long disappeared under an inner ring road and a shopping mall. I strolled around some of the old streets close to the River Nene and found myself wondering what my father would have made of a road sign that enquired, 'Would you be interested in sponsoring this roundabout?' I decided not to linger long on this grey early morning and resisted visiting the ancient cathedral, saving myself for the reputed awesomeness of its neighbour at Lincoln. Back at the station I vitalled with a warming breakfast that Tom might have taken at that time: a sausage sandwich and a scone (pronounced 'skoon'). I boarded the diesel, single-train carriage that shuddered ominously at low speed on the very slow route across the fenlands. The low-lying, hill-less terrain stretches eastwards to the other side of the North Sea and then on through the Baltic republics as far as the Russian Urals. The landscape is completely alien to the west of England that signifies home to me.

It became even more so as we navigated vast fields of luminous green winter wheat and cabbages of every hue. The carriage clicked and clacked over endless level crossings but there were almost no villages, church spires, schools or woodland. We were traversing the polders, drained and reclaimed over hundreds of years. The landscape possessed only two dimensions: land and sky. As, however, we eventually approached Lincoln on a wide curve, the cathedral visible on high and bathed in sunlight, took on, appropriately to a pilgrim, a glimpse of a celestial city.

The Victorians, in their attempt to keep the railway on level ground and as close to the centre of Lincoln as possible, planted it four-square

Lincoln High Street in George Boole's time.

across the long, long High Street that was once part of the Roman Ermine Street heading from London to York. Now part-pedestrianised, the thoroughfare recreates the medieval scrumming of citizens that once would have frequented the inns, merchants' and craftsmen's premises. Here too were held the markets for Lincolnshire agricultural produce. Once upon a time during the April fair, 50,000 sheep could be sold. Wool and the weavers' trade gave rise to the distinctive 'Lincoln green' cloth that camouflaged Robin Hood.

Towards the end of the High Street the Tudor Guildhall building straddles the road marking what was once the southern gate of the Roman city. Beyond it the High Street funnels into not much more than a wide path that steeply ascends the scarp to the summit. This point designates topographically the social hierarchy that George Boole grew up within. Above it lies Upper Hill, crowned by the ecclesiastical pinnacle of the

Steep Hill, Lincoln.

cathedral; below the aspirant world of Lower Hill toilers. Steep Hill, as the route is aptly named, evokes a Bunyanesque ring indicating the difficulties to be encountered by one such as George Boole in attaining a higher social standing and grace with God. The houses that perch on the wayside tantalisingly refuse a sight of the cathedral itself as one climbs. Attaining the summit square at Castle Hill there are fine old buildings. An intrusive CCTV camera protrudes incongruously from the Magna Carta pub. Here at last, through the Exchequer gateway, one appreciates the immensity of the cathedral – visible across Lincolnshire from twenty miles away.

Built over three periods from 1085, it has survived fire and earthquakes. The unusually wide western facade, like a bookend, buttresses the two high and once-steepled towers immediately behind. Overawed and still slightly out of breath, I ventured in. A Morning Prayer in level amplified tones

33

The west front of Lincoln Cathedral.

The George Boole window, Lincoln Cathedral.

was being given from a near pulpit thanking the entirely empty seats below for sharing the occasion. John Ruskin thought the cathedral to be probably the most beautiful of all the gothic church splendours of England. The length and height of it is indeed mightily imposing.

The George Boole memorial window in the north aisle of the nave was installed here five years after his death in 1869 by public subscription. Of the three panels, two depict Boole's role as a teacher. The more telling bottom one is the 'Calling of Samuel' chosen by Mary Everest, his wife, because it was one of George's own favourite biblical passages, echoing the idea that the humble could be drawn to serve God. This is an apposite depiction: Jesus himself was the son of a workaday artisan, self-taught and dying well before his three-score years and ten.

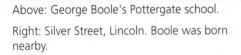

Above: George Boole's Pottergate school.

Right: Silver Street, Lincoln. Boole was born nearby.

Outside, across the cathedral lawn, stands number 3, Pottergate. Here Boole in 1840 ran a profitable, well-established school before he left to live in Ireland in 1849. Relatively opulent, the red-brick, large-windowed building is almost quadrangled in front by a garden of trees. Below sits the Pottergate arch through which he and any wheeled traffic too wide for Steep Hill would have descended to the townscape below.

Boole was born at the foot of the hill on 2 November 1815 and grew up a resident of humble Silver Street. His sister, Mary Anne followed in 1817, brother, William in 1819, my relation to him, and Charles in 1821. The surrounding locality in which George was baptised, played, educated and found supporting friends provided the essential milieu that nurtured his ideas. His aspirations were not those that would have him craning his neck towards the heights of respectability and orthodoxy above. He remained a man of the people.

George's father, John Boole, was born in 1777 in Broxholme, a small village six miles northwest of Lincoln. Not much is known of his background except that he was proud of the fact that his forefathers were literate, master thatchers. In 1791 he became apprenticed to a shoe-maker but either, like my own father, through lack of prospects or an enquiring mind, set off to London around the turn of the century. Here he established himself as a journeyman shoemaker. His work in a damp cellar, however, led to unfortunate subsequent ill health. He married Mary Ann

35

in 1806, returning later to Lincoln to become a master of his own trade. The city, about that time of 15,000 people, was shod by over forty shoemakers. This would probably not have enabled wealth-amassing but John seems to have had wider interests.

He was described as 'a man of thoughtful and studious habits, possessed of an ingenious mind, and attached to the pursuit of science, particularly of mathematics'.[1] John Boole's manual skills and scientific knowledge combined in his study of optics, making complex instruments such as telescopes, even sending a theoretical paper on the subject to the Royal Society in 1825. George Boole was born into a quite particular social environment in which his genius was nurtured. Daniel Cohen suggests that to appreciate his significance 'requires equal parts technical discussion, biography and broader cultural history …'.[2]

Although historian E. P. Thompson in *The Making of the English Working Class* often warns against the perils of generalisation, he clearly identifies a social category of artisans in the first half of the nineteenth century that was highly active and influential during this period. These urban skilled workers were conscious of their separateness from illiterate agricultural labourers, and 'costermongers, drovers, slaughterers of cattle … cads, bricklayers, chimney sweepers, nightmen [and] scavengers …'.[3] He lists their typical trades as 'printers, apothecaries, teachers, journalists, weavers, saddlers, tailors, cabinet makers, blacksmiths, shoemakers and engravers'.[4] Their mental attitudes as well as their occupations set them apart, chief of which was their desire, perhaps more than money or status, for self-improvement through acquiring knowledge.

Thompson writes of that period: 'the towns and even villages hummed with the energy of the auto-didact …', spurred on to learn despite, 'overwhelming difficulties – the lack of leisure, the cost of candles (or of spectacles) as well as educational deprivation.'[5] Their independence, compared to straitjacketed factory workers, coupled with motivated inquisitive minds, encouraged not only sobriety and self-respect within family life, but, taking little on trust, fostered a degree of anti-authoritarianism. A sense of identification with others like themselves led to mutuality and 'to a belief in the active duties of citizenship'.[6] This in turn nurtured a belief in the rational possibility of change, tending towards radicalism, republicanism and utopian dreams.

Many important radical figures of the early industrial revolution followed artisan trades. William Blake's father was a hosier, his son an engraver, Thomas Paine's father a corset maker, Robert Owen's a saddler.

Thomas Spence's father was a shoemaker. John Boole's similar trade hits the nail squarely on the head. Shoemakers seem to figure very often in such historical accounts. Perhaps the high skill combined with a portable small-scale workshop of tools encouraged a craftsmanship and mobility that enhanced a sense of independence.

Non-conformist religious belief was often a strong adjunct, placing believers relatively in opposition to the status quo. Although Methodism in its heyday laid a stress on sobriety and encouraged self-improvement through literacy and Sunday schools, it also endorsed the study of practical disciplines such as botany, biology, chemistry and maths. Thompson avers to 'that peculiar phenomenon of early Victorian culture, the non-conformist parson with his hand on the old testament and his eye on the microscope'.[7] Methodism baulked, however, at intellectualism for its own sake and speculation beyond authorised works. The study of philosophy, politics, poetry or biblical criticism was not endorsed. At its worst, Thompson describes the 'psychic terror' of its concentration on 'a morbid preoccupation with sin, guilt and the saving of souls'.[8] Satan is the enemy within, not so much the social system without.

Other non-conformist denominations were more congruent with the artisan world-view. Unitarianism, with its rejection of the Trinity, laid stress on the oneness of God, on unity. Jesus Christ is allowed a prophetic status only. The Bible should not be treated as literal truth, in particular its notions of original sin, eternal damnation and predestination. Man with free will and reason could combine both faith and science to create a better society. The lack of liturgy and autonomy of the chapels suited the independent status of the skilled journeyman.

George was precocious from the outset; his appetite for learning led him to a preparatory school for tradespeople. He was considered shy and retiring but not so much that, once, in an almost biblical tale as a young boy, he was found spelling out words to a crowd in exchange for coins. His father's tutoring provided the mind-stretching his son needed. George's scholastic learning was tempered practically by helping him to construct optical instruments.

A telling anecdote recounts how his father, having finished making a telescope, placed a notice at 295 High Street in the window of the shop to which they had moved. 'Anyone who wishes to observe the work of God in a spirit of reverence is invited to come in and look through my telescope.'[9] It precisely expresses the product of self-reliant application yet a democratic openness to all to combine scientific enquiry and a sense of devotion. Although the Boole children were baptised as Anglicans, the sentiments of the family were those of Unitarian non-conformism. At 290 High Street, the bookseller's shop was run by William Brooke, scholar, printer and an ardent liberal. He exercised a paternal influence on George, lending him the novels of Sir Walter Scott into which he loved to escape.

In his early teens he found himself at a commercial academy because the family could not support grammar school fees. He set himself to recoup the deficit by learning ancient Greek on his own. Encouraged by his father, George sent a translation of a Greek poem to the *Lincoln Herald* which was printed in May 1830. This resulted in a spat of long-running correspondence with a local citizen who refused to believe the age of the originator. Bright-eyed with enthusiasm, although short-sighted, probably from over-reading, George also learnt Latin, French and German, wrote poetry and played the flute.

Beyond the free-thinking atmosphere of his environment there is little to link George directly to the radical upsurge taking place in England at that time. Both the agricultural labourers revolt of 1830 and the clamour for enfranchisement that culminated in the 1832 Reform Act would have been apparent. An interesting insight, however, into the more explicitly radical craftsman prototype can be glimpsed through George's own cousin by marriage, Thomas Cooper.

Born in Gainsborough not far away, ten years before George, his life pre-shadows him. The son of a dyer, brought up in poverty, his precocious drive to read found him absorbing mathematics, Latin and other languages, aided, like George, with a prodigious memory. His insatiable desire for knowledge led him to astronomy, Shakespeare, Locke, Virgil and Byron: the polymath's pantheon. He too attended local elementary schools and similarly taught as a school assistant. Aged fifteen he had to take on a trade, choosing that of a shoemaker. Echoing his Lincoln cousin's later perspective, his desire for spiritual certainty was bedevilled by rationalist qualms. While recognising the moral beauty of Christ, he could not abide the 'tinselled rags of miracle and imaginary godship … the false idolatrous and enslaving forms in which Priestcraft clothes that glorious Galilean peasant'.[10]

In 1827 Cooper moved to Lincoln where during Christmas week 1829 he met up with fourteen-year-old George and later married his cousin. (George himself was not immune to falling in love: one strong attachment faltered because he could not reconcile a Christian marriage with Church doctrine.) Cooper wrote in his autobiography that George 'overwhelmed me with enquiries about the contents of books I had read'.[11] They worked together socially but their paths diverged after Cooper left Lincoln in 1838 to campaign militantly for the Chartist cause. Cooper's political fervour and egoism, unlike George, resulted in a two-year jail sentence for arson. Much later in London the two met up on friendly terms, Cooper feeling somewhat dwarfed in his presence. In 1872, looking back, Cooper made the highly astute observation that Boole's name would become truly illustrious in some wiser future.

Rather than being apprenticed to his father as a shoemaker, George pursued his education in order to become a teacher. When John Boole's business took a turn for the worse he was required in 1831, aged sixteen, to help the family by taking up a post at Doncaster, some forty miles away, at a Methodist school. George seems to have been a rather lonely outsider there, missing his mother's gooseberry pie, but was by all accounts liked and respected. No Dotheboys Hall, the atmosphere was relaxed but suffused by its religiosity. Fellow teacher, Thomas Dyson, later recalled that this presented problems for George as he was becoming too engrossed with his study of mathematics. Blasphemously, he continued during the Sabbath. Dyson concluded that his colleague had developed a more general faith in a benevolent deity rather than some dogma-based creed.

This must have become more focused during a period two years later when he experienced some kind of deep intuitive flash while crossing a Doncaster field. According to Daniel Cohen, this epiphany revealed that the mind 'has an innate sense of "Unity" that it constantly uses to synthesize its understanding of the world'.[12] More concretely, Desmond MacHale describes it as the notion 'that it might be possible to express logical relations in a symbolic or algebraic form and thus be able to explain the logic of human thought and to delve analytically into the spiritual aspects of man's nature'.[13]

Perhaps George's 'flash' was more than a revelatory *pensée* and something closer to a pantheistic experience. His wife Mary wrote that when her husband worked on differential equations he was in 'a sort of religious rapture'.[14] We can assume that the Doncaster moment had a lingering effect, probably augmenting his drift spiritually towards Unitarianism. With a heart also over-brimming with the humanist stuff of literature and art in a time of political and religious debate, the idea of rationally finding a way to enhance clear thinking would seem to be highly desirable. Did he at this time, one wonders, think of himself as almost a prophet, called like Samuel?

Probably becoming ever more disaffected with his post, George was asked to leave the school in June 1833. A short period teaching in Liverpool brought him back to his native city. His own widening intellectual horizon would have chimed well with the proposed setting up in 1833 in Lincoln of a Mechanics Institute, designed to further adult education on technical subjects with the help of a library, classes and lectures. The petition to the mayor was signed by John Boole, Thomas Chaloner, George and his brothers and sister, Thomas Cooper and Brooke the bookseller. Lincoln was ten years behind the first such foundations in Liverpool, Manchester and London. In 1837 the Institute celebrated Queen Victoria's coronation marching under the banner 'Wisdom is the principal thing'.[15] It displayed emblems of faith in science: a picture of Isaac Newton, a geological map, a globe, electrifying machine, steam engine, telescope, galvanic battery and a chart of the kings and queens of England.

It was duly established on the ground floor of the ancient Grey Friars stone building, still standing just below Silver Street. Periodicals subscribed to included *The Library of Useful Knowledge*, the *Penny Encyclopedia* and *Constable's Miscellany*. George, an enthusiastic supporter, donated four volumes of *Spectacle de la Nature*, a popular work of natural history, plus a stuffed emu and one of its eggs. His father became curator in residence, giving lectures in optics, but later resigned the position, dissatisfied with the Institute's lack of seriousness. Thomas Cooper lectured in Latin and French. George became a committee member and unpaid teacher, taking particular interest in its library. A report of his in 1846 complained of a lack of broad philosophical and economic volumes. Regarding maths, there was either not enough applied works nor more advanced, adding that most of the users of the latter were unlikely to have enough leisure time anyway. The constitution of the Institute banning the holding of works of 'party politics' or 'controversial divinity' gave him greater anxiety as an issue of personal liberty. At a lecture to its members he protested that both controversial areas must be part of enquiry into the gravest and most serious questions with which a human being is concerned.

On the occasion in February 1835 of a marble bust of 'unrivalled beauty and purity' being presented to the Institute by local worthy, Lord Yarborough, George, despite his avowed incompetence to do justice to so great a theme, gave an address in honour of fellow Lincolnshire genius, Isaac Newton. He expounded the latter's theories of optics, light and colour, gravitation and the bold triumph of astronomy. The understanding of the planets, he maintained, freed mankind from the superstitions that had surrounded them for generations. Astronomy above all demonstrated 'the economy of the universe, the capabilities of our own immortal natures and the majesty of the Being who created them'.[16] Boole is trying to unite religious reverence and rational thought. The pure transcendental realm of mathematics – a bridge to the Divine – was fully demonstrated a decade later when Neptune, the eighth planet, was literally divined by mathematical theory and subsequently verified by observation.

George's impecunious family had to be supported, his two brothers being still at school and his sister, Mary Anne, not able to contribute. Being the breadwinner he took a teaching position as chief assistant at the Gentleman's Boarding Academy five miles south of the city at Waddington. This post seems to have been a happier one but in 1834 he returned to his native quarter of Lincoln to begin his own school. Education was now the family's artisan trade. His uncle, William Boole also ran a school in the High Street at that time.

As a teacher George was both creative and successful. The formal side of learning was represented by the usual three Rs, Latin and Greek and the offering of private instruction to gentlemen preparing for a university. Boole, however, offered more than rote learning, employing geometric models, maps and globes and the blackboard to encourage intuitive understanding rather than mere abstraction. Language was to be absorbed by reading and speaking rather than by a crammed grammar. Handwriting must be clear, not just because of discipline but to enable communication.

Charles Clarke, son of the Lincoln Inspector of Corn Returns, attended all of George Boole's various institutions and attested to his 'refined taste, liberal ideas of vast width and depth … to him the schoolroom was his home'.[17] In 1838 when George gave up his own school and resumed a post at Waddington as its head, Clarke, as one of his favourite pupils, followed him there, boarding at £6 per annum. George's whole family accompanied him, all to be involved in teaching.

Extra-curricular activities included running a newspaper, a debating society and mock elections to encourage free-thinking. Some physical recreation was unavoidable but under-used by George himself, preferring to take classes on walks in the countryside. Clarke recalls being taken to a hilltop to watch an approaching storm. George timed the gap between lightning and thunderclaps to demonstrate the speed of sound. On another occasion during a solar eclipse his pupils observed in a darkened room its progress via the shadows thrown by one of John Boole's telescopes.

At totality the class emerged into the dim garden outside 'to see the closing of flowers, to hear crowing chanticleer and to watch jackdaws winging home to roost in the lofty cathedral towers'.[18] One can imagine their teacher rapt in awe at one of God's wonders. Boole had spoken to Clarke on another occasion of how nature itself was as beautiful as the cathedral's vaulted ceilings. Ironically, one boy was beaten by him for robbing birds' nests of eggs. Brother William recalled that, like himself, he could not bear the idea of hanging a live worm upon a hook as bait for fishing.

In 1840 Clarke followed the Boole family back to Lincoln when the school at Pottergate was founded, bringing at last almost a decade of relative financial security. Its syllabus placed emphasis on the moral side of education, of acquiring 'the habits of industry, integrity and mutual kindness … the distinction between right and wrong [and] a reverence for sacred things'.[19] All of these attributes seem to have come easily and unbidden to its saintly proprietor. Robert Harley, a fellow mathematician, noted how he would visit sick pupils and offer them little luxuries out of an 'enlarged sympathy with the joys and sorrows of others … a tenderness towards their faults and a ready recognition of any redeeming qualities in the worst of them'.[20]

George Boole's consistent underlying sentiments are those of the liberal humanitarian. If humans truly desire the best for others, their intrinsic goodness, guided by the God-given reason of science and debate, will provide a harmony of individuals within society.

The crusade to repeal the Corn Laws that artificially raised the price of bread at a time of poor harvests and unemployment must have augmented the latent opposition between Upper and Lower Hill. Boole's pupil Clarke, possibly influenced by him, pitched in with the Anti-Corn Law League, writing youthful pamphlets, organising petitions and promoting demonstrations. When Clarke emigrated to Canada in 1844 George offered him his view that England had 'a too great regard to the distinctions of society', and America 'a too eager pursuit of wealth'.[21] Not possessing Thomas Cooper's active revolutionary passion Boole did not become involved in confrontation but took on the mantle of the earnest radical social reformer. One cause was the Early Closing Association fighting to obtain better conditions for shopworkers.

This national campaign paralleled the fight for the reduced working day of ten hours in factories. George became one of its Lincoln vice-presidents. Its members included old friends such as William Brooke and local left-leaning vicar, Edmund Larken. George was called on to give an address on *The Right Use of Leisure* at the Mechanics Institute in February 1847. Robert Harley supplies a physical description of the speaker: 'A man of middle stature, light-complexioned, slenderly built with a countenance in which both genius and benignity are suppressed, and a manner gentle and modest, almost to womanliness ....'[22]

George repeats his firm belief in God, 'that the laws of the material universe are the expression of His Will, their unfailing certainty is the symbol of His Immutableness; their wondrous adaptation is an argument of His Universal Presence'.[23] The deity does not subscribe to one denomination being known variously as, 'Great Author and Founder' and 'Infinite Intelligence'. The Scriptures he damns with faint praise as 'not unfavourable' to the free development of the human mind in so far as they affirm man's equality.

Far more important is the god-given scientific law of induction that enables our reason to discriminate general laws and understand His world. If this allows us to comprehend the behaviour of tides so should it allow us to discern our social obligations bound by a higher than human law. The implication is that society can be better organised by philanthropic exertion, especially for those less fortunate. To this end, he asserts, rather prudishly, that one should put aside the fading character of all merely sensual pleasures, although he allows the value of music and poetry. Instead, while observing both sides of a disputed position we should educate ourselves in our leisure by reading history and the biographies of great men, volumes 'that unveil to us those secret springs of action ... from which revolution and historic change have resulted'.[24] Reading not so far between the lines, there lies an underlying fighting talk.

One year later in 1848, the Year of the Revolutions, regimes toppled and wavered across Europe. In England the Chartist Movement delivered its petition signed by six million citizens in favour of basic democracy and parliamentary political reforms. The authorities, fearful of violence and very mindful of their responsibility for shooting dead twenty-two Newport Chartists in 1839, imported hundreds of soldiers into London. Queen Victoria was exiled to the safety of the Isle of Wight. Such great upheaval was not to George's liking. He may have felt a strong attachment to the common man, encouraged debate and the employment of reason, but he disparaged anything resulting in serious discord and factionalism.

In his own use of leisure he penned a good deal of poetry, mostly idealistic and noble rather than personal. In 'The Fellowship of the Dead' he raises sentiments praising radical bonds, '... the free unfettered soul / Bows to no enforced control', sharing 'deeds achieved, and perils dared …'. 'All who felt the sacred flame / Rising at oppression's name / All who toiled for equal laws / All who lov'd the righteous cause, All whose world-embracing span / Bound to them each brother man'. 'All that to the love of truth / Gave the fervour of their youth … Bringing starry wisdom down / To the peasant and the clown …'. 'Till the morning come(s) at last / And an Eden blooms again / For the weary sons of men'.[25] George is clearly on the side of an undogmatic egalitarian utopianism.

As we have seen, he is prepared to lend his support to ameliorate social ills in the here and now. That perennial source of fascination and pity, prostitution, was in Lincoln, as everywhere else, in the public eye. George would have witnessed its spectacle in the Brayford Pool dockland area of the city. The warehouses, tenements and pubs were inevitably a home to low living. The sexual vice appalled the Lincoln well-to-do, who complained of 'the torrent of sin that was streaming down their streets'.[26] The local clergy seemed unable to act but separate committees of men and women met and established a Penitent Females Home in June 1847 on Steep Hill, well clear of the murky tide below.

George was involved from the outset and supported similar schemes later in Ireland. Twenty-one women, some as young as fourteen, became occupants, taking in washing and doing needlework to help earn their keep. A larger house was acquired from where, behind the painted-over windows, the penitents would emerge in their lilac dresses and straw bonnets to march on a Sunday, somewhat conspicuously, uphill and out of breath to the cathedral. Many of them later gave up penitence and ran away.

Boole would no doubt have noticed the latent hypocrisy behind all this, aware of the need for more than charity, but stopping short of any formal theories of systematic social change. Creating model institutions is a halfway stage exhibited in the spirit of the Mechanics Institutes (and later as we shall see, of the Settlement Movement that involved other Boole/Hintons). It is consistent that around the time of the Society for Leisure and the Penitents Home, George seriously considered taking up the position of principal at the newly founded People's College in Nottingham. The institution was dedicated to offering a sound secondary non-denominational education for working-class boys. Its prime mover and enabling financier, George Gill, himself from non-conformist stock, apprenticed as a hosier. Boole demurred because of his prospects in Ireland.

This move towards grassroots education was approved by him in the support he expressed in a letter to Mary Anne in 1850 regarding a Mutual Improvement Society in Lincoln. As his father, John, had bemoaned much earlier, the Mechanic Institutes were liable to become too leisurely and middle-class in outlook. The fees and opening hours plus proscription on political or religious debate did not encourage democratic tendencies. The Mutual Improvement Societies that spread from the 1840s, inspired by what Thomas Cooper had called 'knowledge Chartism', gave elementary education at a penny a week to both men and women. Meetings were held in chapels, pubs and houses and were radical in outlook, forming and disbanding spontaneously.

Changing the economic relations vested in property was another avenue for improvement at this time. Boole was a director helping to establish a Building Society in Lincoln to enable the poor to buy and sell land in small lots. According to Frances Hill, it helped a large number of middle- and working-class citizens to acquire their own houses.

George had one particular close friend who was an active initiator and participant in all of his activities promoting social reform. Edmund Larken, almost the same age, was the kind of Christian Boole could respect, even if his pedigree was distinctly Uphill. Eton- and Oxford-educated, cricket-loving Larken had married into the rich Monson family whose estate lay three miles from Lincoln at Burton. Here in 1843 he was given the living of the Church as 'domestic chaplain' to the Monsons at £420 per annum – enough to support his eight children and six servants. As a Christian Socialist, however, radical causes were a mission. His own appearance itself was a bold statement, being the first vicar with a beard to preach in an Anglican church – a clear identification with the male working class.

In one sermon he outlined utopian community schemes similar to the Harmony settlements proposed by Robert Owen. The underlying principle was one of mutual co-operation inspired by the basic tenets of Christ. There is every reason to suppose that Boole would have taken a sympathetic interest in his friend's views: Larken had entrusted his boys' education to George, who had in turn dedicated poems to him. George would have respected his active religious tolerance and adherence to the lived fundamentals of Christianity. Larken's political rebelliousness was no doubt responsible for his lack of preferment at the cathedral.

Throughout the troublesome 1840s George had become immersed in the world of mathematics. Unlike the recipients of private instruction, George, like Thomas Cooper, could foresee little chance to go to Cambridge

University. Despite increasing financial security from the success of his schools, it was beyond his means. From his background, lumbered no doubt with a Lincolnshire accent, he would have been something of a fish out of water. He reminds one, in his shy gaucheness, of Thomas Hardy's Jude. Unlike him, however, George had the support of a local liberal and benefactor, Sir Edward Bromhead, who lent George books and took the budding genius under his wing.

In 1838 George submitted a well-received paper, *On Certain Theories in the Calculus of Variations*, to D. F. Gregory at Cambridge – a major achievement for someone coping with running a school as well. In early 1839 Boole visited Gregory and was no doubt impressed by varsity life making plain a desire to enter. Gregory, however, warned him in a letter of March 1840 not only of the high expense but recognised Boole's unorthodox self-taught nature: 'You must be prepared to undergo a great deal of mental discipline which is not agreeable to a man who is accustomed to think for himself.'[27] In any case George still bore the responsibility of looking after his family and as a theological misfit would have had problems with the orthodox tests for entrance which extended to Jews and non-conformists.

More papers in the 1840s brought him into contact with the select world of mathematics. In 1841 his *Exposition of a General Theory of Linear Transformation* was published in the *Cambridge Mathematical Journal* and initiated a correspondence with the mathematician, Arthur Cayley, who had, unlike George, entered Trinity College aged seventeen. Around this time also he befriended Augustus de Morgan from a middle-class colonial background who had entered Trinity College in 1822 at the even earlier age of sixteen. Their interests, including algebra, overlapped into flute playing, a love of literature and a secularist stance that landed de Morgan in trouble when he objected to discrimination against a Unitarian friend. In 1843 de Morgan helped George revise his paper, *On a General Method in Analysis*, before its submission to the Royal Society in 1844. After initial confusion about what to do with a paper from an unrecognised source, it was awarded a Royal Medal. Despite being born on the lower slopes of the hill, his reputation as a mathematician had been established. Impressed with the scholastic life on a visit to Cambridge in 1845 he penned the doggerel: 'Twas something on the banks of Cam, to see / Men known to science, known to history'.[28]

Boole had another three papers published in the mid-1840s and meanwhile was probably concentrating on running his Pottergate school and the social activities to which he was committed. In relation to his subsequent career, however, ideas were formulating that had been dormant for a long time, dating back to his 'mystic' experience as an eighteen-year-old in Doncaster. Boole wanted to understand the inherent capacity to reason that enabled man to reconcile different interpretations of God's world. Boole wrote to de Morgan, 'I do think that when we know all the scientific laws of the mind we shall be able to be in a better position for a judgement of its metaphysical questions.'[29] This would apply as well to matters of social conscience where moral obligations also needed to be clear-cut. The task would come more easily to someone from his social milieu – Larken was much more of a rebel from his.

In the field of knowledge, mathematics and its intrinsic beauty was God's especial tool to find a way through discord. Even here there was to be no fundamentalism. His wife, Mary Everest, later declared surprisingly that 'mathematics never had more than a secondary interest for him; and even logic he cared for chiefly as a means of clearing the ground of doctrines imagined to be proved'.[30] There is evidence to suggest that she was right. In May 1840 Boole warns his old school friend, Joseph Hill, that maths 'can deaden the imagination and destroy the relish for elegant literature and indispose the mind for everything but the bare pursuit of abstract truth'.[31] Six years later, despite his growing success as a mathematician he writes again to Hill that ethics gives him more 'solid gratification' than mathematics as the subject appeals to both emotion and reason.[32] To a former pupil, Charles Kirk, now at Cambridge, George writes in 1847 that if he likewise attended he would not make mathematics his principal pursuit. A year later, further into professional success, in a letter to Larken he writes ironically, 'In pursuance of my intention long since formed and never to the present time lost sight of, I send to your careful custody my mathematical books.' Keep them three years, he suggests, and should he ask Larken for them back he is to burn them.[33]

In 1847 an inspired Boole wrote *The Mathematical Analysis of Logic*. Where J. S. Mill had searched for a felicific calculus to rate happiness, Boole was looking for a calculus of reasoning to aid thinking. Language itself was full of ambiguity and imprecision. As Leibniz maintained, we need to make our reasonings as clear as those of mathematicians so that we can find our error at a glance and when there are disputes we can simply say: let us calculate. *The Mathematical Analysis of Logic* was a precursor to Boole's seminal work *The Laws of Thought*, to which we shall return.

The Pottergate school was now supporting his family: his aged and infirm parents, his brothers Charles and William and sister Mary Anne. Teaching, alongside an active involvement in social affairs, was a vocation for him, yet he worried about the future. It was suggested that he might apply for a professorship at one of the three Queen's Colleges being established in Ireland to facilitate non-denominational university degrees. Such an appointment, as well as giving financial security, might help further harmony regarding the bitter religious divides there.

In October 1846 George applied for a post at one of the new colleges. To the fulsome testimonials from academics such as de Morgan and Lincoln friends like Larken, he added modestly, '... it ought to be remembered that I have not taken a university degree and have never studied any college.'[34] George had to wait nearly three years before the Chair of Mathematics at Queen's College, Cork was offered him in 1849. Uphill in the Castle Square at the 'White Hart' inn, Lincoln friends gathered for a send-off, presenting him with books and a silver inkstand. George ventured the subsequently forlorn hope that one day he would return to live out his life peacefully in his native city.

Across the Irish Sea lay anything but accord. Continued failure of the staple potato crop because of blight had decimated the population through starvation and emigration. The successful repeal of the Corn Laws in 1846 cheapening the price of cereals had little impact on the Irish population as it was too destitute to buy at any price. Other home-grown agricultural produce fed the richest or found its way into export. George was aware of the state of the Irish nation; in a poem of 1849 he writes of finding life '... clamorous in thy streets, and where the gold / of plenteous harvests waved, the plashy plains / O'er which the bulrush towers, the ragweed reigns'.[35]

To add to the general appalling distress of the Great Hunger, Cork city had been flooded by high tides. He wrote to Mary Anne of the sufferings of the poor living in basement cellars. He made visits: one to the Cork Union workhouse, itself swollen with nearly five thousand inmates, and another to some schools for poor children. The damp in general had caused his walking stick to twist, but did not prevent him taking excursions into the countryside to enjoy the scenery and the air. On occasion he kept company with Dr Ryall, the Vice President of the college, and his visiting future father-in-law, Thomas Everest, whose sister Ryall had married.

Boole's honest empathy is conveyed to Mary Anne regarding the conditions of the peasants: of deserted cottages and how a hospitable woman had cooked them potatoes for dinner. In complete contrast he writes in distaste to his mother in 1849 having attended a dinner

Queen's College, now University College, Cork.

that included truffles and champagne. In a later lecture to the Lincoln Mechanics Institute he gives graphic detail of entering a roadside cabin without windows; the only food visible being a couple of turnips. The invariable story of the famine consists of husbands dead or absent seeking employment. He found he acquired 'an indifference shocking even to myself … to witness the unmerited sufferings'.

Despite the weather and the disturbing social setting, he settled into lodgings with a friend he had met in London, Raymond de Vericour, who he describes as 'an unbounded favourite in Cork'. Boole writes to his mother that he is lonely but finds Cork people to be kind and friendly. His income is secured at £250 per annum plus fees directly from students, enabling him to send money to his family back home. His father had died in 1848 and his mother was ailing. Mary Anne was struggling to run a school; William, also a teacher, had married Eliza in 1846; Charles is living on the Lancashire coast married to Ann and earning a living as a miller. Although George writes that he wishes he could see his brothers, his affection is greatest towards his sister, who comes regularly to Cork. He in turn often takes the boat back to England.

The college is establishing itself; George is liked and his lectures have proved popular. He sometimes takes breakfast with his pupils. Trouble was brewing, however. Despite half of the 1850 intake being Catholic, the three Queen's Colleges were being viewed as colonialist impositions. In 1850 the Primate of Ireland spoke of the 'propagation of error through a godless education'. De Vericour sparked a strong criticism of himself by the male hierarchy of the college following the publication of his history of Christianity that queried papal authority and the role of the Church. Accused of religious partiality contrary to the principles of the College, he insisted that his work outside was his own affair.

George characteristically decided to mediate between his friend and the President of Queen's College, Sir Robert Kane. The discord quietened but Kane's own lack of ideological balance and querying of teaching methods led to an open and unfriendly debate between him and Boole. To add to the discord, an official Vatican condemnation of Queen's College was issued in 1857. George must have felt once again in his life that institutional religion anchored in bigotry was a brake on reason and social harmony.

Regardless of dissension in the Catholic hierarchy Boole was elected as Dean of the Science Division of the Faculty of Arts in May 1851 and gave an address to it entitled *The Claims of Science, especially as Founded in Relation to Human Nature*. It outlines the philosophical context from which three years later his seminal work *The Laws of Thought* emerged.

He begins by re-iterating his conviction stated in his Newton Address of 1835 that immutable laws abound which demonstrate the order of God's universe; the task of science is to apprehend them. General laws exhibited in nature can be discerned by experiment and observation and the use of reason. In this process there exist laws of thinking to which obedience is forced as much as in the physical world. Their discernment is the provision of mathematics, 'not of number and quantity but in logic as universal reasoning expressed in symbolical forms and conducted by laws which have their ultimate abode in the human mind'.

Armed with these tools of thought, it will be possible to interpret other less certain laws of behaviour where there exists a 'liberty of error', where obedience is claimed, not enforced. Against, for example, the evidence of 'social and economical sciences that afford us valuable information as to the general tendencies of society and institutions, there is a wilful disregard of the cause and effect between an undrained uncleansed condition of our towns and the prevalence of fever and a general high mortality'.

Science, informing our ethics, can deliver us from 'that dark prejudice of chance in the physical world, of fate in the moral world, to which Ignorance clings with inveterate grasp'. Those emblematic banners of the Mechanic's Institute lauding scientific progress expressing 'man's dominion over the inorganic world', if rightfully applied, will produce a time when painful toil is replaced, when 'the most prolific sources of disease as crowded cities, undrained swamps, pernicious indulgences … and the extreme inequalities of wealth and the miseries which they entail shall have yielded to a better moral or social economy'.

Boole expressed two sides of his artisan background here: a belief in science and forward-looking material progress, augmented by a strong sense of social justice and change. His utopian aspirations are optimistic:

> When relieved from the offensive bondage of physical wants, man shall be at liberty to accomplish … the higher end of his being: that while the earth shall shine with more than its pristine beauty, the human family shall not only be clothed with the fair assemblage of the moral virtues, but shall add to them that crown and safeguard of knowledge which has been won from the hard experience of ages of error and suffering.

Noble sentiments indeed. The scientist guided by clear thinking will be a key player in effecting this vision: 'The authoritative Moral perception when wedded to the discipline of true science … is favourable to a sound morality. If it exalts the consciousness of human power, it proportionately deepens the sense of human responsibility.'

We have already mentioned Boole's first book on mathematical logic, *The Mathematical Analysis of Logic* published in 1847. In 1854 he published *The Laws of Thought*. This aimed to correct and perfect the earlier book. Its scope is lucidly summarised by Boole himself in the opening chapters:

> The design of the following treatise is to investigate the fundamental laws of those operations of the mind by which reasoning is performed; to give expression to them in the symbolical language of a Calculus, and upon this foundation to establish the science of Logic and construct its method; to make that method itself the basis of a general method for the application of the mathematical doctrine of Probabilities; and, finally, to collect from the various elements of truth brought to view in the course of these inquiries some probable intimations

concerning the nature and constitution of the human mind.[36]

Boole accepted traditional Aristotelian logic but wanted to take it further. He created mathematical foundations involving equations as well as expanding the range of problems logic could treat and the range of applications it could handle. A short exposition of some of the basic elements of these mathematical foundations is justified here because of their profound impact on subsequent advances, most notably in the field of computing.

Boole argued that logic is more akin to mathematics than philosophy and articulated this through two significant contributions. First he demonstrated that symbols, such as X and Y, conventionally employed in algebra to denote numbers, can also be used to represent groups (or 'classes') of real things. For example, we could define Class X to be things that are green and Class Y to be things that are edible. Something which is green and edible, such as spinach, falls into both these classes. The word 'and' in the previous sentence is termed an operator. There are several such Boolean operators, including 'or' and 'not'. Changing the words in a logical argument into mathematical symbols and using operators to combine them in different ways enables logical problems to be solved like equations. Boole called this 'the calculus of reasoning'.

Secondly, Boole asserted that every logical statement must be either true or false. So the question, 'Is spinach green and is it edible?' can only be answered logically either 'yes' or 'no'. Translating the language of logic into the language of mathematics, 'yes' is designated by 1 and 'no' is designated by 0. Using this new form of logical reasoning, seemingly anything can be expressed as symbols, operators and the two numbers 0 and 1, known as binary digits. More than a century after Boole's death, this has become the language of computers, which today process all their information as binary digits, since any number can be represented as a combination of ones and zeros.

Despite the success of *The Laws of Thought*, he questioned, however, whether Queen's College would be the most useful and best-equipped setting for his advance. He considered applying to Manchester and to St Andrews but had reservations about the severe Church of Scotland. In 1854 a professorship in Melbourne was considered, promising a large salary but was too far away from friends and family. Even though they lay across a narrower sea, he possessed an enduring affection for his brothers, Charles and William, and his sister, Mary Anne. George resigned himself not altogether unhappily to his Cork existence, but in 1860 writes that

'Ireland is a country in which I can never feel at home [and] is looking forward to settling in England in some way'.[37] The opportunity he may have had in mind was the prize of a chair at Oxford, but once again religious divides seemed to be an obstacle.

His desire to avoid confrontation was not always realisable given his strong sense of justice. Relations with the President of Queen's College, Sir Robert Kane, worsened to the extent that Boole eventually avoided informal contact with him. Desmond MacHale chronicles in detail the chronology of their disputes, culminating in a public slanging match in 1856 between them, centred on the diligence of Kane's role as President and that of Vice-President, John Ryall. An uneasy truce arrived but not before Boole had shown that underneath his sensitivity lay a steely sense of rectitude.

In 1850 George had met Mary Everest in Cork while she was visiting her uncle, John Ryall. This began a five-year period of increasing friend-ship. One year later, George noted that she was visiting her uncle again, remarking also that she played the piano quite well. In the summer of 1852 George was staying with Mary's father, Thomas Everest at his vicarage home in Wickwar in Gloucestershire. In January 1853 he returned yet again, five months later meeting again in Cork. George writes to his sister, Mary Anne, that he has been giving Mary maths lessons, praising her 'extraordinary quickness of apprehension and solidity of judgement such indeed as I have never seen surpassed'.[38] They continued to correspond on the subject.

In 1855 George crossed the Irish Sea again to stay with the Everests, continuing to be, as Mary put it, 'kind and fatherly to us younger members of the family',[39] and enthusing over planetary motion, Milton and maths. George was nearly twice Mary's age, emphasised by someone who told her that he was the sort of man you could trust your daughter with. Boole's reservations about 'imprisoning a young girl's life'[40] and warnings from her friends about his overworked brain did not prevent their continuing to come together.

Thomas Everest died in June 1855; George and Mary were married quietly in Gloucestershire three months later, the ceremony attended by her famous uncle, George Everest. A honeymoon followed at the romantic site of Tintern Abbey on the River Wye, immortalised in the Wordsworth poem. Mary remembered the period as a 'sunny dream'. George wrote optimistically of their joint future in a poem of his own: 'The rainbow spans the fields and gilds the shower / And sunset glows once more with golden light'.[41] Their illumined union was tragically not to last overlong.

The pair took up residence in October 1855 near to Queen's College, Cork, moving in 1857 and then 1863 to the suburb of Ballintemple as their growing family required. Their firstborn, Mary Ellen, arrived in June 1856, nine months and a week after their marriage. Margaret, Alicia, Lucy and Ethel were added at two-yearly intervals. During this period we learn little of Mary's life, preoccupied as she was with begetting and rearing children. In spite of her strong character, she was in general willingly adumbrated by her husband. Mary reports that George 'had made a thorough study ... of domestic relations and cultivated family peace as if it were a tender plant'.[42]

Fame, it seems, was of no consequence nor did he seek it for his five daughters. Desmond MacHale in his biography of Boole proposes an image of the absent-minded eccentric professor, so engrossed in his thought that, on one occasion, he failed for an hour to notice his class of students sitting patiently. His affability survived even Mary's burning of his poetry, 'to preserve his brain from needless exertion'.[43]

When not working he enjoyed innocent pleasures with his family: 'childish stories, funny rhymes, tame pretty landscapes, ordinary dance-music ...'.[44] Egalitarian in spirit, he mixed unselfconsciously with neighbours, once inviting a town band home to tea. One old lady remarked, 'I never saw a learned man that one could talk to.'[45] He drew the line, however, at letting his children associate with 'those rich people who make a display of wealth'.[46] 'We have a great deal to be thankful for,' he told Mary, 'friends and books and leisure to profit by them and the enjoyment of natural beauty.'[47] For those not so fortunate he had great compassion.

Mary, still in her early thirties, must have become impatient with her sequestered life. Motherhood alone was not going to satisfy her. George encouraged her to attend his lectures at the college but she desisted when told it was 'quite unmaidenly'. She became useful instead as a sort of lay figure to render his maths formulations more intelligible to a reader. College undergraduates assisted and visited their house, leading to student parties 'of glorious fun'.

In late November 1864 George Boole walked the three miles from their cottage to Queen's College in pouring rain, gave a sodden lecture and returned home with a feverish cold. Although he complained of the damp climate and rheumatism, there seemed no reason for this to trouble him unduly. Yet on 8 December 1864, George Boole, aged only forty-nine, passed away. The death certificate cited pleuro-pneumonia as the cause.

Boole was buried in a modest grave at St Michael's church, Blackrock, Cork. The college, in recognition of the late Professor of Mathematics, raised funds for a set of five stained glass windows to be erected in its Great Hall. Celebrating his academic success in the cause of reason and his own contribution to logic, he is pictured in the central panel earnestly writing, overlooked by the figure of Aristotle and Euclid.

The irony of the complementary Lincoln cathedral window lies in that George had found himself 'Uphill', immortalised in splendour by an institution for which he did not care overmuch. As Mary records, on the occasions when George visited the cathedral with a party of boys it seems it was mostly to hear the musical offering of the Anthem, sitting where the sound would be most majestic. The window does, however, play out the theme of the search for truth in God's world and the need to bear witness to it in one's life.

Although George Boole's egalitarian and radical cast of mind inspired his mathematical impetus, one would not want to claim him as an advocate of a particular political philosophy.

Although happy to work with others in the cause of social reform, he was perhaps too rigorously honest and critical to accept easy formulas. His moral standpoint and the individual integrity of his beliefs are further exemplified in his strange relationship with the theologian F. D. Maurice.

During his fatal illness Mary had brought George a favourite portrait of Maurice and put it next to him on the bed. 'Oh, that is

F. D. Maurice (1805–72).

delightful,' said her husband. What inspired this veneration? It most likely came from a long acquaintanceship with Maurice's ideas through the latter's books and sermons and following his life and career. Six months before his death in 1864, Boole had stayed in London and visited St Peter's church, Vere Street, to sit immediately under the pulpit listening with a sense of awe to Maurice preach. For George, it seems, there existed an adoration for the man, yet his shyness and difference in beliefs had prevented him from meeting him in person.

Ten years older than Boole, Maurice had been born into a radical Unitarian family that shared the same rational faith in a benevolent God directing mankind towards a tolerant and just society. Unlike Boole,

Maurice went to study at Trinity College, Cambridge in 1823. Residing afterwards in London, he became a progressive literary romantic, valuing an idealistic spiritual search for brotherhood and truth to reveal God's 'vast unity of plan, of laws affecting the least and greatest of the creatures who surround him'.[48] This humility before God's universe will lead to the recognition of His laws, typified by examples that George Boole would have recognised: the laws of steam that run a railway or the navigation of a ship by the stars.

For Maurice, the key to understanding God's world lay in God's word made flesh in the life of Christ. He had come to accept the fundamental notion of his suffering and atonement, the Trinity and the ritual practices of the Anglican Church. These were, however, to Boole a matter of idolatry. While recognising God's splendour, it is through doubt and the pursuance of the Unseen God – science – that the workings of His world will be revealed. The one aspect that they did share deeply was the moral obligation, whether indicated by the teachings of Christ or the inferences of science, to ameliorate the social conditions of life. Maurice states, reminiscent of Boole's lecture, *The Claims of Science*, 'God does care for the sanitary condition, for the bodily circumstances of the people of my land and of every land.'[49]

After the scale of the political unrest of 1848, in which Maurice supported Chartism, he and others felt the need to address proletarian needs. He chose to implement the idea of mutual co-operation in the Bloomsbury area of London. Here with Charles Kingsley, of *Water Babies* fame, they set up as Christian Socialists. In July 1848 they opened a night school, holding Bible classes and printing a paper, *Politics for the People*. There was a revolutionary zeal that similarly underlaid George Boole's feelings: 'The world is governed by God, this is the rich man's warning; this is the poor man's comfort ... Liberty, Fraternity, Unity ... are intended for every people under heaven.'[50] Property and excessive competition were decried. The aim was to make Christianity socialist and socialism Christian. Thomas Cooper reappeared on the Bloomsbury scene but moved on.

The social unit was, unsurprisingly, the community and artisans' working associations combining together and distributing profit. It developed first among tailors and then builders, printers, bakers, needle-women and, of course, shoemakers. 'How much nobler it is to make shoes than seek for principles,'[51] declared Maurice. The association gradually failed, however, for want of sufficient profit. Maurice subsequently put his energy into creating a working man's college, launched in 1854.

Boole even at a distance can hardly have failed to appreciate and admire the social commitment to local mutualism that he had himself supported while in Lincoln. What Boole could not agree with was the complete veracity of the Bible and the nature of Christ. It was for this reason that George could not face Maurice in person and create dissension towards someone whose social creed he clearly admired. For Maurice, theology came first. This was at a crucial time when the scientific evolutionary theories of Darwin had been supplemented by the works of David Strauss and Ernest Renan, arguing that the Bible was based on myth and that Christ was a historical figure like any other. His opposition to such a drift influenced him to resign from his post at Vere Street. He was talked out of it, however, enabling George to languish later below his pulpit. Boole would no doubt have been deeply impressed by Maurice's risking of his position for the sake of his beliefs.

Shortly before he died, Boole characteristically acted in like fashion to defend his own principles. On 30 June 1864 he received a letter from Herbert McLeod at the Royal College of Chemistry, London enclosing a copy of a *Declaration to Students of the Natural Sciences*, asking for him, like the others declared, to append his signature. It asserted that 'it is impossible for the Word of God, as written in the book of nature and God's Word written in Holy Scripture to contradict one another …'. At present, it maintained, 'our reason only enables us to see through a glass darkly …'.[52] The authors looked forward to a complete agreement from all those sent the statement.

On 21 July George Boole replied that as a matter of duty he had to decline. The authority of a list of names to influence opinion was 'absolutely wrong', especially if signing allowed the temptation of 'standing well with other men, of pleasing the great and powerful … It would amount to moral injury.' Boole would not have taken exception to the essential idea that the Word of God is 'written in the book of nature', what he objected to was the forcing of scientific men to 'take sides and abandon that ancient tradition of the fitness and the lawfulness of argument alone for the support of Truth, which they have received as a sacred heirloom …'.

When the *Declaration* was published the following May (in 1865) after his death, 717 scientists had signed up. Boole's friend Augustus de Morgan had mounted a campaign against it, calling it a 'vote of censure on free enquiry'. Boole himself stood at the crossroads between faith and science, unable to cast himself adrift from the sense of mystery and wonder at the universe, to which he was happy enough to append the signature of God but unable to ignore the blessing given to mankind to speculate honestly and openly. One hundred and fifty years later he would no doubt be astonished that in some fundamentalist 'creationist' circles we are still seeing the substance of the *Declaration* through a glass darkly.

George Boole's desire to facilitate open communication using a binary system of algebra lay dormant for nearly a century. Not until Claude Shannon in the 1930s in the US was looking for an efficient means of transmitting information did it acquire practical use. He noticed an essential similarity with telephone switching circuits that mimic logical functions. All computers that have since developed rely on the use of one and zero and the rules of logic set out by Boole. Thomas Cooper was right when he foresaw that his kin by marriage would have a deep effect on the future.

One wonders what George would have made of the modern world that he helped spawn. The ability to reason clearly and pursue truth that he valued so highly depends on access to 'objective' information concerning the world around us. The internet revolution has brought more of it than we know what to do with. In the end it is probably unreasoning faith that informs the moral problems that still beset us, even more than in Boole's day.

Boole and Hinton descendants at George Boole's grave, Cork 2015, during the 200th celebration of his birth. Marni Rosner is to the right behind the headstone.

# THE MISSUS

Probably none of George Boole's five daughters ever visited the small market town in which their mother, Mary Everest, had been raised, or the church in which she and George were married by her father, the vicar, Thomas Roupell Everest. I equally would have had no impetus to travel to Wickwar in Gloucestershire but for my own distant connection with these Victorians. I do, however, enjoy putting flesh on the rattling bones of biography. In July 2010 I made my way to the village some twenty miles south of the cathedral city of Gloucester.

The place has a slightly French feel atop a hill outlying from the imposing Cotswold ridge to the east. The wide High Street, as at Lincoln, served as a market place, although the market hall has now disappeared. The wool trade by the end of the eighteenth century was on the wane. The parish church of the Holy Trinity, a tall-towered, dignified building of yellow stone, is sited, unusually, not close by but to the north, reached by a raised path over what once was a pond. It has stood, nevertheless,

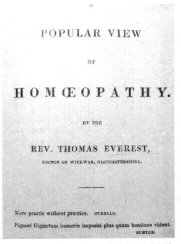

Thomas R. Everest (1801–1855), title page of *A Popular View of Homeopathy*.

Wickwar Church, Gloucestershire.

Mary Everest Boole (1832–1916).

since the twelfth century to justify the order of peasants, artisans, burghers and yeomen farmers, all under the patronage of the Earl of Ducie.

The church's separation was appropriate perhaps to the independent thinking of their vicar, Thomas Everest, despite his inclination to want to be at the heart of parish life. His name is now engraved among a list of other incumbents on a sloping stone sill beneath the large east window. As a token of a small moment in the unfolding church's history I added my own name to the visitors' book. Preceding me was someone who styled himself as 'Smurfpiss', another who wrote, 'Thanks for a moment of pease', and another asking for prayers for their loved ones and friends 'in Afghanistan at the moment'.

In 1832 at the time of the birth of Everest's only daughter, Mary, England was beginning its transformation from a relative feudalism towards a modern democracy, symbolised by the Reform Act enfranchising the new industrial middle classes. Thomas would have had the vote by virtue of his Cambridge education, the benefits of which he was keen to pass on to his flock. England had been enjoying 'pease' since 1815 with the defeat of the upstart Napoleon and would continue to do so until the Crimean War of the 1850s. What disturbed his sleep most was his chronic asthma, for which he had taken up the revolutionary practice of homeopathy.

This is not surprising given his general free-thinking disposition, 'his passion for liberty and individuality in all matters of spirituality and intellect',[1] as Mary described it, characteristics that would have bonded him with George Boole. He promoted a sense of equality with and service to his congregation. Why shouldn't the church stay painted pink inside if they liked it so, and the church band be kept instead of a modern organ? Why shouldn't country girls curl their hair like their betters? Thomas had no great love for the seemingly everlasting landed gentry. Mary reports him, somewhat oddly, as declaring, 'we are no nearer … to getting rid of aristocrats … than of roses and symphonies'.[2] Blushed pink like his church walls, he preached that his duty was to provide 'an object lesson in Christian Socialism'.[3] This same tinge coloured the foundation of Mary's own thought.

Thomas Everest's homeopathic interest arose as 'a learned occultist',[4] aware of such practices as mesmerism and clairvoyance. Mary doesn't tell us how he came to know Samuel Hahnemann, the founder of the pseudo-science, but her father's book, A Popular View of Homeopathy, published in 1834, was well ahead of its time. It was not popular with medical hierarchy. How is it, Thomas asks in a letter of 1851, that we can have come to understand by scientific observation a part of God's 'adorable order and unswerving harmony',[5] as wild as a Caribbean hurricane, and yet have no laws for the cure of disease? A one-shilling book of the time seems to bear him out: Sir Astley Cooper's treatment for cancer merited salt water and senna pods; paralysis a whipping with stinging nettles. The use of bleeding was based on medieval ideas of humours in the blood.

Hahnemann's experiments to discover medicines that in echoing the symptoms of a disease would effect cure, must have seemed rigorous in comparison. By exciting fever with Peruvian bark (quinine) an underlying fever can be resisted, much as pruning a tree incites growth. How infinitesimal quantities could cause major reactions is as much a mystery now as it was then. Everest talks of 'molecular friction'. He was not afraid to practise on himself. Writing in 1834, 'I am at the moment suffering under the action of two globules *Dulcamara* 11 … it is beautiful and well worth it … to watch the characteristic symptoms of a medicament.'[6] Thomas took his family off to France in 1837 to study with Hahnemann, then living in Paris. Mary was aged five.

Their stay at the draughty Chateau de l'Abbaye at Poissy near Paris was highly formative for her later intellectual ideas. Bats flew up from the uninhabited ground floor allowing Thomas to demonstrate the marvel of their flight, able to navigate the strings he had rigged across the stairs. The bitter cold in winter was only kept at bay by the two children running up and down corridors. To add to this they were put to an unyielding regime to build resistance by walks before breakfast and cold baths, the ice on which Nurse broke with a stick. Mary's response to the treatment, perhaps unsurprisingly, took the form of almost cataleptic depression alternating with fierce explosions of anger. The globules of *Ignatia* for grief and *Sambucus* for quarrelsomeness that her father administered can't have been of much help. Fortunately in the summer the shrubberies of the overgrown garden presented a delightful wilderness for a child to become lost in. More importantly, Mary read avidly anything allowed by her father and became warmly immersed in her education and strongly attached to her teacher, Monsieur Deplace.

After a culturally awkward period in a Catholic school, Thomas's almost rabid hatred of its teachings led to private tuition in French and arithmetic with the Frenchman, who seemed to awake excitement in Mary in more ways than one. Although she doesn't remember anything resembling a 'caress or an affectionate word',[7] he clearly aroused in her a hero worship through simply appreciating and understanding her. She recalls the keen pleasure of giving her tutor a red rose from the garden. That she still remembered him and sought him out again twenty-five years later suggests the profound effect that he must have had on her after, as she puts it, 'the dawning of the physical passions'.[8]

Coupled with this crush surfaced an abiding lifelong intellectual pleasure with mathematics, made accessible by Monsieur Deplace's method of teaching: asking questions and eliciting answers rather than preaching. This approach stayed as the basis for her subsequent theorising about learning. Being 'alone in peace by a smoky fire' with her master led her to the revelation, 'I know nothing about God except that he made Algebra ... sufficient reason for loving Him with all my soul and heart and doing ... whatever I thought He wished.'[9] God was an ally and an equal, 'when He said six times seven is 42, He was not afraid to allow me ... to ask, "Then what is seven times six?"'[10]

Mary's precocious worship of God's algebra was, however, to be overshadowed by 'the blackest horror of all my life'.[11] Overhearing her father talk of Cambridge University, she learnt that, as a woman, it was a holy place to which she would be denied entry. God's calling was to be stifled by the arbitrary rule of men. Her logical and enquiring mind was made emotionally turbulent by the contraries she experienced during her years in France. She writes of 'the antithesis and antagonism our lives were spent in ... homeopathy against allopathy, French against English, Protestant against Catholic'.[12] Add to this the chasm of men stunting women and it is easy to understand why she felt a need to make sense of the opposites of self-realisation and prohibition, joy and pain. It led her to proselytise a universal rule of life, augmented by her late husband's ideas, that became her mission to impart to others.

George Everest, (1790–1866), Mary Everest Boole's uncle, surveyor of Mount Everest.

In the spring of 1843 the Everest family returned to England to resume parochial duties at Wickwar. In the same year another of the important male influences in Mary's life – her uncle George Everest – returned from his own long stay abroad in India. Born in 1790, his love of adventure had taken him there as a sixteen-year-old cadet in the Bengal Artillery. With his engineering skills he became assistant to William Lambton undertaking the Great Trigono-metrical Survey. This vast enterprise to map the whole of India with two geodetic arcs from east–west and north–south to Nepal required not only mathematical skills but derring-do and fortitude of the most extreme kind. The work involved lugging heavy equipment, siting and building observation towers from which to triangulate co-ordinates.

George told her about his exploits jumping into a cactus hedge to escape wild cattle and having to leap off a precipice to hang on a cliff-side tree. He not only learnt how to cope with tigers, swamps and tropical weather but developed a great respect for Indian culture and religion that Mary later incorporated into her worldview. The greatest bar to his work was illness: fever, convulsions and abscesses took him back to England to recover in 1825. Five years later, having returned to India, now as Surveyor General, he was at death's door again and put to the cures his brother so despised. On one occasion 1,000 leeches and thirty cupping-glasses were deployed. Mercury tablets were swallowed in doses proportionately, to Thomas, as big as the mountain that was later in 1865 named Everest in George's honour. His qualities of individuality, free-thinking and tolerance matched the instincts of his brother living more delicately in the foothills of the Cotswolds.

Mary recalled later in life the parishioners' respect for her father, who as one put it, would 'run all down street arter I, to give I his arm in the frost or hold his umbrella over I in the rain'.[13] He enlisted Mary's support as a teacher at Sunday school, while she continued her own education in mathematics with the aid of his books. The villagers held her in esteem too: when Mary offered payment for sewing parts of her trousseau they refused to take any money. The garment in question arose from her own journeying afield. As we have seen, in 1850, aged eighteen, she had crossed the Irish Sea to Cork to visit her mother's brother, Dr John Ryall, Professor of Greek at the newly formed Queen's College. It was here that she met George Boole. What must have attracted her to him would have been, in part, his traits similar to those of her admired father: an enjoyment of the simple pleasures of life, a lack of religious dogma and a respect for and equality with others.

Mary's own personality shared some of these traits. She was open to a search for truth but intolerant of others who did not share her discoveries. Her cantankerousness was perhaps driven by a sense of the injustices of the male world. Her forthrightness, plain speaking and writing could often alienate others. It was most likely her intransigence that was responsible for her husband's death.

In line with her acquired homeopathic views, George's shivering from cold after his drenched walk home would best be treated, 'like with like', by wrapping him in wet sheets. Had the infinitesimal dose been part of the prescription she could have as well assured his life by waving a moistened handkerchief from the bottom of the garden. George's fever was in any case a state that Mary believed could 'produce a season of wonderful intellectual and moral progress'.[14] Ethel Boole, then only six months old, was brought to her father's bedside. As in some Victorian melodrama, 'My little angel; she is a vision', were almost George's last words. Ninety years later it was she who wrote to her nephew Geoffrey Taylor, 'The cause of father's early death was believed to have been the Missus' belief in a certain crank doctor who advocated cold water cures for everything. The Everests do seem to have been a family of cranks and followers of cranks.'[15]

There was another more positive side to the blinkered vision that hovered over George Boole's deathbed. It was her intuition that resolved his anguish over the rift between himself and F. D. Maurice. It happened while they had both been considering the biblical passage from John 3:8 that reads, 'The wind bloweth where it listeth, and thou hearest the sound thereof, but cannot tell whence it goeth: so is every one that is born of the Spirit.' It occurred to George that Jesus was referring to the 'geometric figure of the dust-whirl ... with its system of tangents and normals'. This suggested to him that 'the historic Jesus must have been at least a mathematical psychologist of great brilliancy and power' – someone he could regard favourably without the associated dogma.[16]

Mary took this even further, suggesting that in a whirlwind, spiral forces are blowing contrary to each other – but the still centre moves forwards. Opposition does not necessarily deny progression. The insight may have sprung from her father's own mentioning of the hurricane. George, encouraged by both their accounts, wrote to Maurice as a result, asking him to come and stay. It was not to be; according to Mary, 'he went to bed very tired, was taken ill in the night, and never rose again'.[17]

The spiral of Jesus, as sanctioned by her husband and secularised to 'Boole's Law', was the 'clue' that persuaded her to devote her life in his memory to empower logical thinking. *The Laws of Thought*, George's legacy, was meant to throw light on the nature of the human mind. The formula of the spiral expressed the idea that to find unity in God-given truth (however 'God' is defined), one should utilise the natural power in the mind of man to 'rhythmically' pose opposites and extremes in relation to propositions, searching for a synthesis at a higher resolution. It adds up to the literal 'divination' of scientific method by falsifying hypotheses, or in daily life a kind of rigorous empathy. There may ultimately be one possible truth but there are many interpretations in uncovering it. Blake expressed the underlying idea too in his *Proverbs*. 'Without contraries is no progression.' 'The road of excess leads to the palace of wisdom.' Mary had sanctified a law of thought that George had expressed in algebra.

Sounding rather like a washing machine trademark, this mode of thinking became labelled by her as 'pulsation'. F. W. Daniels, one of Mary's disciples, asked how many of us employ such a thinking tool when disputing the spheres of religion, theology and politics. 'What pacifist,' for example, 'could bring himself to listen sympathetically to the best that can be said for militarism and vice versa?'[18]

Charged with a mission in life, meanwhile, Mary Everest Boole, at sixes and sevens, was left with five daughters aged from eight years to younger than one. Some of George's friends rallied to her support and secured a Civil List pension of £100 per annum. Despite relatives in Ireland, she seems to have had no desire to remain there. Hoping for employment in education, she took up an offer, made appropriately by F. D. Maurice, to become librarian for £60 a year at another Queen's College – in Harley Street, London.

Maurice was one of its founders in May 1848. Its object was to provide education for governesses: the only respectable occupation open to a widow or unmarried woman without private means. Classes in basic subjects such as maths, languages, geography and history were given by men, chaperoned by Lady Visitors. The standard of teaching was high and far removed from the rote learning typical of the time.

Mary saw her role, apart from earning a living for herself and her five daughters, as an opportunity to attend to the moral problems of schoolgirls. In an era of Darwin and growing secularism she ideologically sought to reconcile the value of science and mathematics with her basically pantheistic religious belief. Such a quest had been instilled in her by the Everests, her husband and Monsieur Deplace. Maurice, wedded to a mystical gloss to his religion, although committed to a notion of social justice, could not make the leap.[19] The question he put to Mary, 'Were you never afraid of God?' meant no more to her than whether Christ was divine. She wrote that in discussion, Maurice's shy sincerity towards her produced 'exquisite tolerance and affability', but essentially, 'we were like a bird and a fish talking together'.[20] This separation at least allowed her for a few years to create her own sphere of influence in the college.

Mary was appointed Tutor in Arithmetic and Geometry from 1871 to 1874 but had also over the years held informal 'true logic' (i.e. 'Boolean' logic) classes for resident pupils on a Sunday evening, designed to encourage the 'free play of their minds … free expansion and unification'.[21] She no doubt also dabbled in their 'moral problems'. Attending these meetings on occasion, probably to the disquiet of the management, was James Hinton – an aural surgeon with a passion for polymathic thinking like herself. (We shall meet him soon in detail.) The two were introduced in the late 1860s although the Boole family's acquaintanceship goes back to her father, who had persuaded Hinton to experiment with homeopathy. The connection between Mary and James must have been profound; their ideas indeed interpenetrated. Both were looking for frameworks to uncover the secrets of God's world of nature and His plans for us. Mary had already formulated her key idea.

Where might we look for examples of the divine pulsation at work? There are indications available to us in the form of 'notations'. As metaphors they allow us to see 'spiritual law revealed in physical fact'.[22] Mathematics, as she intuited as a child, is a pure way to reveal God's truths, but there are other sources. Besides His holy word as revealed in the Bible are His holy works in nature. One of the purest examples that combines both is the rainbow, such as the one that had entranced both George and her at Tintern. As an essential phenomenon it gives 'A temporary explosion of colours which soon fades into the Unity of white light, leaving No-Thing behind …'.[23] What attracts her is the way the rainbow appears after the stress of a storm, entrances us aesthetically, gives us a lesson in the science of colour and disappears, leaving no trace to be owned and idolised. It is an 'all' or 'nothing': a visual representation of the binary one and zero that lay at the basis of Boole's algebra. This may be poetry, but of an insightful kind.

Other notations she suggested are more eccentric and personal. The old religion of Cornwall (where she took her children, generating a lifelong passion in Ethel) evoked another notation: the logan stone. Its delicate rocking balance and poise between opposing forces would have been revered, she believed, by ancient worshippers seeking truth. Similarly, an ancient healer holding the stem of a forked branch cut from a sacred tree could demonstrate polarity. Its reversal holding the fork, provides a unifying tool enabling water to be divined. The healer would later burn the wand to avoid idolatry and fixed thinking, showing that its powers came not from him but the Unseen Father. To a twenty-first-century cynic this may all seem to be mystical mumbo jumbo, but Mary at least, in true Boolean fashion, put the hypothesis to test by dowsing in Kent using wands from various species of tree. 'Some were like the horns of beetles, oxen or stags; several would serve for Phallus symbols ... one ... bore the most extraordinary resemblance to the hind-quarters of an ape in a very nasty attitude.'[24]

Lest all this seems merely quaint, it is perhaps worth remembering that the desire to search for signs of order and beauty in a disordered, unstable world is a constant historical quest. It dominates today in a materialist world where science reigns supreme. Spiritual 'messages', however, are still found in nature's toys such as crystals and pendulums. The countryside is tramped over searching for geometric revelations in ley lines, zodiacs and crop circles. Many of these searches for succour and certainty bolster their efficacy with 'scientific' claims of proof often employing as yet unknown 'energies' and 'forces'. Human personality, astrologers tell us, is formed by the action of planetary rays on the head of a baby emerging from the womb.

If the above point of view seems a little excessive, might I suggest that if you had spent several months wading through the dense, 1,500 pages of Mary Everest Boole's *Collected Works*, you might feel forgiving. She at least understands a basic rule of science. Certainty can only come, she maintains, when questions are answered upside down and inside out with what Thomas Huxley referred to as 'methodical doubt'.

Mary attached great importance to the idea of genius. She no doubt saw her husband as one, alongside Luther, Cromwell and Wesley. The genius individual is not just a visionary poet but a pulsating thinker of power with a mission to 'always bring freedom ... to break ... the fetters of mere mechanical conventionality'.[25] He (rarely a she) appears to be isolated and often mixes with the lower orders to learn the ways of the world, as Christ did. The Prophet is ahead of his time, misunderstood and often scapegoated, put into a madhouse or killed. As a member of a priesthood he looks to explain God's world by experimentation, observing heavenly bodies and erecting sacred stones or planting a grove of trees.

These tools of enlightenment, however, come to be regarded by the masses as symbols to be venerated in themselves thus impeding fluid growth. Eventually these will be overturned as, 'Jesus, High priest of pulsation offered himself to death rather than support any ... fixed doctrine; yet having been slain by one set of idolaters he was made into the object of sensational worship by another ... .'[26] Viewed as a crude sociology of religion we're not so far from Marx's notion of religious alienation. For Mary, God's loving world will be realised through the power of his agents of genius moving men towards liberation. There is one collective agency, however, that has a special role. True to form she ventures beyond the stereotypes that ran deep in conventional society.

If the Bible is to be viewed as one of God's aids to understanding His purpose, it is to the Pentateuch, the first five chapters, that we must look. It is there, for example, that the rainbow appears symbolising God's covenant with Noah and mankind that after the Flood there will be harmony. Its inability to be idolised is part of the inclusiveness of Mosaic Law with its 'hygienic and ethical discipline', that makes Jews part of a 'hereditary nation of priests'.[27] Adherents to its essential beliefs, by being misunderstood, 'act as a solvent in world history'.[28] Not especially given to sensuality or the need to convert others, Judaism has produced intellectual individuals of genius (Marx, Freud and Einstein might serve as examples). George Boole too was drawn towards the monotheistic unity of Judaism.

Mary as usual puts her beliefs to the test. At Queen's College where anti-semitism was not unknown, she and the Lady Superintendent took pupil Jewesses aside and talked up their inheritance. Their resultant 'delightful air of calm repose'[29] had a discernible effect on the school. One of the pupils was later to write that Mary had given them power to think for themselves.

She was not dogmatic enough, however, to fail to recognise the gentile genius of Charles Darwin. She extols his work with fulsome adjectives, 'beautiful ... refined ... delicate and hopeful',[30] somewhat surprising for a theory that rocked authorised views of God's world to its foundations. Not so for Mary – with a sprightly hop, skip and jump in thought so typical of her, Darwin's message demonstrates that, as the genius Nazarene carpenter put it, the meek shall inherit the earth. A reference not to the weak surviving against evolutionary odds but to Darwin's own undogmatic spirit remaining an agnostic in relation to the Creator.

In a letter to him in December 1866 she asks hopefully that if he could allow that the Spirit of God influenced the brain of man as well as the forces of natural selection, there might be a missing link between religion and science. Darwin replied, in meekness, that his opinion was not worth much and that he was not responsible if the reconciliation of the two realms of thought should be far off. In any case Mary neatly accepted Darwin's message, that the apparent misfortunes of one species demon-strates the way in which 'the Creator develops its power and perfects its type'. It was the purposive 'unseen chisel', after all, 'that gave the fox his cleverness and the bee her geometric instincts'.[31]

Although her work as librarian at Queen's College had been applauded, her intervention in school affairs and attic free-thinking sessions must have irked its hierarchy. Her natural spirit of opposition began to chafe its symbolic apex in the person of F. D. Maurice, despite his liberal disposition. He expressed disquiet over a proposed biography of her husband and disapproved of a series of lectures that she gave to groups of churchwomen, eventually published in 1883 as *The Message of Psychic Science to Mothers and Nurses*. In the book she dips in and out of a medley of current scientific exploration to help carers in education and health put her grander theorising into practice. One portion of God's world in nature, for example, is the force of magnetism and animal magnetism.

Franz Mesmer, a German physician (1734–1815), sought at first to relate individual health to astrological influences on a magnetic fluid that was part of the human constitution. He claimed that by his personal contact, or the use of magnetised water, blockages in its flow could be overcome, effecting cures for various ailments. A French Royal Commission that included Benjamin Franklin and Joseph-Ignace Guillotin observed cures but suggested the power of auto-suggestion. The idea lingered, however, that through mental power and magnetic force people's thoughts could be influenced. Mary writes that 'every mother unconsciously mesmerises her children, every nurse her patient, every teacher his pupils as the nerves receive and transmit the life force'.[32] Children possess a direct way of thought-reading through magnetism until around the age of six when there is a mental closure and external modes of reception take over. For some, especially among the early Hebrews, this process is retarded, giving rise to the exceptional magnetic powers of Jesus. Such influence, however, has to be used carefully, as employed among a religious crowd it can lead to epidemics of insanity and crime.

Ever concerned with more specific aid for the ill, Mary counsels that non-pulsating religious bigotry should be left outside the sickroom door: the patient should be treated with ordinary pleasures. In the schoolroom gifted children may fall prey to the over-magnetisation of a teacher. The temperamental child needs a correcting pulsating influence, perhaps when she 'catches as in a vision a glimpse of something gloriously lovely in the wonders of creation'.[33] When grown up, this type is a 'favourite with clever men' but not much liked by the ladies of her family and suffers 'rare but violent paroxysms of temper usually connected with perceived injustices. She may become a pessimist writer or a pantheist poet.'[34] One wonders if Mary is here presenting aspects of her own life or has anticipated some essentials of her youngest child, Ethel.

Phrenology was another aspiring but doomed branch of the emerging science of human behaviour. With the aid of callipers and reading bumps on the head, individual personality could be assessed. It enjoyed a vogue throughout the nineteenth century and was used variously, as a diagnosis for finding partners and employees for example. Mary is most interested in that part of the brain designated 'moral and religious', distinguished by its capacity of easy magnetisation and voluntary self-renunciation of pleasures. Its capacity to energise the whole system, however, lays it open to abuse, such as the dogmatist religious practitioner who visits the sick and encourages a fuss about their souls rather than immerse them in 'the common business of life'.[35]

Mary's tenure at Queen's College was somewhat mysteriously terminated. Her outspokenness and eccentricity must have irritated her superiors. Elaine Kaye in her official history of the college noted that Mary was 'eventually regarded with some suspicion by the College staff'. Her 'unstable character' and 'dangerous ideas' led to the termination of her lease.[36] There is also a hint in the biography of her grandson, Geoffrey Taylor, that Mary had an illness in 1874 which later had 'assumed the form of temporary derangement'.[37] I decided to consult Queen's College archives for myself. Strangely, one other enquirer about Mary Boole and the college in the 1980s noted that the librarian at the time had clammed up completely regarding any disquiet a century before. There was also the collateral need to search records to find out about her children's attendance and education there.

Unlike many London city-centre educational institutions, Queen's College has not decamped to the suburbs and still occupies its row of Georgian terraced houses in Harley Street. Ex-alumni over the years have included Gertrude Bell, Kathleen Kennedy and Christina Onassis. Permission was kindly given to me to rummage around its archives at the top of one of the houses where Mary Boole and her five daughters had lived from 1865. Up aloft in the two tiny cramped rooms I found logged in the round-shouldered volume of *Council Minutes*, 10 March 1873, details that Mrs Boole intended to retire from her tenancy at the end of the summer. The *Council of Education Minutes* recorded 15 February 1875 that she had been incapacitated '... as Lady Teacher in mathematics since October 1874 and been given leave of absence'. A month later the Council Secretary, establishing Geoffrey Taylor's assertion, indeed recorded her 'temporary derangement', and that letters 'received from her led him to think that she was in too excitable a state to return to the college without risk'.

This view had been confirmed by her medical advisor – one James Hinton – who 'had begged that no further action should be taken for fear of bringing on a relapse'. I searched diligently for evidence of her 'excitable' letters but found nothing. I also found myself wondering whether the meeting of minds between Mary and Hinton might have led to physical minglings. Some accounts indicate that she became his secretary, suggesting possibly that she moved into Hinton's household after leaving Queen's College. Mary was certainly broad-minded enough to entertain risks.

Certainly she was not bashful about sex in her theorising. Like Blake, she believed that 'God is in Nature and whatever is natural is holy'. Sexuality is sacred: it creates 'a power of sympathy and a development of … altruism'.[38] From her pantheon of names for the Almighty she selects the libidinous Jewish Adonai, Creator-Inspirer and Pan, the 'authoriser of all freedom and condoner of all licence',[39] who descends in Boolean fashion 'at the moment of contact between differentiated polars …'.[40] This is a rather prosaic way of describing sexual congress but shows that she is no prude. Despite her full-frontal mention of the pulsating dowsing rod, she writes reservedly about sexual nitty-gritty: 'implements of racial preservation, organs of generation';[41] although the 'pole of his nerve-battery'[42] has a little more zing. Sex is sanctioned as long as it is an honest creative impulse to propagate the species.

It is not a matter, however, for jesting – nor for moralising by religious bigots who regard it as evil and preach abstinence. She compares their meddling to creating a train crash but admits that 'sometimes the very physical nervous structure of the man and his posterity is wrecked and sets up for itself a variety of anomalous sensations, desires and lusts such as no wild beast can conceive of'.[43] She adds that anyone 'who is going to try extra-legal experiments should not entangle … any woman *who would not have associated herself with him had she known it*'.[44] She also has, of course, down-to-earth advice for superintending mothers. Boys should learn that their organs are most delicate and not to be experimented with until grown up. Order and sequence is sacred; early meddling is rejected.

Mary has views on prostitution – as on almost everything. For her it is a matter of social policy. She shared the later stance of W. T. Stead, the journalist, that subjecting women to physical examination is not the answer to men's immoral needs. Writing in 1893 about 'authorised' prostitution in the Indian army she calls for restraint, duty and care of mutual welfare.

Mary emerged from the troubles at Queen's College to continue her mission to advance the more important ideas of her husband, George Boole. One in particular had become her over-riding interest: education. Well ahead of her time, she had developed a theory of the unconscious and its relation to learning.

If A is the conscious and B the unconscious, according to Boole's law, the pulsation between these two provides an individual with an accession of mental force C when he returns to the thought of unity after attention to contrast. Rational reasoning can be enhanced by thought processes just out of reach in daily life. The still centre of inspiration comes from many states of mind: religious meditation, conditions produced by morphia, chloroform, alcohol and other drugs, hypnotism, coma, trance, clairvoyance, automatic writing, delirium, catalepsy and epilepsy. The one esoteric area she leaves out is spiritualism. Although she intimates that she has been acquainted with séances, it seems unlikely that she would have been a fully paid-up member. Her use of the word and 'mediumship' equates with the generalised psychic power of magnetism and thought transference, less with the hereafter. Since the first decade of the twentieth century when she suggested the above list, some states remain hallowed, some have been proscribed, some exposed as chicanery or seen as evidence of higher states of consciousness. The lower-level notion of productive day-dreaming or empowering the liminal mind while driving a car or performing monotonous activity is well known.

An over-dwelling in C, however, can lead, she warns, to a spiral of deviance. The individual cannot distinguish between 'lemonade and brandy, an innocent sweetmeat … and some nauseous drug', and rather more seriously, 'the sensuous sting of a self-flagellation and the sensuous sting of a fleshly lust, between his own wife and other's men's wives … other men's sisters and his own, between a woman and a child, or a man and a beast'.[45] Asceticism however is not necessarily an answer, 'the most pious saints … who habitually mortified the flesh, have been amongst those to whom Pan appeared in his vilest forms'. Such influence in society at large may produce 'some ghastly tragedy'.

The vital force of imagination must be put to sound pedagogical use. Mary was not completely averse to the normal practice of rote or habitual learning at that time; grammar would be acquired by laying down rules. The classroom, however, should be a place for expanding minds, both those of the pupils and teacher. In science, in particular, right conclusions will emerge from the disorderliness of the circular storm focusing together the 'rays of Truth'. Experimentation and even the temporary pain of failure should inspire learning if it is given time to accumulate in experience. Early education, in her estimation, should be 'hands-on' and leisurely, allowing the conscious mind to alternate with periods of 'going slack' to allow gestation and inspiration. The Bible, she asserts, tells us that the purpose of the Sabbath was precisely this, to allow a calm in the storm to generate synthesis.

Parents too or teachers out of school can add to formal lessons by playful stimulation 'out of bounds' with familiar objects. Nails and magnets will demonstrate polar attractions; sealing wax and a flannel: static electricity; bath water: hydrostatics; the spinning top, swing and the sling: tangential motion. In the kitchen the natural world can be encountered by growing cress, watching gnats' eggs in a bowl of water, tending gardens and caring for animals (if not caged or over-indulged). From the time when an infant begins to stroke a cat, she asserts, the child should have access to geometric solids or ornaments as toys. Back in the kitchen, the sharing of a cake usefully invokes fractions. 'A child should not see a multiplication table till he has made one,' she declares.[46] Out in the wild lies the possibility of collecting seaweed on the beach, shells become money.

Early education should excite children's wonder. It is somewhat ironic that her grandson, Geoffrey Taylor, who professed little interest in her writings, was the exemplar of the scientist 'playing' with ordinary objects to advance his own ideas. Taylor loved to be with children engaging in direct exploration. It was her own experimentation when young that led her to develop curve stitching, consisting of a card with holes punched around a circle through which threads can be drawn allowing geometric shapes such as parabolas to emerge. These sewing cards were later sold commercially and put to the test in 1904 on children under ten at Bedales School in Hampshire. After a brief training in their use the pupils were left to express themselves by free association. Mary loftily claimed that if the principle were extended to subordinated women such as 'servants, factory hands and shop assistants … to express their own freaks of fancy freely …', it would provide a means of a truly national evocation of creative and organising power such as inspired the women needle-workers of ancient India.[47]

Although the essence of her educational ideas now seem no more than common sense, they were strongly at variance with the 'idle hands make devil's work' ethos of the Victorians. 'Slacking' was not a word that found favour. Only in middle-class enlightened families could her ideas be realised. For the rest, elementary education was limited by poverty, child labour and the religious dogmatism Mary loathed. Not until the 1870 Education Act was a beginning made to systematise secular state provision. Other theorists such as Montessori came later to stress the importance of self-directed education, but Mary Boole was the first to develop a link to the importance of unconscious learning. Moreover, the child-like receptivity of the unconscious should not be allowed to disappear but in later life harnessed for individual and social good. The fairyland that children often inhabit is the same domain that inspires a grown-up sort of fairyland, 'the world of scientific hypothesis ... that Unseen World ... No mortal has the right to deny the truth of what another sees there.'[48] The formal education that shuts it off, dedicated only to competitive exams and commerce, only prepares pupils who leave school 'with brains mutilated like the feet of Chinese ladies'.[49]

It is very difficult to get a 'human' picture of Mary Everest Boole, this tireless person ever diagnosing, analysing and proselytising. *The Collected Works* are exhausting; reflective but never in repose. One imagines an ardent, mercurial nature, worrying life like a dog with a bone, hardly settling to relax with something unworthy and un-advancing like a novel or going to a concert. Her living room would not be graced with artworks, one suspects.

There were always more causes to rally to. Another, as we shall see, was a life-changing matter for Ethel, the fifth daughter. Mary became interested in the Russian tsarist exiles having corresponded with their leader, Sergei Stepniak, in 1885 asking to meet and offering help. She was impressed by what she regarded as the mutualism and liberal principles of the Russian community in London. Their bravery, atheistic high intellect and dedication she contrasts with the idolatry of those who speak of spirit, faith and future life.

In the 1880s she was attending a London asylum with Henry Maudsley, the pioneering psychiatrist, to study mental illness and lecturing still on the forms of mental aberration in girls' schools. She was also trying to help teachers in the US by correspondence. Throughout the 1880s and 1890s Mary wrote many articles for the Jewish press, various secular journals, and delivered lectures at meetings of the Christo-Theosophical Society in Bloomsbury Square (held appropriately on alternate Tuesdays). One of her *bête noirs* was vivisection. This she saw as 'the eruption ... of a disease ... infecting all the intellectual and educational life of the world',[50] in that it confounded pulsation by emphasising dissection, i.e. analysis at the expense of synthesis. Should it become a habitual mode of thinking it would deny the all-important elasticity of mental processes. Vivisection did not take the position of the suffering animal into account.

About 1903 a group calling themselves the Cranks met in a London vegetarian restaurant. Their eponymous magazine published her writings. One is not be surprised to learn that she tended towards a diet of vegetables, fruits, nuts, pulses and grains. From an address in Bryanston Square not far from Harley Street she had moved westwards in the early 1890s to 16 Ladbroke Grove, in Notting Hill. Lucy was the sole daughter living with her by this time, occupying the first floor. Mary was ensconced in the back parlour of the ground floor separated by a curtain from the sitting room, resplendent with a case of ferns. Ministering to Mrs Boole was her housekeeper, Agnes Musk, known, after Dickens, as 'Barkis' who had been with her since 1886 at least.

According to E. M. Cobham, one of Mary's devoted adherents, Barkis would join her sitting in the small back garden under the Virginia creeper to feed the sparrows who, ungratefully, would in spring eat the primroses and crocuses. At that season Agnes would be left with vacant possession to clean the whole house while the Missus went visiting in the countryside. Mary Boole's small abode must have been crowded on occasion. 'Thither

Above: Mary Everest Boole in old age.

Left: Mary Everest Boole and 'Barkis' her housekeeper in London.

came people from many quarters each with his or her question to ask,' writes Cobham.[51] One visitor would probably have been H. G. Wells, whose novel *The New Machiavelli* of 1911 referred to a Mrs Boole. The disparaging likeness, however, it was said, was put in to annoy her.[52]

One of Wells' lovers, Dorothy Richardson may also have been another. Elaine Showalter quotes Richardson, that 'the feminine mind is capable of being all over the place and in all camps at once'.[53] Fortunately perhaps, its higher power of intuition and harmony affords shelter. Richardson allowed her own discursiveness full reign with her ground-breaking stream-of-consciousness novels. Mary Boole, who wrote with 'cabalistic intensity', is invoked in one as the unprecedented advocate of the psychology of polar opposites. 'In fifty years' time her books will be as clear as daylight,'[54] says a proponent.

Cobham details some of Mary's other diverse attendeés at the soirées including members of higher echelons.[55] Dalliance with the aristocracy may seem out of character for Mary but her singlemindedness made contact wherever there was a convert to be had. A revealing picture emerges from meetings in the 1880s and correspondence kept by Victoria, Lady Welby (1837–1912), an almost exact contemporary. Like Mary, Victoria did not receive a formal education but as godchild and maid of honour to Queen Victoria she lived very much on the other side of the tracks. Welby's strong intellectual curiosity made them allies, however, both concerned with reconciling religion and science. Mary seems almost to have had a schoolgirl crush on Victoria. She writes of her as 'one of the people whom one does love … the longing to be friends with you and … give you a few lessons in the art of thought integration, although I tremble at touching so exquisite an instrument'.[56] At the same time in her characteristic, often blunt, style she states, 'I do not like fine ladies one bit.'[57] Despite the mission to enlighten Welby, it falters over Mary's strong beliefs: the humanness of Christ, her anti-trinitarianism and the fringe dabbling with spiritualism – Mary had been in touch with the Lady's spirit-guides. In one letter she makes a rare reference to reincarnation.

Mary cannot resist lecturing from her sturdy platform: 'just get that clearly into your head; you and I have a lot of work before us'.[58] Their mutual friend, Julia Wedgwood, far more refined and poetic, declares of Mary that she is 'as noble a being as I have ever known though streaked with … extraordinary arrogance'.[59] The noble being at times knows this too, writing, 'I am a bear,' and rather sorrowfully, 'I can't think what makes an exquisite bit of china like you want to sail down with an iron pot like me.'[60] Victoria abandoned the voyage and went on to develop theories on semiotics that must have left the pot unstirred. They did share, however, one perspective: the emergence of feminism.

Mary had a deep-seated distaste for the supremacy of the male world, although her father, uncle, husband, Monsieur Deplace and James Hinton were people she greatly admired. What she hated was the institutional discrimination of universities, for example, although this had begun to slowly change. But Mary does not want simple equality: women should not compete with men in tasks they themselves find difficult but reserve their strength for doing well those that men find impossible.

Unsurprisingly, the pulsation of opposites lies at the bottom of her assumptions regarding roles. Her own anthropology, common to many, compares the male hunter who relies on muscle tension and constant wariness of opposing prey, to the female brooder exercising slacker muscles suited to the nurturing of the young. Teased out by analogy, the former is represented by the authoritarian policeman's grasp, the latter by the mother's caress. A favourite phrase of hers expresses this: 'Indolence is the Mother of Philosophy, Reason the Father.' Family life involves the sacrificing woman at its centre, until post-menopausal, when she becomes more fertile as a prophetess or seer. Otherwise, 'it is men's business as a rule to decide on facts and to make theories'.[61]

This may all seem woolly thinking and clash with today's perspectives on gender but there are echoes of the re-assessment of roles that took place as feminism emerged forcefully a century later in the 1970s. The male world was questioned in terms of its cold, mechanical, scientific analysis favouring hierarchy and competition, compared with the female universe embracing intuition, feeling and consensus. The wise-woman's empiricism as herbalist was compared to the diploma-carrying, drug-wielding doctor. The destructive products of science such as phallic nuclear weapons were compared with the sacrificing but resolute gatherers' womb homes at Greenham Common. Yet true to Boolean tendencies these oppositions have more lately resolved in a synthesis allying with women: that they can dwell equally well in both camps of science and creativity.

Mary Boole would probably have been happy with this. Women exercise logic just as well as men, they just do it in a different way, 'not while herding in gangs or competing against each other … but while pouring out our hearts and lives in the shelter of our own homes'.[62] At times of conflict such as during 'the wars of men and, the extraordinary state of confusion that affects even the cleverest of them ... women rise out of their dollish idleness ... and their slavery to fashion'.[63] Their sphere of power may be domestic, but it is not submissive; 'every utterance of opposition ... is the outcome of some personal suffering or of revolt against some injustice'.[64] Like a cat when matters concern her brood, 'her claws come out ... and are savage, so very, very sharp'.[65]

Mary was writing in 1868 after the Crimean war, with heroic Florence Nightingale in mind, but she had also lived through the Boer War which had politicised her. It represented what she termed 'swindler's algebra': the pursuit of symbolic values. The nation at that time was urged on by a 'rag on a stick', the Union Jack, 'made sacred by hypnotic suggestion of falsehood', to support a horrible 'torture of men and horses, sorrow to relatives of killed soldiers and heavy taxation for war expenses',[66] all for gold and diamonds, material to use as token values compared with the real world of 'corn, coals and wood, muscle, nerve and brain'.[67] Empire means 'extensive land grabbing, wasteful expenditure, shameless ostentation and a steady pitching downwards to maniacal confusion'.[68] Money itself is a token value when pursued for luxury and idleness; a financier is 'the croupier of a gigantic gambling table'[69] whose 'trade is to sell insanity'[70] and live in luxury on the fruits of other people's suffering and toil'. It escapes all auditing until it is audited by national disaster'.[71] With the Iraq war and the recent banking calamity in mind, her analysis is timeless.

Religion comes in for a bashing too; children are inculcated into Church by 'a dramatic representation, picnics, recreative excursions into some stratum of an outgrown past, visits to the tombs of the Mighty Dead'.[72] She discounts the certificates of education awarded to children whose brains are no more than thinking tools. Art too becomes a totem when fashion, not aesthetics, is worshipped. Perhaps disappointingly from a feminist standpoint she does not support the campaign for another token: the vote. She forecasted the current growing disillusionment with politics.

'The Franchise ... means the privilege of electing a representative who "stands for" them in the House of Commons. Whoever England is governed by just now, it is not governed by *voters*.'[73] She offers her own sociology here. Voting would be a 'practical and honest manner of conducting the affairs of a tribe small enough for all the members to sit round one camp fire ... By the time we have arrived at organised differentiation of labour ... voting is a mere game.' Mass society lets in 'the exploiters whether they call themselves priests, schoolmasters, college dons, political leaders or organisers of syndicates and trusts ...'.[74] Mary reserves special wrath for the working-class leader tokenised by the establishment when offered a peerage.

Her exegesis rings bells. Hush, hush, whisper who dares: Mary Everest Boole had a strong anarchistic leaning,[75] stepping out in later life from a wardrobe of old coats and moth balls wearing a *bonnet rouge*. She even changed the name assigned to her articles to Virginia de Mericour. Virginia perhaps may have signified the tenacious creeper in her back garden, red-tinted in its autumn days. 'Mericour' may have been a tribute to Raymond de Vericour from her Cork days, but also suggests Anne-Josephe Theroigne de Mericour, a fiery, enigmatic French revolutionary who harangued mobs and demanded that women should have the right to arm themselves. Like Mary, she kept a political salon.

This fits well with a letter Mary wrote to Victoria Welby in 1885 distancing herself from her aristocratic set, declaring, 'I was born mentally and morally on the *Place de la Bastille*. My father took me there when I was a tiny child and told me it was the most sacred place in the world.'[76] She envisages women dancing the *carmagnole* in streets that are running in blood, possibly even in England. At that time her mixing with Russian anarchists must have coloured her worldview. It certainly inspired her daughter Ethel to journey to St Petersburg where real blood was spilled by revolutionary women.

Like many an anarchist at heart, optimism and faith in humanity (whether inspired by God or not) is tempered by an angry compassionate sense of justice, caught perhaps by the title of her 1910 book *The Forging of Passion into Power*. Mary's arrogance may be forgiven in this light; her 'passion' is almost physically violent at times. Those drugged by the 'conceit of masculinity' should be given a slap in the face, not enough to pierce the skin but in a spirit of science, one that 'combines the largest amount of momentary stimulus with the smallest risk of future injury, e.g. a bunch of stinging nettles'.[77]

In a letter of 1908 to a female friend protesting about state support of denominational schools she inveighs that nothing can be done about it until the working classes refuse to send their children to school and 'receive the School Board official when he calls with a pitchfork or a red-hot poker'. 'Once mothers have begun to defy law for the sake of their children, property won't be safe in the district.'[78] In another of 1905 she writes that, 'hatred, contempt, desire for revenge ... have as good a right to live as rattlesnakes or tigers ...'.[79]

In her mid-seventies her health waned somewhat: her sight was failing and her freestyle knitting (*à la Tricoteuse*) numbed her hands. Never at a loss, she turned to crochet. Her enforced idleness, however, given Boole's law, created a stimulus to think laterally. She was still having visions and inspirations but she drew the line with theosophy, questioning an advocate point blank about 'spooks, shells, astrals, etc', dismissing it all as 'a philosophy concocted by people who know hardly any mathematics …'.[80] Her headquarters became a quiet haven of refuge where one could gain a respite from the war of 1914. Those of her daughters who remained – Margaret, Alice and Ethel – continued to visit but expressed little affection for her. In 1914, aged eighty-two, she was still writing snippety items including a retrospective childhood amble around Wickwar, *At the Foot of the Cotswolds*.

Mary Everest Boole 'passed on', as she might have liked her death described, on 17 May 1916. 'Passed her tests' might be another description, as she once declared that it would be like 'being let out of school … and allowed to go home'. On her deathbed Ethel reports that her mother announced, 'This is interesting. I never died before.' The intellectual inheritance that she bequeathed has not been generally recognised. This is partly due to her meandering, over-ardent, discursive writings that oscillate from grand theory to small observations as her mind takes her. Some of her views are repetitive and off-the-wall but esoteric theories were popular then among many. Her enquiring mind is perhaps too open to influence.

Her legacy has suffered from the sexism of being not only considered an appendage of George Boole but of having misrepresented the great man's thought. Her *Dictionary of National Biography* entry states that she 'came to confuse the eccentric [James] Hinton's disordered views about mental processes with her husband's work in logic'. Or in other words, her psychological interpretation of it was disordered. This has been standard fare for decades but has now been questioned.[81]

Her lack of recognition in a feminist hall of fame presumably rests on her failure to advocate the franchise. This seems to entirely ignore her elevation of women in many ways and the amazing strength of her bringing up five young children while doggedly pursuing her writing and campaigning. The main area where her legacy has been treated with respect is education, one in which women found a certain leeway. Mary added to the idea of child-centred learning the notion that the unconscious mind is the best teacher of the conscious. This was long before

Freud was being read and dreams interpreted in Britain.

To the last, Mary had not lost her touch. The suitably entitled magazine, *Plain Talk*, of September 1913 carried her piece, *The Shelter of the Vatican*. In her 'humble opinion' she agrees, ironically, that the Catholic Church is indeed a refuge from conflicting ideas backed up by 'the learned syndicate of men, the Jesuits, as honest even as the House of Commons'. She would rather remain outside the Church, 'perverse in old age, amongst those who would prefer to collect wild herbs rather than kneel under any roof, those who believe in … no holy water except the spray of the thought-storm and who take refuge from the warring elements in the calm centre of the whirlwind itself'.[82]

## 5

# TADPOLES INTO FROGS

The four volumes of Mary Everest Boole's *Collected Works* had been lodged in my family for some years. I suspect my Aunt Muriel had acquired them hoping to position herself somewhere in the wake of the famous Boole dynasty; she had after all been given the middle name of Everest. How this happened I have no idea; her mother Henrietta Boole and her husband Sam never spoke of any Himalayan connection. Having yourself named after mountains and scaling algebra would have been in any case far too rarefied and 'booshwa' for him, a communist plumber on the Great Western Railway. On Muriel's death Mary Everest's tomes had been handed on to her sister, my Aunt Doreen. It was at her funeral that I first heard of the *Voynich Manuscript*.

I was not alone on my side of the great Boole divide. Opening Volume One of the *Collected Works*, out fell a scrap of paper in Aunt Doreen's hand. Ethel Voynich's novels were listed on it as well as a recipe for scones: 4 oz flour, 4 oz sugar, 4 oz marg, 2 tsp ginger, ungreased tin, 40 mins. I doubt whether they had been inspired by reading Mary Everest Boole even at her down-homiest. (Do try Boolean scones for yourself; it's not a bad formula.)

More to the point, what had their eccentric widowed mother, Mary Everest, passed on to her five daughters allied to the genius of her husband, George Boole? We have very little information about their early lives to go on. The life of the youngest, Ethel, was fortunately relatively well-documented because of her celebrity status in later life. Even then she was hardly forthcoming about her childhood, which seems to have been for all of the sisters something of a nightmare. We don't even know for sure where they all were after their father's death in 1864.

The records of enrolment at Queen's College, Harley Street, where Mary Everest was librarian and later teacher, had shed some light. Mary Ellen, Margaret and Alice Boole were enrolled there from autumn 1865 to July 1873 when their mother gave up the tenancy on the top floor.

Lucy only attended classes from September 1866 to July 1869 and Alice was absent over a period 1866–67. Lucy's leaving, aged seven, and Alice's interrupted schooling give credence to Ethel's (and others') assertions that the children variously spent some time with their grandmother, Mary Ryall, in Cork. There are strangely no records for Ethel herself and no information came from her; she might too have been in Ireland when very young. She could also have been, from one year old, with her mother in London and looked after by a minder of some sort. Either way it must have troubled her later life.

Ethel told her (unpublished) biographer, Anne Fremantle, that Mary was seriously overworked and as a result the children only saw her for one hour a day. Paying £80 per annum for rent and eight guineas per term for tuition must have left the family on an extremely tight budget. This would not have stopped her insistence on education; apparently she sold George Boole's Gold Medal from the Royal Society to buy a harmonium. Ethel mentioned to Fremantle that they were 'hideously poor'. This was

Margaret, Alice and Ethel Boole as children.

borne out by her reminiscing that Mary had bought a bolt of brown material with large white blotches, and had dresses made from it for all the sisters until it was used up. She added that, in any case, Mary disapproved of prettiness with clothes as it reminded her of the 'ugly faces of people who were starving', entirely in keeping with her moralising attitudes. Mary would have given them a diet of plain food even if she'd been rich. Her menu included bread, plain-boiled vegetables, solid meat and raw apples. Nevertheless, the girls were attractive children as evidenced in an undated studio portrait of Ethel, Alice and probably Margaret.

It is difficult to assess what influence their father, George, might have had on the eldest three sisters who really knew him: Mary Ellen, Margaret and Alicia (known as Alice). Commentators tend to stress the genetic inheritance. One imagines Mary overseeing her children, when able, with a detached concern, obeying her own rules for their development. She no doubt followed George's advice, discouraging baby talk and competition but encouraging duty and work. Childhood would not have been nursery-cosy with fairy stories; their fantasies and imaginings would have been treated rationally. Mary recounts that Mary Ellen, the eldest, brought her 'birdie', a beetle, for her to kiss, and did so rather than destroy the child's illusion and show bad faith. One wonders how many kisses and hugs there were when knocks and bruises occurred. Someone who, however, could write, 'all genuine inspiration means being drunk with joy',[1] must have conveyed this somehow to her offspring.

It is highly likely that Mary practised what she preached about stimulating the unconscious of her own children. When a child 'begins to realise that a tadpole grows into a frog … a seedling into a flowering plant it's time to acquaint them with the flow of geometry'.[2] A night-light, for example, can project shapes onto paper, a suspended ring will describe ellipses and circles. Hang up a corkscrew wire and one can evoke the path of the planets or her favourite allusion to the whirling stormwind or the coils of snakes. (Ethel later professed having a special fascination with them.) Mary would no doubt be pleased to see that a box of bricks is still a favourite toy today.

The influence of their mother's teaching and milieu in their later childhood years would have been both bewildering and a boon. Her emotional collapse in the first half of the 1870s must have been traumatic. Ethel remembers her at this time as not being in her right mind. Mary Everest seems to have regularly overtaxed herself in her efforts to bring up a family and pass on her ideas. She once wrote to Victoria Welby that she had been 'ill, just like a woman after childbirth, prostrate …', having had a 'thought-child',[3] in her case one of a large progeny.

The children seem to have been relatively isolated from others but to teach them 'independence' and 'fearlessness' they were made to take daily walks around the nearby Regent's Park. Ethel remembered them with fear and had nightmares. Some sort of compensation may have been gained from her girlhood remembrance of 'holiday tramps round the wild Cornish coast and of talks in sanded kitchens with poor folk'. The visits were absorbed into her later life as a novelist. Yet offsetting this romanticism she recalls her mother's house-meetings full of stimulating but weird guests debating religion and philosophy. The sisters were made to sit in a passage-way, 'where they heard discussed all sorts of subjects which were unsuitable for children'. She added, 'before I knew where babies came from I had my nose rubbed in the facts of prostitution'.[4] One of Mary's guests was James Hinton, her collaborator in the college's attic, extra-curricular sessions.

Some luck did come the family's way in 1874. Her uncle, Reverend Robert Everest, died and left the large sum of £3,000. This was very fortunate coming soon after the termination of her tenancy at Queen's College in 1873. It also must have helped at that critical time when Mary Everest was suffering from 'temporary derangement'. It is most likely that it financed the training of Margaret, and possibly Alice and Lucy, to be nurses in Cork. They would have been able to live there with their relatives, the Ryalls. Mary Ellen, now nineteen and Ethel, eleven, would have stayed on in London.

After James Hinton died in 1875 contact with the Boole family was maintained through his son, Charles Howard. He had been a regular visitor to the Boole household, sufficient to woo the eldest, Mary Ellen, and marry her in 1880. Mary Everest would no doubt have approved of his instructing the girls with a cube of coloured bricks to which he'd given Latin names. The aim was to memorise their positions as an aid to visualising other dimensions. Ethel must have found it completely pointless; Lucy apparently tried to engage out of a sense of duty. Alice was the only one who seems to have been inspired.

This may have been because Mary had singled her out of the five daughters as most possessing a synthetic, intuitive ability. Born in 1860, she was only four when her father died. Most references to her follow articles by H. S. M. Coxeter, the famous geometrician, a colleague of hers in her later life. He maintained that she stayed in Cork until she was thirteen or so, claiming, therefore, that her ability must have been hereditary. This ignores the role of her mother. Alice later became inter-nationally recognised for her skills in demonstrating the fourth dimension.

The talents of Mary Ellen, the eldest daughter, drew her towards music. Her Queen's College reports showed promise with the piano. Margaret, like Alice, developed a visual skill to be able, before photography, to reproduce eye and ear pathologies in watercolours as a diagnostic aid. Lucy, of whom we know least, struck out towards a pioneering career in medical research. We shall return to examine their progress in Chapter 14.

The most notable and singular of the brood was the youngest, Ethel.

The author with 'Treacle Baby', a portrait of Ethel as a child.

It was she, sharing her mother's determination and single-mindedness, who tried to pursue an artistic vision prompted by both her parents ethical sense. A portrait of her aged about six, 'Everybody's Darling', shows a serious redhead fingering a coral necklace. It was dubbed 'treacle baby' by her mother so as not to excite the child's egoism. (I took on the task of tracking the painting down and taking it to a permanent public home in the US). Somewhat spoilt by her mother, Ethel felt that she was resented by the others. She told Anne Nill, her companion in later life, 'I alternatively quarrelled with my mother and adored her.' She described herself as 'a horrid little termagant with a vile temper', adding that she was 'unhappy, high-strung, exasperating, selfish, conceited and difficult'. None of this was helped by a saga worthy of Dickens.

At about aged eight she developed erysipelas, a bacterial skin infection, sometimes known as 'St Anthony's fire' owing to its redness. Her mother, unable to cope at that time, dispatched Ethel to the care of her uncle, Charles Boole, the manager of a coalmine at Rainford near Liverpool. If separation from her sisters and illness was not trauma enough, Charles it seems was a hell-raising religious bigot and a frustrated pianist. One episode is recounted that possibly sets an Oliver Twist to the rest of Ethel's life.

Falsely accused of stealing a lump of sugar, she was locked in a dark room pending her admission of guilt. Her uncle threatened even to administer some chemical to vindicate his assertion. In retaliation, Ethel vowed to drown herself, and obviously single-minded enough to be believed, was set free. This highly dramatic incident, recounted in her old age, underscored one of the driving forces of her life: an absolute pursuit of justice. It perhaps also contributed to the other characteristics remarked about her: a certain obduracy and over-seriousness. Ethel told Anne Fremantle that returning home aged ten in 1874 she suffered a nervous breakdown. This is hardly surprising given that it coincided with her mother's period of 'derangement'. The events in tandem must have left a deep scar in her life.

There was, however, a more inspiring event that contributed to Ethel's desire to put the world to rights. She told Anne Fremantle that about aged fifteen in 1879 she went to Ireland to stay with the Ryalls, by all accounts better hosts than Charles. There she read Emilie Venturi's biography of Giuseppe Mazzini, the Italian patriot who had inspired the 1848 revolt against Austrian occupation. His stormy life involving secret societies, arrests and exile in France captured her imagination and came to influence her own first novel. It also complemented a strange story with an epic historical narrative she had heard from her mother.

In 1859 a group of Italian patriots were being transferred to Naples from the notorious Spielberg prison near Brno, then within the Austrian Empire. Part of the journey involved a ship. Somehow the prisoners managed to bribe its crew to steer to England. It ran aground, however, after a storm near Cork, where the Boole family were then living. Two of the fugitives were housed by them. George Boole and his wife Mary obviously harboured romantic and underdog sympathies. Ethel's eldest sister Mary Ellen, only three years old, apparently charmed the escapees and invented the story that they'd been imprisoned in their attic and that she had set them free.

Of all the five Boole daughters Ethel's life was the most colourful, partly owing to her literary talent and a sense of mission, but also, like Lucy, she never really took on the potentially excluding role of motherhood. Coming across her story whilst researching *The Voynich Manuscript* fired my own imagination. As a result she and Wilfrid take a central place in the first part of the book. I became hooked on both the *Voynich Manuscript* and the congruence not only of Ethel's political views to my own, but the strange biographical trajectory that we shared.

# THE WIZARD

James Hinton (1822–75), the 'Wizard', aural surgeon and philosopher.

*Seat 58, Rare Books and Music, British Library. 24 November, 2009.*

Sir Samuel Wilkes in his 1875 *Lancet* obituary of James Hinton noted that his mind was occupied by 'the ideal, speculative and metaphysical …', so much so apparently that it adumbrated his profession as an ear specialist. Wilkes adds, 'I feel astonished that he was ever in it but being *in it*, was never *of it*.' Hinton was one of a rare breed: an aural philosopher, highly skilled but with his own head in the clouds. To this strange mix was added the fact that he was known as 'The Wizard'. Fascinating, I had to find out more.

I ordered up some of the published works of James Hinton; a tattered copy of *Life in Nature* (1865) arrived. I waded into the tome and took notes despite understanding almost nothing. Chapters included *Vital Force, The Living World, How the silkworm illustrated the Law of Admiration and Loathing, If Nature is Living, why is it perceived as Dead*. I've always hated that capitalisation of nouns like gravestone headings and the pompous Victorian way of explication. I gather only that James Hinton is trying to merge our understanding of the scientific world with the spiritual, but the verbosity is beyond me. Mary Everest Boole in comparison reads like a DIY manual.

So I called up *Life and Letters of James Hinton*, edited by Ellice Hopkins (1878). At the front there is a brown-stained, thin leaf of transparent paper, like an old-fashioned toilet paper sheet, over a photogravure of the man himself. He's very handsome: almost bulbous full nose, full mouth, strong forehead, wild side-parted hair and a straggly but not biblical beard with a cleanshaven upper lip. He looks sad and weighed down. At least I like the look of him. Reading Hopkins' account, it's all rather disappointing: we never get to find out why, most importantly, he was labelled 'The Wizard'. I decide wimpishly to go back on the internet and follow up his son, Charles Howard Hinton.

An essay by Mark Blacklock, *The Fairyland of Geometry*, came up concerning the family. Interesting names populated the screen, including Olive Schreiner and Havelock Ellis. James Hinton, I read, had shocked many people with his writings about women, prostitution and polygamy, and even further by his apparent goings-on with a number of them. And then, Blacklock writes, 'At the time of his death [he] had been living with his wife, Miss Haddon [his sister-in-law], a spinster and Mary Boole the widow of the mathematician George Boole. Of these he had shared physical relations with his wife, Miss Haddon and Mary Boole …'. WHAT! WHAT?

WHAT?? Impossible. This is news to me, but at least wizardy. The conventional photo of Mary Everest portrays a Queen Victoria lookalike, leading commentators to dwell on the fact that she remained a widow for fifty years after her husband's death. An earlier one (in Chapter 4), however, shows her still serious but rather good-looking. She would have been forty-three in 1875 when Hinton died, not exactly a frisky spring chicken, rather a mother hen with a brood of five orphaned chicks. James was ten years her senior. No one had mentioned cusp-menopausal romps. I was stunned but girded on to persevere with the life of the Wizard.

James Hinton was born in Reading in 1822, the second son of eleven children, to an outspoken Baptist minister, John Howard Hinton, known as the 'weeping philosopher' but also a geologist and naturalist. His mother, Eliza Birt, was something of a visionary herself. In James' early development it seems he derived a love of nature and an emotionalism yet rationality from the former and a compassionate generosity of spirit from the latter. The relative poverty of their family gave him an empathy with the underprivileged and the underdog.

Although blessed with an outstanding memory, James didn't exhibit any particular academic excellence and was put to work as a cashier, aged sixteen, in a draper's shop in Whitechapel. This complete change of environment from a relative backwater was accentuated by the grimness of life in the squalid courtyards of a notoriously decadent area of London. The gin alleys and destitution were made more shocking to him by the brutality handed out to beaten wives and prostitutes. The preacher's son was horrified to find himself confronted by two women asking him, 'Which one of us will you have?' The impression, he maintained, never left and came back to haunt him decades later.

Not long after he was assigned as a clerk to an insurance firm in the City and began a fervent period of self-education in languages, history, maths and metaphysics. The urgency of his later thinking shows the characteristics of the self-taught polymath from the same dissenting mould as George Boole. Aged nineteen, he met and fell in love with Margaret Haddon, a printer's daughter from a non-conformist background herself. Hinton, somewhat *gauche*, failed to make an impression. He turned his attention to a career in medicine, and after a brief voyage to China as surgeon, gained his diploma from the Royal College of Surgeons in 1847. Unable to reconcile his doubting religious views with marriage to the more pious Margaret, he escaped to sea once again, this time as medical officer on board a ship carrying 'liberated' Yorubas from Sierra Leone to Jamaica to work on the sugar plantations.

Margaret Hinton (1822–1902), wife of James.

His year there seems to have convinced him of the universality and intrinsic worthiness of people corrupted by the society he saw around him. In particular he was struck by the pride and hard work of the women harshly dominated yet again by their menfolk. Writing after his return to London in September 1850 to the still-waiting Margaret, he asserts that, 'in the byways and alleyways of this city and your town … the divinity of human nature is not extinguished there'.[1] To his parents' distress this faith in humanity could not be coupled with a simple acceptance of the literal truth of the Bible. His travels, his wide-ranging learning and rational study of the body via medicine had made this impossible, yet not at the expense of a faith in God.

Inspired by the restoring of his mother's hearing by a routine syringing of her ear, Hinton resumed his medical studies at St Mary's Hospital, London. In a typical spirit of enquiry he also began to visit regularly the recently founded London Homeopathic Hospital. He writes to Margaret in May 1851, 'I have become wise and discovered that I was a fool … what everyone has known from the remotest antiquity. Fancy kills or fancy cures.'[2] He recounts how a doctor, twenty years previously, was walking through a field of peas absent-mindedly rolling some between his fingers. A woman in bad health saw him and, assuming the peas were pills, asked for some. The next day she was cured. Hinton asserts that what was demonstrated was the influence of the mind on the body, a conclusion that Mary Everest would not have shared regarding homeopathy.

The episode spurred him, however, into other investigations in his insatiable way: why women blushed, how headaches could be cured by 'hope' and how tears in a child must not be disturbed because of their beneficial effect on the eyes. Hinton asked further: if the brain is just an organ like capillaries, tear ducts and muscles, what is it that makes it think? Hinton answers that we must suppose a spiritual substance: a notion that sidesteps the metaphysical divide between the religious who say that it is the spirit of God that thinks and wills and the scientist who declares more mundanely that only the brain does so. 'What if the brain thinks and the spirit wills?' he suggests. 'Matter can only obey the forces that act upon it.'[3] The brain is like your piano, he writes to Margaret in August 1851, passively capable of producing sweet music or random sounds as when a cat walks over the keys. It is God, he declares, in tones Mary Everest would have agreed with, who reveals to us his divine plan of harmony and love for us to realise in our actions. The avenue to our understanding lies in interpreting His material world in nature through our scientific faculties.

Hinton's musings on faith and its power to influence the body had renewed his own faith in the Bible, but not in any slavish way; his Christianity was not of the institutional kind. He tells his sister-in-law, Caroline Haddon, of his doubts about a God who sends little children to hell and values Sabbath days more than human lives. Hinton had come to have faith too in the usefulness of his own work in medicine. 'Will it not be delightful to aid the afflicted, to soothe the distressed, to impart heavenly consolation to the sorrowful,'[4] he writes to Margaret in March 1852.

He is also down-to-earth enough to want to earn sufficient for 'tables and chairs and house, butcher meat and bread, a good coat and a gold watch'.[5] In the same year his prospects, working now as an aural surgeon, enabled his marriage to Margaret Haddon at last. They produced two sons: Charles Howard (known as Howard) in 1853, William in 1854 and two daughters, Margaret in 1855 and Ada (known as Daisy) in 1856. Such worldliness, however, was not at the expense of a continued intellectual enquiry. He writes to Caroline in August 1859, 'I am under an incurable fate to think … it is deeper than a passion.' Prophetically he adds, 'I could conceive it growing into a raging torture, or a madness.'[6]

He did indeed already evince an obvious eccentricity in his outward appearance and demeanour. Margaret commented, 'He was such a trouble-some man, it was so hard to have his hair cut and have his photo taken … He would never be got to a dinner party.'[7] A colleague described him 'steaming along the street', the nap of his hat brushed the wrong way, gloveless and with a book under his arm. Agnes Jones (the 'spinster' from Hinton's amorous quartet) reported on his impulsiveness: how he once went barefoot and shabby down Fleet Street to try to understand the likely feelings of a beggar; how he once got drunk to see whether he would attack his wife. As with Mary, 'contrariness' is the key to understanding.

There is a mystical drift to Hinton; like Blake, all of life is holy. Unlike Blake's concise aphorisms and visionary art, however, Hinton's glorying in creation is suffused with the disciplines of the mathematician, scientist and practising physiologist looking reverently for correlations in all that he sees in the world around. He could never write dispassionately as some scientists were wont, regarding subject matter mechanically as 'dead'. The conviction that the sensible world manifests a divine impetus brings with it a search for processes and movement in nature and growth and becoming in Man. It is coupled though with the everyday profession of the surgeon empirically diagnosing and effecting cures for humanity. As with Mary his intellect is at the service of its usefulness. His observation, for example, that spirals are a basic form in nature resulting from deflected motion may have been drawn from his close acquaintanceship with the cochlea in the inner ear. Mary's interpretation was, as we have seen, somewhat wilder and more grandiose.

Typical of the development of his ideas, the spiral figure returns in many contexts. In his *Selections* Hinton adds geometrically to the notion: a spiral constitutes three turns at right angles within a 'flowing cube', formed by 'alternate action in rectangular directions'.[8] A few pages later he moves on in tangential fashion to note how force and resistance produce tensions in architectural forms that require resolution. This is an axiomatic theme akin to those of Mary Everest – what he calls the process of 'nutrition' leading to 'function' – a storing up and liberation leading to a higher state found, for example, in the life stages of a frog.[9]

The idea reappears in a collection of essays, *Life in Nature*, 1862. Plant growth is a continual process of motion taking the least line of resistance. This is true, for example, of the unfurling of a fern frond, 'because the central part grows, while the ends are fixed, with the increase of the plant it becomes free'.[10] Once again the spiral is cited as an essential construction, 'manifested from the lowest rudiments of life upwards through every organ of the highest and most complex animal'.[11] In a letter to Caroline Haddon he writes how the spiral as a helix is 'nature's hieroglyph for the fact of life which meets us at every turn in the animate and inanimate world ... each round of the corkscrew coming back to the same point but on a higher level'.[12] (His son, Charles Howard, later made use of the idea in his own fourth-dimensional musings.) A century later James D. Watson, co-discoverer of the helix structure of the DNA molecule noted how the knowledge would have been a great satisfaction to James Hinton. Compared to these rational observations, however, with Blakean innocence, he writes to Margaret, 'The world indeed is wonderful; it is divine, spiritual, eternal. This is heaven. We do well to be intoxicated, ravished with its beauty ... and wonder ...'.[13]

In 1860 after the successful publication of *Man in his Dwelling-Place*, in order to devote more time to writing, he took his family to a cottage in Tottenham, north of the city, then a village. Once again a garden proved revelatory to Hinton and most appositely among the spiralled tendrils of a bed of peas. Getting his face scratched while picking them led him to muse that greater ends are sometimes the result of pain, and since causes are not always mutable they must necessarily be endured, even welcomed. Pain can be a 'nutrition'.

A profounder universal example, he maintained, might be found in the weariness and toil of a mother, who, in ministering to her child, nevertheless brings a wider joy. For Hinton, if our pains are unendurable it only shows that we are sick (perhaps like Blake's rose), unable to embrace them although they are part of 'our proven nature and destiny ... an essential element of the highest good, felt as evil by want in us'.[14] Altruism consists in the serving of others and acting gladly regardless of the cost. For James the highest example of such a principle was shown by God's sacrifice of His only son and His sacrifice for us in turn. The world we experience, according to Hinton, like all its phenomena, is not what it seems immediately to us; there are higher laws, an essential part of which is God's gift of pain. As in Christ's life, such an experience is one of passion. His thoughts led to the publication of *The Mystery of Pain* in 1866, which proved popular with a Victorian public married to the idea of renunciation. It was passion of a very different sort, however, that ultimately brought him and his followers into social disrepute.

Hinton and his family moved back to London in 1863 and set up house in George Street, Hanover Square in order to practise as an aural surgeon at Guy's Hospital and at a private practice in the West End.[15] Until about 1870 he embarked on a period of 'hybernation' from his frenzy of thought. The respite was to pay off; he was no enemy to Mammon, able to charge high fees for his specialism and investing enough wealth to buy an estate in the Azores to grow oranges. He was not averse either to the bourgeois trappings of professional life; as a regular concert-goer and lover of Mozart and Beethoven, he was often observed with his head in his hands listening in ecstasy. His emotional delight was inevitably accompanied by theoretical musings. Music, while being 'the highest mode of the soul's affirmation that the universe as a whole is absolutely beautiful', also embraced 'pains' as discords, 'things evil in themselves', but necessary, of 'service' to the whole.[16]

He also began to collect paintings, being a great admirer of Turner and the rural scenes of David Cox, both modernists in their time. On the occasion of a medical meeting in Birmingham, a colleague found him enraptured in front of a misty sunrise by Turner. Hinton observed that the daubs of light were of a distinctly spiral form. In 1871 he staged an exhibition of his most valuable paintings in his old haunt of Whitechapel and proposed to donate them eventually to the poor.

Hinton was inclined to see poverty not with Victorian righteousness as a punishment but as a kind of nutritional pain storing up martyrdom and a force for change. 'The sufferer works true actual good; his sufferings take away sin; his life is a ransom.'[17] Any reward, however, is not postponed to an afterlife but to be taken in the now. Unable to forget his early east London experiences he remained keenly aware of its human degradation. He despised 'wealth fed on poverty … virtue that was buttressed by vice … a law that created crime …'.[18]

While, like George Boole, he did not elaborate any systematic theory of changing society, he did consider specific amelioration. In 1871 he published his *Thoughts on Health*, a collection of articles in a popular style advocating health education. Agreeing with Florence Nightingale, he believed in nursing as a proper profession for women, advocating both higher pay and social status. In 1874, contributing to a volume of *Physiology for Practical Use*, like Boole and Maurice he elaborates on the general poor health arising from overcrowding, long working hours and unemployment. He examines specific trades where dust, cramped conditions and poisonous materials are used. Hinton actually proposed an early form of a Settlement community in the East End where the working class could receive hands-on education. It was later realised as Toynbee Hall.

His 'hybernation', however, had not precluded intellectual pursuits; he established philosophical soirees. Ellice Hopkins writes, 'Many date the beginning of a higher life from these meetings.'[19] In 1870 at the prompting of Tennyson he joined the Metaphysical Society that included, as well as the poet laureate, eminent names of the time such as Gladstone, Ruskin, Maurice and Huxley. They discussed such issues as *Has the frog a soul?*, *The Absolute* and *Hospitals for Incurables from a Moral point of View*. R. H. Hutton recalled that 'the wistful and sanguine … almost hectic idealism of James Hinton struck me much more than anything he had to say'.[20]

Now financially secure, his 'frenzy' returned to his philosophical speculations. There was to be no snap, leguminous revelation to inspire Hinton towards the aspect of his philosophy that scandalised Victorian life. To use one of his own organic analogies, his body of thought had like an egg morphed to a grub and after a chrysalis gestation emerged gaudily for all to view. The fluttering beauty Edith Ellis described as able to float in a region undreamt of by the merely sane and solid. It spiralled out of control. To a modern ear the issue has perhaps only the weight of a flight of fancy but to Victorian society, outwardly secure but inwardly reeling from the latent impiety of Darwin, some things could not be said.

According to Hinton, as part of the majesty of creation, God has given us impulses that need not only to be obeyed in so far as they serve others but also enjoyed as pleasures when they serve oneself. Acting for oneself can combine with acting for others. The suckling mother (a regular mammary theme for Hinton) derives pleasure not just for herself but the species. When we eat we can take pleasure in the act and its similar service to humanity through its resulting perpetuation.

Unselfish pleasure is a natural force, not a social vice; 'Service stands as the other powers of nature, as fire, steam, electricity … waiting for man to use them, hurting him till he does.'[21] Taking the maxim further into risky climes he suggests that our needs are actually a cause for excitation, for 'men are good, wonderfully good and loved by God with a passion of delight, to shadow which forth He made the love of man to woman … Sexual passion is the great spiritual power of human life; the regenerator of man; the means by which his life is raised to its true height.'[22] He writes in 1870, 'How utterly all feeling of impurity is gone from the sexual passion in my mind! It stands before me absolutely as the taking of food.'[23]

Despite his intimate knowledge of the human anatomy, he rarely ventures in his writings into the nuts and bolts of sex. He mentions masturbation briefly and remarks upon the parallel between the nose and the vagina: both are in part physiologically organs of excretion. He suggests that if we do not cover up the former why should we the latter? One needs the sight of a woman's body, 'only then will her soul be seen also for God has so made her that she bears Paradise in her body'.[24] Women similarly he maintains do not feel repugnance and abhorrence physically for what he bashfully calls 'the thing' but for the way relations are undertaken. 'May it not be,' he suggests, 'that women are truly sensuous … even more than men …'.[25] Women's sexual passion he suggests is repressed.

At a time when female orgasm was diagnosed as 'hysterical paroxysm' such ideas were scandalous. James Hinton saw intimacy as an act of beauty compared to praying or regarding flowers. Sexual impulse should be enjoyed; restraint for its own sake is a sin as is its twin evil, overindulgence. Hinton had nothing but scorn for the ascetic and a wagged finger for the glutton.

Problems arise, however, in this optimistic declaration. Pain for the individual, we have learnt, is also heaven sent and to be taken with passion and a pleasure for service to others – but who is to decide the justice of their taking? Not for Hinton the utilitarianism of John Stuart Mill and the confluence of self-interest and the greatest good. For Hinton, because of the wonderfulness of man as God's creation, aspiring to the good with a pure heart is sufficient to ensure harmony. The purity of motives is all important. Even if they're not always easily discernible, intention is everything. Pleasure is possible to the utmost when one has 'cast out the self'.

This is, of course, hippy love, and all you need; it's the golden vision of the guru and also the poetic, but godless, philosophical anarchist. Hinton concurs: 'Society is organized necessarily upon the assumption of a certain amount … of organised goodness among men.'[26] We don't need legislation either: trying to get a 'just' law, he maintains is 'like trying to bring a sharp point to a pencil that can't be made sharp'.[27] As we shall see, however, the mismatch of good intentions and others' sacrifices in Hinton's libidinous life would rebound seriously to damage his moral legacy after his death.

One social evil was to become his mission in life. Hinton had never forgotten his early days in Whitechapel and its most evident symptom: prostitution. For him it was a clear example of how the undue restraint of passion led to criminality and vice. If within 'respectable' marriage, he argued, wives, inhibited by the Christian slur on sexuality, withheld its joys, men, given their nature, would turn to other women to satisfy a god-given impulse. Men, he maintains, are constructed to need more sex than women as they do food. Stimulated desire would bring its reward in increasing affection and delight within the home and thereby banishing prostitution. Lost women could be saved; Jesus himself recognised this. At the Resurrection scene it was a woman like Mary Magdalen he chose to greet. 'Prostitution is dead, I have slain it,' Hinton declared, and noting that Christ was a saviour too, added, 'and I don't envy him a bit.'[28]

If Hinton's worldview was based on the simple axiom, 'serve others and do what you will', he certainly didn't need socialism, for example, to provide a guide to justice. His appeal was essentially to the individual. Hinton suggests that if society represses itself a force within it will build, waiting to be expended, as the seed stores up growth for eventual harvest. Great individuals will emerge to act momentously at times of historical pressure and institute change. Jesus, for example, arrived on the historical stage at a point of least resistance. Laws that had become ritual barriers needed to be overthrown as Christ did opposing the Sabbath, as St Paul inveighed against circumcision and Luther against celibacy. Society often reprises with punishment but the suffering and martyrdom are gladly born. Mary Everest saw genius in a similar light.

James Hinton probably felt that a mantle of prophecy had descended on him, as some saw it, obscuring his vision. 'Why is it,' he writes, 'that every shame of womanhood makes my soul wither as if it were my shame?' And further, blasphemously, 'O God, have I not borne the sins of the world?'[29] Although he loathed the idea of founding a sect of disciples, his writings take on a more exclamatory and lecturing tone, full of yea-saying and avenging. In an egoistic journal entry of 1870, quite unlike Mary Everest, he writes, 'I am like a man climbing a mountain, every limb strained to the utmost, every nerve tense, and he or she that would be with me must accept life so, must climb the mountain or be content to keep upon the plain.'[30]

Hinton believed that marriage as a social institution must change. The laws that enshrine monogamous marriage, that 'funeral hearse', as Blake declared, must be amended. 'Surely what we call marriage,' rails James, 'is the cruellest idol ever known? Is it not lust embodied?'[31] Victorian marriage and the family, he maintained, is sanctified above all else, but in reality is mostly a hypocritical hell of property relations, financial settlements, competition between women and the creator of a hierarchy of enslavement, from the prude wife to a class of mistresses – and beneath all, the common prostitute.

Although he recognises that marriage can work well in some unions, the solution to its evils might lie in a greater variety of forms including polygamy. His writings on the subject were not circulated widely, but the Havelock Ellis collection of letters in the British Library makes it clear that he spoke on the subject urgently, especially to women listeners. Although Mormon and other polygamous societies in the US had been publicly discussed, to the stuffy Christian orthodoxy of the day it must have appeared very shocking. But why not enact polygamy, asks Hinton. More women would be married, competition between them lessened, the frustrated single celibate woman would disappear and women could have careers if they were not solely responsible for the family. He points to the practice in Islam as effective, rationalising even that animals are polygamous, as will humans be in heaven.

There is more at stake, however, than re-engineering a sacred social institution; there is a fundamental justice involved. Women have been in 'a state of coercion, subordination, suppression all these ages', he asserts. 'I want one law for men and women, but a law of the spirit – one law, the absolute desire for good in both.'[32] This realisation of nature will be according to men and women's own respective natures. Women to Hinton are essentially altruistic in their caring for others, most especially as mothers. It is this that gives them intrinsic purity but men, whose lives are out in the world, must learn to give too, recognising women's need to be loved. Mutual passion, not self-seeking but 'in service', will provide a bond, especially for the man. 'Embracing a woman is the most spiritual of things', to be compared to prayer and music.[33]

Women will be a nutrition, the regenerators of future civilisation. 'A woman's age is coming for the world, a time in which the woman's power shall be felt throughout it and her nature fulfil itself in all its doings.'[34] But she must make a sacrifice: 'And what has she to give up: this, nothing more: a claim on him to be a hypocrite. If Man so loves, so desires, unconsciously longs for the passionate sympathy of women … their surrender of exclusive marital rights and embracing of passion will persuade men to cease their philandering.'

Edith Ellis (wife of Havelock Ellis), a disciple of Hinton's writing forty years later, while recognising the authenticity of his feelings for women has a more modern take. For her, economic independence is the answer. She reminds us that his polygamy was never polyandrous; men would be

Havelock Ellis (1859–1939).

the chief gainers of a liberated sexuality. It would also require a comradeship between women that the world has never yet seen. Hinton's appeal to the new band of polygamous wives for them to be 'heroic' and put aside jealousy was highly optimistic. Did he extend the idea into his own life, as Blacklock suggests, incorporating his sister-in-law Caroline Haddon, Agnes Jones and Mary Everest Boole? He had, it seems, entertained the notion of a 'spiritual wife'. Margaret, his legal wife, told Havelock Ellis that they'd contemplated 'marrying a girl'. Edith writes of it, however, only as an obsession that absorbed him towards the end of his life as a principle but not with any special woman.

Despite his enormous energy, exhausted by the strain of all his various endeavours, in 1874 Hinton resigned his position at Guy's Hospital. The following autumn he set sail for the Azores. He died there a few months later on 16 December of 'inflammation of the brain'. The property seemed not to be all he had wished for and this increased his melancholia. In a late letter, however, to his son, Charles Howard, he indicates some notion of hope about the intense wrong in society: 'I dashed myself against it … it was too much for my brain; but it is by the failure of some that others succeed, and through my foolishness, perhaps, there shall come a better success to others.'[35] Subsequent events proved that this was not to be the case.

One can easily imagine the shared empathy and earnest excitement of the meetings between James Hinton and Mary Everest Boole when they energised a group of Queen's College young girls at 68 Harley Street. Both poured out their intuitions, sparking in all sorts of directions but aiming, whatever their source material, to stitch an embroidery showing God's harmonious world. Unsurprisingly, given that they were both the offspring of preachers, they sermonised and invoked parables.

Their views overlapped considerably: the importance of growth by opposition, the role of women and the necessity to translate theories into practicality – for Mary through education for James in relation to prostitution. Although his interest in the subject was aroused by a genuine sense of injustice, it is also fairly obvious that James, unlike George Boole, entertained a sensuous relationship with womankind, part of a more generalised aestheticism and love of beauty in music and art. Mary hardly refers to either subject. Hinton's views on prostitution and polygamy however are too much for Mary. While sharing his notion that good and evil are not opposites but different aspects of the unity of all things, she could not agree with his joy of pain from sacrifice. Suffering enables growth and development of both a spiritual and intellectual kind. The appropriate organ is the brain, not the genitals – veneration, not generation. Furthermore, any advance must be subject to a reversal according to the laws of pulsation. As the heart and lungs contract in 'rhythmic alternation of contrary motions', so must the mind. Hinton's mistake, she suggests, was not to recognise the principle and instead advocate continuous altruism by which she presumably means his hedonism and monomania. Mary, nevertheless, would have been drawn to James, firing on all cylinders, compared to the quiet saintliness of her husband.

The 'Wizard' apparently had many female admirers attracted by his ebullience. In *Three Modern Seers* Edith Ellis remarks, 'He was nature's child … his visions not those of the drawing-room but of the heavens.'[36] One might suggest further: women wanted, on the one hand to take care of the childlike genius with his magnetic personality able to cry with abandon and be sweetly eccentric, yet one who was also able to suffer willingly in a manly, Christ-like way. The martyr can be an attractive persona for women (as Ethel Boole/Voynich was to find in her life and novels).

Hinton is also a professional healer as a surgeon and a prophet, a 'Knight of Love' who boldly professes the powers of earthly delights, of mutual passion between men and women. The receptive woman finds herself bowled over at the feet of the lovable rogue, hoping for release, subliminally or sexually, yet also wanting to transform and rescue the child, troubled in mind, needing a mother's rescuing touch. Mary Boole would have been probably less attracted to Hinton's glamour, but one rambling letter in the Havelock Ellis collection in the British Library dated 11 February 1870 is illuminating. (Reading it under the gaze of a bust of George Bernard Shaw added to the experience.) It is addressed to 'My Dear Friend' (presumably Hinton).

She recalls one of his central tenets: the oneness of humanity. She agrees that she has accepted the idea – but hates it. She must bear it, however, with joy: 'I rejoice in finding out what God pleases to do with me.'

Bizarrely, she goes on, 'If I wasn't God's child I would like next to be a good man's dog … the thing the man had under his foot, that thing that had no ideas beyond doing the man's will (that is if I liked the man of course).'[37] There is an undeniably obsessive character to the epistle, of being unwilling to accept higher aims in life and wanting to abase herself. It doesn't fit remotely with the way she is considered in any extant accounts of her life. It suggests that there was a compulsive and unstable relationship between the pair.

One further letter, written by Hinton, takes this much further in a dramatic way. While delving into the Bristol archives of Howard Hinton, the entomologist (James' great-grandson), I found a letter lodged by an American researcher, Esther Blanc, who had been working on a PhD about James. She had written to Howard in the early 1970s hoping for help with any relevant material. She mentioned that she had by 'fortunate accident' come across a 'mass' of his documents at the Dr Williams Library in London – the HQ of non-conformist archives.

Intrigued, in 2011 I visited the library and spoke to the present curator. He professed no knowledge of the collection and implied that was the end of the matter. I persevered and in 2012 was told that it had been located but, being unfoliated, was not available. In January 2014, after more correspondence, I finally sat down to yet another assortment of Hinton's ramblings: 135 folios from the period 1870 to 1875 that had been presented to the library in 1934 by Ethel and Ursula Nettleship. Was the Dr Williams Library's reluctance for them to be seen because of any still-remaining notoriety?

The manuscripts are peppered with references to 'Mrs B.' – obviously Mary Everest Boole. James notices in one, dated April 1870 when they were meeting regularly, 'the marks of an external bias on restraint or coercion preventing her from what she naturally would'. It is present too regarding his wife, perhaps suggesting that there had been experiments with the concept of the 'spiritual wife'. Margaret was wrong too, it seems, in relation to some sacrifice, 'which nature does not demand …', adding that she did not tell him 'that Fanny was personally distasteful to her'. This may relate to some other context entirely but he continues, 'looking for duty in any form or way is only a perversion and hindrance … liking good, the duty is gone, all she has to do is what she likes … when the self is cast out doing what we like must be the absolute rule; the right must be in the form of pleasure'. Hinton's sexual aim is to persuade women that they do his bidding not from duty but because his needs and motives are honest.

One undated ten-page letter to his son Charles Howard was detached from the rest of the collection and, for once, written in a clear, determined hand. It serves as a sad and tragic farewell to James Hinton's life. Although undated, it mentions a proposed returning after Christmas, suggesting that it was sent from the Azores shortly before he died in December 1875. There is a reference also to financial problems that occurred about that time. Hinton is bitterly repenting.

> My boy I have erred, do not walk in my steps, nor listen to my voice … about human life and its laws and methods … my thoughts are utterly one-sided, partial, blind. My boy I did not read the Bible. Do not trust the heart it is deceitful … and desperately wicked. Care not for the bodily wants … even though this should be strong and quick ….

Even further, seeming to doubt everything in his life, he writes, 'Do not trust intellectual power … neither your own nor others …'. This leads to a heavily underlined appeal, 'If you know of any copies of *Thoughts on Home*[38] burn them or get them burnt. Be sure of this.' One other appeal is even more heavily underlined. 'Dear Son, do not go to Mrs Boole … do not spend any time more in her company … absolutely and totally avoid her influence for ever. I lay this on you as an absolute charge.'

What had poor Mary Everest done to earn such draconian disfavour? Had she dropped her 'external bias on restraint or coercion' at last? Had they had the affair that was to be claimed later in the 1880s? It is he who seems to be at least complicit, referring to 'My terrible act … oh how the agony is terrible to me'. Was it part of Mary's mental breakdown of 1874/5 that had led her to 'temporary derangement' while at Queen's College and the dispersal of three of the Boole girls to Cork? Something scandalous and destructive must have taken place for Hinton to be so adamant.

He also adds, however, that Charles Howard should be 'kind as ever you can to her children'. This wish his son evidently did comply with, meeting Mary and the family again to introduce Alice to the world of the fourth dimension and before long to satisfy another appeal from his father: 'Choose a good quiet amicable pious wife as much like Mamma as you can.' In 1880 his son wedded Mary Ellen, the eldest Boole daughter. James further advises, 'Above all … reverence the purity of women.' As we shall see, Charles Howard singularly failed to do so. Had he listened, his and Mary Ellen's exile from England in 1887 might never have happened.

# TRAPDOORS AND VELVETEEN

It's not easy to explain the fascination that things Russian can hold. The picture postcards are compelling: snow on the gilded onion domes, straining horses pulling a troika, wooden *dacha* cottages in deep birch forests – snapshots akin to Dutch windmills and Swiss alpine pastures. The vastness and emptiness of the Steppes, of man pitted against horizonless nature, strike a profound chord too – but Canada has its wintry plains that conjure up, it seems, mainly images of mounted policemen. More than anything what appeals, to some people, is Russian culture. Its beautiful language fleshed in an elegant otherworldly script lies on the surface of a history written on large canvasses portraying sadness and suffering, the extremes of the human condition in an extreme natural environment. Religious mysticism, political tyranny and rebellion have provoked the turbulent artistic and philosophical world of Dostoyevsky's *The Brothers Karamazov*, Tolstoy's *War and Peace*, Eisenstein's *Potemkin* and Pasternak's *Dr Zhivago*.

Maybe these parallels could be found if one were knowledgeable enough about China's impressive history and culture. Yet the word that comes readily to explain the less obvious attraction is 'inscrutable', meaning completely foreign. 'Enigmatic' adheres to Russia – enticingly different but familiar. Churchill's famous aphorism that Russia is 'a riddle within a mystery inside an enigma' compels, as does the *matryoshka* doll with its layer within layer leading to some secret homunculus essence. We might add that the use of the word 'enigma' clings as thoroughly to the *Voynich Manuscript* with its beautiful script and images that, even more so, demand comprehension. As with the manuscript, there is an enticing familiarity recognisable to a European worldview, but also something distant, verging on the absurd and mad.

If someone – anyone – had mentioned in my youth our family's connection to Ethel Boole and her Russian revolutionary dealings, rather than to her father, the logical Mr Boole, I'm pretty sure my life would have taken a different turn. The politics of my father had disposed me anyway towards radical views, which at university in the 1960s, augmented by studying sociology, had morphed via Marxism into an anarchist leaning that still informs me. My desire to help change society would have gained a strong extra impetus from knowing of the Boole family link with Russia and what was happening in the Soviet era. As we shall see I only belatedly became directly involved in its affairs in the 1980s during an intense period of the Cold War

At the risk of resorting to pop psychology it is tempting to suggest why the darkness of tsarist Russia might have become such a magnet for Ethel. She knew the injustice of poverty and distress in her own life from the outset. She would have been influenced by the idealised moral beacon of the father she never knew, urging her to redress wrongs. Their scale in Russia at the time exceeded everywhere else in Europe.

Ethel's passion for life came over in her love of music, Bach and Chopin especially, and her love of literature. Piano lessons were taken when money allowed. She read copiously Shakespeare, Milton, Shelley and Blake, his *Songs of Experience* being a favourite. The romantic element of oppression, nurtured by Mazzini's life, stayed with her. Like many a 'teenager', its outward reflection emerged in a style of dress. Fremantle was told by Ethel how Mazzini's 'Dark hair curling down to his shoulders, his melancholy beauty and distinction, remained with me. I determined as soon as I was allowed to choose my own clothes to copy him. I dressed in black until my marriage, mourning for the state of the world.'[1] More lastingly, the revolutionary Italian context was to provide the setting much later for her novel *The Gadfly*, inspired by her experiences in Russia

The inheritance her mother had received enabled her to travel to Berlin to study piano at the *Hochschule der Musik* for three years. We know little of this time, although Ethel was to tell Fremantle another story that epitomises her extraordinary, almost obsessive, single-mindedness, reminiscent of her mother. 'Professor Spitta, a great expert on Bach was lecturing to us and explained … that in tuning, the third and fourth notes of the octave had to be a little off, or the octave wouldn't fit. From that instant I hated God and despised this Almighty Creator of all things visible and invisible who couldn't make even eight notes fit.'[2] She later in old age admitted that this verdict might have been a little bit hasty. Obviously there's nothing 'even-tempered' about Ethel; the ingredients of a majestic, cosmic petulance seem to be in place. The physical humiliation meted out by her uncle Charles and the intellectual imperfection of God the Father must have left her with an early disillusionment about life and the male principle in general. Fortunately it didn't dent her passion for finer things and a keen moral sense, rather than complete despair.

Her unambiguous disavowal of the Almighty may have brought a swift divine reprisal. At the end of her study in Berlin, aged twenty, she became ill. Evgenia Taratuta, Ethel's Soviet biographer, reports that 'incomprehensible cramps convulsed and twisted her hands …'.[3] The doctors said there was no cure. This must have been a horrible blow to her ambitions. She did continue to play but presumably the dexterity required for a concert standard had gone. She was obviously at a crossroads in her life. Her response was to take off on a post-dated 'gap' year travelling in the Black Forest, Lucerne, and finally settling in Paris for several months, then, as now, a destination favoured by the soulful traveller.

We have little information about her stay there save for Taratuta's charming but 'heroic' account of Ethel's visits to the Louvre. Clothed in her long black dress, Ethel strode swiftly past the works of the great masters, the Giaconda even, to stand transfixed by a portrait of a young man by the painter Franciabigio. Like Ethel he is clad in black and exudes in his eyes 'an infinite grief, the sign of long, profound and hidden suffering. His whole figure is full of hidden strength, great pain and reticence.'[4] It clearly encapsulated a mirror image of her own searching for the perfect, romantic, idealised figure, noble in thought and deed and prepared to withstand physical suffering to realise just ends. Probably tinged with sexual sublimation, it summarised her life so far. Ethel acquired a copy of the painting, which hung in her New York flat until she died; a touching photo shows her, white-haired, with the painting in the background (see page 267).

It's difficult to imagine this young figure strolling carefree in the Jardins de Luxembourg, or idly sitting in the Tuileries watching old men playing *boules*. She's too earnest to savour the *joie de vivre* on offer. I see her tut-tutting at life's frivolities, waiting impatiently for some rousing to conviction and action. Back in London in early 1886 the clarion call came from a distant country that suited her mood, one whose sufferings had been in the back of her mind for several years. To understand her later engagement with revolutionary upheaval in Russia one needs to place its nineteenth-century history in context.

Living in London within her mother Mary's intellectual circle, the assassination of Tsar Alexander II in St Petersburg in March 1881 must have been a major talking point there as everywhere else. It was but one

Mikhail Bakunin (1814–76), Russian revolutionary anarchist.

of the latest in a series of violent strikes against autocracy on mainland Europe. After the failed revolutionary attempts of 1848 to bring about constitutional democracies, radicals trod more far-reaching avenues including the rise of socialism. In Russia the grosser inequality of the ossified autocratic tsarist regime nurtured radicals supporting terrorism, such as Mikhail Bakunin (1814–76). His maxim, 'the passion for destruction is a creative passion', argued for uprisings, violent if necessary, against capitalism and the state. His life followed the almost obligatory fate for a Russian rebel: in 1849 Bakunin was arrested, endured six hard years in jail and was exiled to Siberia. From there he escaped, travelling east through Japan and the US, passing through Liverpool in 1861.

Although he accepted Marx's class analysis, Bakunin as an anarchist proposed a society without government, fearing, and accurately forecasting, that a revolutionary workers' state might lead to new despotism in the name of a 'dictatorship of the proletariat'. (In later chapters some of our *dramatis personae* will become involved in the staging of one in Maoist China.) Society must instead organise itself federally from below based on the collectivist notion of the village commune.

Disaffected aristocrats like himself questioned all of the repressive institutions of monarchy, religion, the family and property. Reason should govern individual and social life, not faith or tradition. In Turgenev's 1862 novel, *Fathers and Sons*, the character Bazarov declares himself a nihilist, someone who bows to no authority and accepts no principle without examination. Bourgeois art, fashion and sexual mores were to be rigorously challenged; women should consider themselves free to shake off conventional restraint.

The negative side of nihilistic anarchism became introverted into the kind of existential detachment found in Dostoyevsky's character

Raskolnikov in *Crime and Punishment* who stood outside the law. In *The Devils* he used the real-life example of Sergei Nechaev, who in 1869 had murdered a fellow conspirator, declaring that a revolutionary should have only one purpose – to destroy society. But for many of the educated elite, a positive faith still remained in the notion of 'going to the people' in towns and countryside. As ascetic educators they would learn from their lives in turn. These anarchist missionaries were known in Russia as *narodniks* under the loose organisation of *Zemlya I Volya* or Land and Freedom.

Sergei Kravchinsky (1851–95), Russian anarchist, known as 'Stepniak'.

The members of one particular group in St Petersburg, circa 1872, initially engaging in study and self-improvement, became known as the 'Chaikovsky circle' named after its founder, Felix Chaikovsky, another aristocrat. This was the group that Ethel Boole came to know personally twenty years later after its members' exile to England. One of the most important to influence her was Sergei Kravchinsky, who later adopted the nickname Stepniak, or 'Son of the Steppes'. His book *Underground Russia*, eventually translated into English in 1883, became a worldwide bestseller.

The book itself was not an ideological call to action as such but a number of sketches and profiles of heroic life under Romanov tyranny. Resigning an army commission in 1871, Stepniak took up forestry as a practical skill with which he could aid the peasantry. His experience with 'the people', however, was probably typical. He wrote how, clothed in peasant dress with another *narodnik*, they met a peasant driving a sled. They told him that he should not pay taxes, that officials are robbers, and that the Bible preaches the need for revolution. The peasant merely urged on his horse. After a month, detained by a village elder, they had to flee the police.

Peter Kropotkin (1842–1921), Russian anarchist c. 1900.

Another later personal mentor to Ethel was Prince Peter Kropotkin. Born in Moscow in 1842, his father owned estates across Russia with 1,200 serfs attached. Despite an elite military education in St Petersburg, his twenties were spent in Siberia engaged in geographical survey work. Exiled from Russia for his subversive activities, during a stay in the Jura Mountains of Switzerland among watchmakers, George Boole's artisan class, he became convinced of the value of small-scale localised industry. Returning to Russia, disguised as a peasant, he mixed with workers in St Petersburg until his arrest.

Stepniak narrates Kropotkin's escape from the prison hospital of the Peter and Paul Fortress in 1876. Noticing on his prescribed daily exercise walks in its courtyard that an outside gate was open for periods guarded by only one man, he decided he could make a surprise run to an awaiting getaway carriage. A red balloon would be raised above the prison wall when the coast was clear. Strangely unobtainable, an improvised india-rubber ball filled with gas was tried but merely fell to the ground. A violin playing from a room overlooking the courtyard became the next signal. The moment eventually came, as he recounted, 'If I remained in prison I was certain to die. "Now or never" I said to myself.'[5] By a straggling whisker one of the most influential thinkers of the nineteenth century effected his clownish escape.

The noble but naïve rural *narodnik* diaspora was not successful; the authorities replied with mass arrests in the years 1873–74. Thirty-seven provinces were affected by the 'socialist contagion'. Yet Stepniak's faith in the peasantry remained, writing that it only needs a spark 'to make its hatred burst out into an immense flame which would destroy the entire edifice of the state …'.[6] The aristocrats paid heavily for renouncing their privileged background. Many died in prison, were exiled or fled.

The 1877 trial of the 193 *narodniks* detained in prison for up to four years had awoken some public sympathy for their cause. This sentiment extended to the case of Vera Zasulich, yet another aristocrat's daughter, aged thirty-nine, who, acting alone in January 1878 had shot and wounded General Trepov the tyrannical Governor of St Petersburg. Tsar Alexander's reaction to the outcome of the trial was to annul a petition for pardon that had been granted to many of the 193. With Dostoyevskyan logic Stepniak declared, 'against such a government everything is permitted.'[7] Grassroots non-violent education was replaced by militants advocating 'propaganda by deed'; insurrections were plotted both to avenge repression and to trigger support and revolt in the wider society.

In August 1878 Stepniak returned from exile in Geneva to St Petersburg with retribution in mind. In broad daylight he attacked General Mezentsev, the Chief of Police, and struck him in the chest with a dagger. He fled in a *droshky*, an uncovered carriage, pulled unbelievably by the same steed, Varvar, that had whisked Kropotkin away two years before. In *Underground Russia*, depersonalising himself, Stepniak writes, 'The Terrorism, by putting to death General Mezentsev … boldly threw down its glove in the face of autocracy.'[8] Fleeing the city, he spent some time yet again in Switzerland, before arriving in London where his book was already well known.

Despite the failure of Land and Freedom, a second, smaller and more single-minded group, back from the countryside, labelled itself *Narodnya Volya* or The People's Will. With a hierarchical Executive Committee it actually began to look like a political party with a programme of suffrage, liberal freedoms and socialist control of land and industry. But it was terror that emerged as the chief weapon of advance. Among the People's Will, regicide became more of an obsession than a goal.

The ideological distinctions, often used interchangeably, between socialist, anarchist and nihilist mattered little to the state when violence was in the offing. On the 13 March 1881 Tsar Alexander II was routinely travelling in his carriage back to the Winter Palace followed by two sleighs containing his guards and secret police. Along the embankment of the Catherine canal a member of *Narodnya Volya* threw a bomb under the carriage. The Tsar, unscathed, determined to inspect the damage but was felled by a second assassin's bomb. The loss of blood killed him within the day. It was the final attempt of eight; only two years before the same group had tried to blow up his train. The result was to abort the Tsar's plans to establish a first-step Assembly, and for his successor, Alexander III, to launch another period of repression.

The trauma reverberated around Europe. *The Times* in London labelled the nihilists 'apostles of the dagger and nitroglycerine', another editor ranted against 'idiots and scoundrels'.[9] Henry James, the novelist, was led to declare that society was 'dancing on the lid of a gigantic trap-door'. Yet even *The Times*, expressing a popular underlying distaste for tsarist despotism, noted the heroism of those brought to trial in St Petersburg. 'There was a refined, educated lady, an inquisitive chemist, a clumsy peasant and a pale arch conspirator … prepared to give their own lives for the life of the representative of authority.'[10] Stepniak provided details of the lady, Sofia Petrovskaya, in one of his sketches in *Underground Russia*. She appeared in court, 'tranquil and serious, without the slightest trace of parade or ostentation, endeavouring neither to justify herself, nor to glorify herself; simple and modest as she had lived'.[11] His admiration extended not only to her but to all revolutionary women, who he believed were more richly endowed with this 'divine flame' than men.

*Underground Russia* had a profound effect, for example, on Leo Tolstoy in Russia and in England on William Morris, who read it with 'a heady mixture of terror and excitement'. Morris recommended it to others, 'if you want your blood to boil'.[12] Young and impressionable, one can easily imagine Ethel Boole intensely re-reading its pages evoking sacrifice and loyalty to an ideal, morbidly fascinated and inspired.

These characteristics would be plainly evident in Ethel's first novel. Far, however, from such revolutionary aspirations, in England lay another strand hoping for social transformation. Here the desire to nurture a simpler life might be said to be a version of the *narodnik* impulse not so

The assassination of Alexander II, in St Petersburg, 1881.

distant from Tolstoyan leanings towards renunciation of wealth and the superfluous. This anarchistic strain led to the founding of the Fellowship of the New Life in London in October 1883. Among its aims were 'the subordination of material things to spiritual things … simplicity of living … and the introduction of manual labour … in conjunction with intellectual pursuits'.

Havelock Ellis, the pioneer sexual psychologist, the South African novelist, Olive Schreiner and Edward Carpenter were members. Carpenter combined gay sexuality with a Whitmanesque rurality, sandal-making and utopian socialism. Another New-Lifer, Hubert Bland, husband of Edith Nesbit, the children's author, wrote later, 'We felt that we had had the misfortune to be born in a stupid, vulgar, grimy age … so we turned to a world within a world … of poetry, of pictures, of music, of old romance, of strangely designed wallpapers and of sad-coloured velveteen.'[13]

Some Fellowship members felt that questions of how to organise a more equal society at large – the socialist element – were missing and decided to form another faction. George Bernard Shaw quipped that this led to 'one to sit amongst the dandelions the other to organise the docks'. It led to the beginnings of the Fabian Society in January 1884, whose outlook gravitated towards the idea of slow social change and the provision of general welfare through parliamentary pressure. Its members included Sidney Webb, Shaw, Graham Wallas and Karl Pearson (who we shall meet later in the famous scandal involving Ethel's eldest sister, Mary Ellen). Among the artists was pre-Raphaelite painter Ford Madox Brown. Feminism was represented by Annie Besant, scandalous promoter of contraception and Charlotte Wilson, who had attended the recently founded women-only Newnham College, Cambridge.

It was at Wilson's home in late 1884 that the group met for their radical discussions – a rural Chaikovsky circle. Despite being able to live comfortably on her husband's stockbroking income, Wilson wanted to pursue the rural 'new life' direction, moving in 1884 from Hampstead to the isolated shiplapped Wildwood Farm on the edge of Hampstead Heath. She renamed it 'Wyldes' and prettied it up. Another group member, Edith Nesbit, described it: 'The kitchen is an *idealised* farm kitchen, where of course no cooking is done – but with a cushioned settle, open hearth, polished dresser and benches … a delightfully incongruous but altogether agreeable effect.'[14]

In this sylvan setting the ideologies of anarchism and socialism were discussed, by St Petersburg standards somewhat twee. Revolutionary

Wylde's Farm on Hampstead Heath, home of Charlotte Wilson.

commitment did not require the renunciation of aristocratic entitlements or the courting of solitary confinement in a cell. British middle-class radicals sheltered behind guaranteed constitutional rights. Papers published and edited by Shaw as *Fabian Essays in Socialism*, however, emerged directly from the debate there. It was a formative time when political philosophies were being tried on for size as if handed out from a well-apparelled wardrobe. Charlotte Wilson declared herself an anarchist and proceeded to make contact with two of the formidable figures straight from the pages of *Underground Russia*. Both would greatly influence Ethel.

Of these, Stepniak, in London since 1883, was perhaps the most important. He had gained wide sympathy for his anti-tsarism, although no one was quite clear whether he was the assassin of Mezentsev. At Wyldes in his stumbling English he spoke of his experiences to an audience 'listening breathlessly to the sensational disclosures from his own lips of that wonderful narrative of daring events which took place during the Russian terror'.[15] Stepniak himself had a personal allure as a man of action.

Even in London he enjoyed physical work making furniture and managed to install gas lighting in his Alma Square apartment. The dagger that had killed Mezentsev was being used to make kindling.

Ethel Boole had arrived back in London from Germany and Paris in early 1886 at something of a crucial period in her life. She became friendly with Charlotte Wilson's radical set via her mother, Mary Everest Boole. Ethel must have been thrilled to mix in heady new intellectual circles. She too would have been attracted to Stepniak's personality, combining the heart of an affectionate child with a powerful extroversion. His exact political philosophy was less clear, using nihilism as a term partly because it had a certain exciting resonance in the West. He had only a vague notion of socialism; anarchism amounted to a form of federalism. While nurturing hopes for a democratic constitution in Russia, violence would be used if necessary.

The other figure was Peter Kropotkin. After his escape from the Peter and Paul Fortress he found himself in 1883 imprisoned once again, this time in France, for his supporting of illegal strikes. Charlotte Wilson was instrumental in securing his release in January 1886. Kropotkin as a person was widely respected and liked for his humility and sincerity. His general benevolence earned him the title 'The White Christ' from Oscar Wilde. Kropotkin took a cottage in Harrow, then a village well beyond London life, not far from another old-time *narodnik*, Nikolai Chaikovsky. This may account for Ethel not meeting Kropotkin personally until the early 1890s, but his ideas would have been well known. He shared Stepniak's love of manual labour, making his own furniture, growing vegetables and cultivating a vine that produced fifty pounds of grapes a year. His lack of ostentation and ego impressed all, despite his often risible English pronunciation. 'Own' rhymed with 'town', 'law' with 'low'; listeners were confused by 'the sluffter fields of Europe'.[16]

Stepniak did not share Kropotkin's intellectual grasp. Despite the latter's deep humanism, he believed his social theories to be as scientific as the geology he had once studied in Siberia. Like Bakunin he was strongly critical of capitalism and the need for socialism but similarly opposed to Marxist political methods and the idea of a workers' state. He did not believe, however, unlike many of the 'new life' persuasion, that industry and mechanisation were in themselves detrimental. *Fields, Factories and Workshops*, the title of one of his best-known works, suggested the anarchist view of their harmonious relationship under a socialism without government. Also contrary to Marx, he did not believe that historical class struggle was the dynamic force behind progress. Kropotkin posited that in the animal kingdom each species' survival depended on co-operation. He was fond of quoting Darwin's own example of the blind pelican supplied with fish by his comrades. For mankind too as a species, the optimistic, not utopian but rational, possibility lay open that mutuality could flourish in society. In his examination of primitive societies he found the very evidence of social coherence without government.

Charlotte Wilson and Kropotkin were in substantial agreement in their worldviews. Together in London they jointly founded a new anarchist group, *Freedom*. The first edition of its monthly paper of the same name came out in October 1885. It nailed its banner to the mast: 'We are Anarchists, disbelievers in the government of man by man in any shape and under any pretext. We dream of the positive freedom which is essentially one with social feeling … now distorted and compressed by Property, and its guardian the Law … of free scope for the spontaneity and individuality of each human being.'

One can imagine Ethel and Charlotte getting on well. Both were too serious-minded to frequent 'society' even if they had the means, which the former did not. A photo of Charlotte shows her as rather dour and forbidding. Charlotte's public face was evidenced by her explicit radicalism and public stance on a neglected equally important question – the role of women in society. Her outspoken feminism viewed women as equals rather than as class members and creators of labour. Ethel's political compass still wavered, although she did contribute as a writer to *Freedom*.

A composite image of the two women is caught perhaps by an anonymous *Daily News* review in March 1887 of a meeting to commemorate the Paris Commune, at which Wilson spoke. It portrays 'a slender person … dressed in black … the type is the South Kensington or British Museum art student, the aesthete with views'. Ethel would have been dressed in identical sombre fashion. Gemma, the heroine of Ethel's novel *The Gadfly*, was to be modelled on Charlotte similarly attired. Both women had compassion in common. They wanted to see the world change for themselves as free women within a less unjust society.

Even if Ethel had known about Stepniak's murderous role in the Mezentsev affair it is highly probable that she would have excused it. Stepniak and his wife, Fanny, seemed to take to Ethel immediately and she became a regular visitor to their home. They nicknamed her affectionately as *bulochka*, a pun on her surname meaning a sweet bread roll. With determination, she set to work to learn Russian from them and

'Farewell to Europe' by Alexander Sochaczewski. Prisoners being sent to Siberia.

made amazing progress. In only a short space of time after returning from Berlin, Ethel Boole had resolved what to do with her life for the foreseeable future. Her contact with Charlotte Wilson and her entourage had persuaded her that she must visit Russia herself and experience its woes first-hand.

1886, the year of her first association with Stepniak was a year that sharpened many people's awareness of oppression internationally. In May, in Chicago, a bomb went off at a rally to protest police handling of workers striking for an eight-hour day. Four innocent anarchist activists were hanged and civil liberties curtailed. Members of the Wyldes group signed a telegram to Chicago pleading for clemency. Meanwhile an American, George Kennan, who had set out for Siberia in 1885 to prove that accusations about tsarist oppression were untrue, reported back to the Freedom group about his complete change of mind. A series of articles for *Century* magazine in the US aroused much indignation in the public on both sides of the Atlantic. Ethel's decision was something of a whirlwind response. In a letter to a student friend, Irene Hale, in February 1887 she declared, 'my new life will really begin in a few weeks' time at Easter probably. I will cast off the London fogs, both physical and moral and set

The hanging of four Chicago anarchists, November 1887.

forth upon a sea greater than the Atlantic. And what if there should be storms? What else is there?'

Someone else casting off the same London weather and under a black cloud of scandal was Charles Howard Hinton and his wife Mary Ellen, the eldest of George Boole's and Mary Everest's daughters. Ethel continued,

> My sister's life is developing in the following manner. On the 13th [April 1887] she had a baby [Sebastian], another boy. [Charles] Howard has accepted a post in Japan, and is probably going the next week, Mary and the children will join him in the Autumn. It seems as if everyone is setting off somewhere. Of course that is as it should be. Our duty is to go our separate ways and build our own lives.

Mary Ellen and Charles Howard Hinton eventually built theirs in the US after living in Japan, transporting the Boole/Hinton dynasties far from their English roots. Ethel's sister Lucy, however, was going nowhere. Ethel continued, 'I should very much like her to accompany me to Russia but, of course, there can be no thought of that especially when one takes into account Mother's present state of health.'[17] Thirteen years after the Queen's College episode it seems Mary Everest was having another anxious period; whether physical or mental again, we don't know. Lucy, nevertheless, built her own inspired career.

Meanwhile in Alma Square, Ethel and Stepniak were preparing introductions for her visit to St Petersburg. Fanny had two sisters living in St Petersburg, Praskovya and Alexandra, although in difficult circumstances. The former's husband, Vasily Karaulov, had been locked away in the Schlisselburg Fortress not far away for three years, awaiting trial for political activities. His brother, Nikolai, an activist too, was living in the provinces under police supervision. It was going to be a gruelling time for Ethel, aged only twenty-three. Although used to misfortune and relative poverty, she had lived in a free society. Travelling to Russia must have been like visiting another planet, a domain of 'rude hordes' living in 'semi-barbarism', 'half in Asia' and 'intellectually in the middle-ages', as its citizens were cast in one English newspaper.

In April 1887 she set off for St Petersburg. If actions speak louder than words, Ethel Boole more than made up for her lack of expounded convictions with her courage. I have nothing but admiration for this golden-haired figure peering nervously out of her carriage window through the steam, en route to Paris as a first stop, dressed in black as ever, her suitcases in the string rack above.

# CITIZENS' DIPLOMACY

I too know the anxiety of an uncertain journey ahead. Exactly 100 years later, bar one month, in May 1987, I was peering out of the window of an Aeroflot plane at Gatwick airport bound for Moscow to make political contacts in the Soviet Union, tsarist Russia's inheritance. It was the fourth time in that decade I had tried to involve myself. I was not as alone as Ethel Boole must have felt. As a member of European Nuclear Disarmament (END) and its Soviet Working Group I'd been co-responsible for initiating a delegation to Moscow to visit the Soviet Peace Committee and various government institutions. E. P. Thompson, the historian and tireless campaigner against the Cold War, was the most prominent member of the group but did not want to come with us, to avoid compromising his independent stand. The well-known faces on our mission of six were Mary Kaldor, economist and one of the founders of END, Jonathan Steele, *Guardian* journalist and Ian McEwan, novelist.

Left to right: Mary Kaldor, Jonathan Steele, Kate Soper and Ian McEwan.

We were housed, wined and dined in the Stalin-gothic monster Hotel Ukraine, with its creaking lifts and hospital-like rooms that we all immediately scoured, behind pictures and under beds, for bugs of the electronic kind. We had no intention of compromising ourselves either; but I was lucky to be there at all. A few days earlier my visa had been refused for misdeeds earlier in the decade and it was only because of a threat of 'one out all out' that Soviet officialdom recanted.

It was a time of fulcrum changes in the USSR under Gorbachev, who had been embarking on a process of unilateral disarmament that the Peace Movement had been advocating for Britain for over thirty years. Domestically, the introduction of *perestroika* (rebuilding) and *glasnost* (freedom of speech) were two new and exciting concepts for the Russian people. However, the monolithic centralised bureaucracy that was the Soviet Union, vastly more wealthy and industrialised, was essentially as intolerant as the Russia that Ethel knew. We were latter-day Kennans making our own reports on the treatment of dissidents who dared challenge the system, but trying to build bridges with communist orthodoxy.

END was launched in 1979 as a new Cold War began to re-freeze and solidify the world's two superpower blocs: the eastern, Soviet-led Warsaw Pact and the western, US-led NATO. The former's act of refrigeration was the invasion of Afghanistan in 1979, the latter's the announcement that Pershing and Cruise missiles were to be installed in western Europe. Each act reciprocally reinforced the polar climate that dominated the times. It was Thompson's acuity that highlighted the economic and ideological banners that motivated the rivals as one of 'Peace' and 'Freedom' respectively. It was a monstrously foolhardy contest, mission statements are one thing, arms races another. Post Cuba 1962, nuclear war by accident or design seemed to be a grotesque, calamitous possibility.

Across the US, western Europe and Japan the claim to 'Freedom' was rigorously put to the test as marches, demonstrations, protests, acts of non-violence and peace camps were instigated by ordinary people of every possible denomination: as voters, by belief, profession, gender, sexual orientation, age and ethnicity. It was a period of extraordinary imagination to counter the dearth of thinking of the powers-that-were. Every lie and half-truth from both sides asserting histories, moral superiorities and weapons statistics were exposed in public meetings, press campaigns and even parliament.

In the borderlands of Shropshire, where I was based, a very active peace group had been formed, one of hundreds around Britain that met together at the Hyde Park demos that attracted protesting multitudes not seen since the nineteenth century. In all its guises and formations the basic unit of political activity was the individual citizen coming together with others in communities within an overall movement organised and directed from the grassroots. Its ethos was essentially anarchist. It paralleled, for example, social developments in the nineteenth century such as the Settlement Movement that had involved members of the Boole/Hinton families, anti-segregation in the US, the women's movement of the 1970s and the spirit of 1968 a decade before. It entirely accorded with my own view of political change.

Looking today, a half century later, I see not much has changed. The blatant deceptions that justified the war in Iraq rile me as much as those rolled out then. I joined the million-and-a-half plus who marched in February 2003. Sadly our principles do not seem to have had much effect on the international scene. We still have a ridiculous number of nuclear missiles that don't make any sense as so-called deterrence; nuclear proliferation is still a major risk to world peace.

Our delegation, however, in 1987 was putting forward END's manifestly sensible notion of a 'reasonable sufficiency' in warheads in lieu of complete disarmament. At various institutes that we visited this was listened to seriously by opinion-formers. Many years later it became clear that our views had actually filtered through to high levels and possibly even influenced Gorbachev's thinking. We met with prominent players in the media world such as Yegor Yakovlev, the editor of the influential *Moscow News*, and visited one of the discussion 'clubs' that were forming to test the boundaries of the new licence to debate.

If we were playing at politics, there was a caveat; we reserved the right to also engage with the unofficial Soviet peace activists, our 'cousins', who were under constant pressure from authority. This prerogative we termed 'citizens' diplomacy' pursuing 'détente from below', a logical extension of the right to control one's foreign policy as much as one's neighbourhood. We were 'going to the people' internationally. This led to our support for the minuscule but important independent peace group that was trying to find purchase within the Soviet leviathan. END supported other eastern European groups too: in East Germany it was allied to the Protestant Church; in Czechoslovakia, Charter 77, the civil rights group, was making a stand; others existed in Poland and Hungary. As Jonathan Steele was to argue in 1999 in the *Guardian*, it was these beginnings in both Prague and Berlin that shaped events in the Great Thaw of 1989.

We also played at being tourists. Our status bestowed certain privileges in return for our valuable hard currency and the opportunity to visit

Russian institutions. One such is the Bolshoi. Tickets at only a few roubles were available to the public if you were lucky. Ian McEwan and I went to see the opera *Blood Simple*, a Spanish affair by Lorca sung in German to a Russian audience as uncomprehending as we were. The interval drinks of champagne and a pinch of caviar in the plush surroundings were, however, rich 'guilty-pleasure' reminders of tsarist days. McEwan was later in the week invited by the Union of Writers to their HQ as a guest of honour. They plied him with champagne toasts one after the other; he later said that he'd never been so drunk in his life.

In the USSR any fledgling peace group was up against stiff opposition. The Moscow Group to Establish Trust founded in 1982, mainly by intellectuals, tactfully did not try to criticise Soviet foreign policy but to independently implement the constitutional obligation of every citizen to rally under the 'Peace' banner. This they construed in terms of exchanging correspondence and meetings with western peaceniks, holding art exhibitions devoted to peace and conducting self-educating seminars. With *glasnost* in the air, their harassment by officialdom had eased in comparison with earlier days. Then, the enraged Soviet Peace Committee declared its historically destined right to speak, 'on behalf of the peace-loving peoples of the world' and eradicate the usurpers.

Drawing of members of the Moscow Trust Group by Ann Pettit.

In January 1983, Jean, another peace contact, and myself were the first Brits to meet with the group. Not wishing to elevate historical coincidence to some magnificent parallel, I was enacting almost exactly, without the faintest awareness, what my relative Ethel Boole had been doing in the 1880s. It was the first stage on my globe-trotting peace activity that took in the geographical centres of the Boole/Hintons: Russia, China, Japan and the US. Ironically, had I known of her history, I might have gained some partial exemption from the nervousness I felt at the time. In the person of Ethel Voynich, novelist, she had acquired a saint-like beatif-ication in the communist world owing to its veritable worship of her first novel *The Gadfly*. As a relative of hers doors would have opened magically, but, like her, part of my role was to smuggle in and out banned literature. Later this proved to be my undoing. Brief diary extracts record my impressions of that visit as a member of an ordinary Intourist party, the only way at the time to take a peep behind the Iron Curtain.

SATURDAY 1 JANUARY 1983

At Gatwick the Aeroflot plane, a Tupolev-154 was waiting; apparently unusually close to the terminal – normally they park in batches away from the other airlines for security reasons. The Cold War intrudes on day one of the tour.

SUNDAY 2 JANUARY 1983

First visit to Red Square, it's bitterly cold, minus 25 centigrade. Many police aimlessly herding crowds. At one end, St Basil's cathedral is a marvel of intricate, barley-sugar, pineapple and onion shapes of orange brickwork and painted render. After tea, off by coach to the ballet. Expecting 'Romeo and Juliet' at the Bolshoi, as we've been promised, we're driven instead, to our guide's embarrassment, in a coach that banged and spluttered like a firework, to a very minor theatre. Here, to records, six dancers are performing. We decide to leave, to the amazement of the garderobe attendants.

In the afternoon we popped into GYM the State Universal Store on the square. It's a bit like a glorious railway shed, but the goods are very expensive and of poor quality. You have to queue three times to get what you want. One of the group asks for an LP of Swan Lake – sold out, as was the main Melodiya record shop. In the USSR supply and demand don't operate: if something's in demand and you run out, so what, no skin off your nose.

On the Metro down the bottom of the escalator in a little kiosk sits a woman fast asleep as Moscow passes by. There is much under-employment here, but at least everyone has a job – especially as it's illegal not to.

MONDAY 3rd
In the grounds of the Exhibition of Economic Achievement we queue for a troika ride, very evocative on a short trip through woodland. The three horses are lovely and their bells jingle à la Prokofiev. Back in the exhibition pavilion for food – we had stuffed pancakes, eclairs, orange caviar and samovars of good tea with lemon. An electrified pseudo-balalaika band strikes up. They play woodenly and refuse to smile even when we take to the floor looking silly. They are doing their job, no more; smiling is not part of their productivity package.

TUESDAY 4th
Trip to the Tretyakov Gallery. Large queues of Russians outside like steaming animals waiting to be milked – we go straight in, of course. Inside, school parties of children, their white/sallow skins never sport ruddy cheeks. This is heightened by their clothing of very poor quality, man-made materials and without design. One can spot a westerner a mile off because of their bright clothes and individual accessories. How important is the freedom of self display?

Jean and I have arranged to meet the Trust Group. We travel south to one among hundreds of apartment blocks. I feel jumpy at times and wish I was not involved. We're welcomed, however, into Sergei Batovrin's cramped flat. He's an artist, frail, bearded but speaks good English. Yuri Medvedkov is a distinguished-looking political geographer of international standing. His younger wife, Olga works at the same institute. We talk from 8 till 1am. Their bravery amazes me, they are so unassuming, humble and resigned to an almost futile potential martyrdom. They seem to know little of what's been happening in the West, despite getting news of Sergei's recent release from a bogus stay in a mental institution through Voice of America. They desperately want support to mitigate their imminent fates. Yuri says our visit has been like a 'breath of oxygen'.

THURSDAY 6th
At breakfast it was pointed out that my pullover was inside out.

I reversed it and someone noticed a pin in it. I joked about it being a KGB plant and licked it. One hour later on the coach to visit a museum I started to feel my mouth dry out as the 'deadly poison' on the pin began to work. The power of paranoid imagination. It's what fuels the Cold War.

FRIDAY 7th

Take Metro to meet Jean and Valodya Brodsky. A charming man, late 30s, Jewish and a dedicated doctor. He speaks little English, so Jean translates. The KGB have set light to his flat and kicked in his front door plus making an official search-report. He expects to be put away. He's very impish and humorous. Outside the flat KGB are waiting for us in two 'taxis' half-parked on the pavement (not normally permitted). They watch us at the bus stop. On to the Moscow Conservatoire and a piano trio of Beethoven. We discuss where to pass on an article to be smuggled out. I ask him, 'What do we do when both sides believe each other to be responsible for hostility?' He says, 'Dance.'

SATURDAY 8th

Off by Metro to hunt the avant-garde. Finally find a dingy basement of an ordinary civic block of flats. One artist is obsessed by fluffy, budgerigar-like birds, another by Magritte-like still-lives with red-lipped couches, pomegranates, elephants and occasional women. In the west I'd tend to despise this stuff as individual bourgeois fantasy, but here I feel pleased to see it. Later at midnight we take the 'Krasnaya Strel', the Red Arrow train to Leningrad. Brodsky turned up at the station, he and Jean have been having an affair. As the train pulled out he signalled 'V' for victory. I sleep fitfully for 6 and a half hours until we arrive in the dark.

WEDNESDAY 13th

Off to the Hermitage Museum. Rushed around several miles of exhibits, but remember some fleshy Rubens, da Vinci Madonnas and acres of tsarist paraphernalia – thrones, pottery, beautiful inlaid tables, vast chandeliers and malachite vases. Up to the Impressionist gallery, a stunning collection: a magnificent van Gogh, two incredible Munchs. While we are watching a scruffy worker in overalls comes up to an empty frame with a Renoir canvas under his arm and slots it into place as if he were changing daily newspaper posters. Under a Matisse picture a little babushka 'guard' sits fast asleep.

Afterwards three of us walk up the Fontanka embankment to meet the Vice-President of the Leningrad Peace Committee in a large gilded, be-mirrored room. Bottles of Pepsi are arranged for our consumption. After a long introductory spiel about the USSR and World War II and the Soviet need for strong defence, Jean asks about the independent peace group and is told they are 'hooligans'. A general air of hostility is frozen in diplomatic nicety. On return to our hotel Jean tells me that Valodya tried to come to Leningrad but was thrown off the train by KGB and held for 3 hours.

The author outside the St Petersburg cemetery, 1983.

SATURDAY 15th

After a dish of fried eggs, we visit a cemetery in a big monastery opposite the hotel which contains a collection of monuments to Soviet heroes. Eventually find burial area of Russian artists. A notice-board gives the number of their graves and table of inmates. Like some weird board-game we spot Dostoyevsky, Tchaikovsky, Glinka and Moussorgsky. Some snow has fallen overnight, the graves look enchanting with a weak sun behind.

SUNDAY 16th

(Back in Moscow) Still feel terribly ambiguous about the USSR. Wish I could see some non-conformity, private fantasies, deviance somewhere. Did notice two people laughing this morning and some kids playing. No sex anywhere except on packets of tights – saw one man drooling over them in GYM Store. Is sex subversive, does it blunt productivity?

Certainly there is plenty of sexism, the whole 'heroic' ethos of the USSR is male with only token be-spannered women. They in general have less important jobs. Must stop complaining about the country! Underneath the grim exterior I suspect most citizens are very proud of Soviet achievements – new houses, cheap and warm, new roads in the cities, cheap transport, Sputnik, free phones, an international high standing as a superpower – all in 60 'Glorious Years'. Their image of the West is derisory – nobody can get a job, the military duping everyone – taxi drivers have expressed both points of view.

Phone Valodya and go to meet him under the statue of Sverdlov by Red Square, warily eyeing other waiting figures. We go down escalators and come up again as an anti-KGB ploy and head off by Metro and taxi to the usual dingy, cramped flat where a group of shabbily-dressed professors congregate. The meeting is a seminar this time on the medical aspects of nuclear war. Two papers are given and I talk about alienation, haltingly as I'm translated. Later we walk back to my hotel passing Pushkin Square where once a year on his birthday people doff their hats as a mark of respect and defiance.

MONDAY 17th
I meet Mark Reitman outside the Lenin Library. He is bear-like and speaks English very slowly. We walk the streets of slush discussing anarchism and nuclear politics. He has lost his job teaching civil engineering. He is reconciled to prison and will go on an insulin strike as he is diabetic and presumably end his life. What can one say in face of such bravery? How can people live like this? He tells me of a Polish sci-fi story of people who get used to living in a land where stones continually fall all day. He occasionally looks around nervously to see who is listening, but basically believes his goose is cooked – mine too he thinks – he expects they will accompany me on the Trans-Siberian to Peking. We part at Gorky Street. Back in the hotel in bourgeois comfort I phone to say goodbye. 'You good man,' he says. 'You're a lovely bloke,' I reply sincerely. To bed in comfort for the last time in 6 days before the train to China.

Ethel's train arrived in St Petersburg on 17 April 1887; she was met at the station by Stepniak's wife's two sisters, Praskovya and Alexandra. It must have been a long and frightening journey. Her arrival, pursuing her own form of citizens' diplomacy, coincided with a final dramatic episode in the long preceding history of political turmoil associated with terror. Three weeks later on 8 May five men were hanged for their planned attempt on the life of Alexander III, armed with shells packed with dynamite and lead bullets filled with strychnine. Before the court Alexander Ulyanov defended their intention: 'Terror, this form of struggle devised by the nineteenth century, is the only means of defence to which a spiritually strong minority, convicted of its righteousness, can resort against the physical power of the majority.'[1] A statement that could easily represent the views of today's Al Qaeda.

It would certainly have had a powerful effect on Ulyanov's younger brother, Vladimir Ilyich, aged seventeen, soon to be known as Lenin. The repression focused the forces that later toppled the regime. Ulyanov's small group, the third incarnation of the nihilist/anarchists, called itself the *Terrorist Faction of the People's Will*. Its rounding-up inevitably ensued; as in previous times there would be no insurrection by the people themselves. On the contrary, at the site where the previous Tsar's legs had been blown off, a curlicued, multi-domed cathedral was being built during Ethel's stay. The Church of the Saviour on Blood housed a shrine to God's appointed son, garnished with topaz, lazurite and other semi-precious stones.

Most pre-industrial cities evolved slowly in relation to their agrarian hinterland; Peter the Great's edifice, St Petersburg, was raised in the early eighteenth century out of the swampy shores of the Gulf of Bothnia, blue-printed to be a magnificent shop window for the country's trading potential. Its grand streets and squares stood as testament alongside the impressive embankments of granite fronting the River Neva, from which tributaries were channelled into Amsterdam-like canals. The Winter Palace completed in 1817 contained over a thousand rooms and cost the equivalent of forty-five tons of silver. These emblems to St Petersburg's grandeur were extracted from the toiling, subjugated peasantry. By the 1860s, industry had begun to establish itself. The original city was becoming, according to Solomon Volkov, 'surrounded by a ring of grim, sooty factories littered with hovels and ugly tenements ... threatening to become a nightmare'.[2] This was the backdrop for Dostoyevsky's intense stories of exalted and low-life side by side. 'I'm sorry I don't love it,' he declared, 'windows, holes – and monuments.'[3]

The lack of material extant from Ethel's own pen at that time makes it difficult to chart her stay in St Petersburg. Fortunately, however, her Soviet biographer, Evgenia Taratuta, with official endorsement, inspected government archive material. In her introduction to *The Fate of a Writer and the Fate of a Book*, published in Moscow in 1964, she states that she

Ethel Boole, aged about 24, in St Petersburg.

had 'access to tales of people … who knew her personally. The name Voynich [as Ethel became], like the magic word *sesame* opened up hearts.'[4] Taratuta also later corresponded with Ethel. The 'heart-opening', however, inevitably seeded a gush of admiration that has to be picked over by a western biographer with care.

Taratuta doesn't give any details of where Ethel lived in the first few months of her stay but presumably she took work teaching English or music. She met up with the Karaulov sisters, who would have introduced her to the shadowy circle of revolutionaries. Their husbands, the brothers Vasily and Nikolai Karaulov, were absent elsewhere paying the price for their People's Will allegiance. Another suffering friend was the poet Peter Yakubovich, a member of a young section of the same organisation. Arrested in 1884, he had lived with the Karaulovs in the same block of houses built around a grey courtyard. After three years in the Peter and Paul Fortress he was sentenced to death but exiled to Siberia instead where he spent eighteen years. His inspirational book of poems, even though published under a pseudonym, could only contain veiled heroic aspirations. One poem spoke of an eagle bravely hurtling into the face of a storm. Praskovya was struggling to bring up her son, Sergei, and become a doctor. Ethel found herself on the edge of a revolutionary pool without many ripples at the time because its subterranean denizens had been mostly caught and landed.

In complete contrast, in the summer months she took up an appointment with an aristocratic household as tutor to the seven children of the Venevitinov family at their estate in the black earth country near Voronezh, 300 miles south of Moscow. (This connection may have come about through someone from the Hinton circle, Caroline Haddon, who ran a private school in Dover.) Ethel's duty was to teach English and to play the piano in the evenings when guests came. The family were aristocrats of the highest rank; the patriarch had been Chamberlain at the Royal Court, one of the sons had the Tsar as godfather. The children were, unsurprisingly, spoilt and arrogant: 'We hated each other,' she reported,[5] and left after only a few weeks.

Footloose now, she must have decided to continue to explore this strange country. Journeying on by cart along dirt roads, she was given an equally jolting introduction to the extremes of Russian life beyond the big cities. At six in the morning on 7 August (or 19 in England – Russia was behind in everything, including its calendar) a total eclipse of the sun was due, an event as traumatic as the putting out of Alexander II six years earlier. The superstitious peasantry feared the end of the world and bought all the oil they could to keep their icon lamps aflame. Ethel had been invited to a manor house near Kostroma, 400 miles southeast of St Petersburg, to watch the event. At a posting station to change horses, an old woman asked her where she came from. Moscow, Poland, Germany the crone supposed in turn. When 'England' was announced, Ethel was asked to remove her gloves. On doing so she was rebuked for making fun of an old woman seeing that her hands were perfectly normal and not 'clawed' as was typical of the English.[6] The glove incident must have been only one of many revealing to her the depth of backwardness in rural Russia.

Continuing her great tour, she travelled eastwards, joining a boat voyaging down the great Volga river as far as the ancient town of Nizhny Novgorod. When she finally returned to St Petersburg via Moscow, Ethel took up residence with the Karaulov sisters for the rest of her stay in the city. A Police Record unearthed by Taratuta locates Flat 17, block 7, 7th Rozhdesvenskaya Street (now Sovietskaya) in the Peski district, a crowded area of grid-like streets close to the Moskovsky station from which Ethel had first disembarked. Dated 26 June 1889, Section number 4546 of the Security Department, the report lists the Karaulov address home to 'persons known to have associated with politically unreliable persons including the music teacher Ethel Boole and Peter Yakubovich'.[7] The enforced isolation of the Karaulov police-watched household must have pushed Ethel out to wander in the streets. A stroll away would have taken

130

Nevsky Prospekt, St Petersburg's main street, in the nineteenth century.

her to the delights, if she could have afforded it, of Nevsky Prospekt, St Petersburg's Oxford Street.

One of its main attractions was the vast, arcaded Gostiny Dvor building containing hundreds of little shops. Elsewhere, dance halls, fashionable restaurants and theatres catered for bourgeois tastes. With her love of culture she might have found her way into a gallery showing one of Repin's paintings. Realistic, but with romantic subject matter, they would have appealed to her – portraits of composers and poets but peasants too. His 1887 painting of Tolstoy as a ploughman in the fields combined both aspects. She could have heard the Frenchman, Camille Saint Saens play the piano or she might have gone in 1888 to a performance of Tchaikovsky's latest symphony, the fifth. Ethel's personality, prone to the morbid, drew her to visit the Peter and Paul Fortress, 'where the dead Russian emperors lay in luxurious tombs, while next door the best people of Russia languished in grim cells, buried alive'.[8]

After all her travels her use of the language would have progressed enormously. According to Taratuta she was able to immerse herself reading Dostoyevsky – somehow a natural choice, and similarly, Vsevolod Garshin. His grim short stories included *The Scarlet Flower*, the tale of a madman who believes that the three poppies he can see in the garden hospital

Vsevelod Garshin (1855–88),
Russian writer.

Michael Saltykov Shchedrin
(1826–89), Russian writer.

represent all the evil in the world. Managing to pluck them, he dies of
mental exhaustion. Garshin himself, an acquaintance of Praskovya's,
clearly disturbed, threw himself from the fifth floor of his flat in April 1888.
Another favoured writer was aristocrat Mikhail Saltykov, who wrote under
the name of Shchedrin. He had been internally exiled for his satirical
writing in 1854 and later became a civil servant. He began a series of
*Provincial Essays* sending up the bureaucracy and setting them in a town
called 'Glupov', translatable as 'Stupidville'.

In the summer of 1888 Ethel quit the city briefly to experience life in
the countryside once more. She accompanied 'Pasha' and her small son
Sergei to the Karaulovs' family home 300 miles south in a remote village
near Velikiye Luki. The area is one of low-lying, lakeland swamp described
by her in the later novel *Olive Latham* as 'cursed alike by nature and man,
frozen half the year, malarial the other half, neglected, chronically famine-
stricken …'.[9] She nevertheless remembered the beauty of the 'untrodden
tangle of alder swamp, golden with iris, blue with forget-me-nots and the
circling of great hawks by day, the howl of many wolves by night'.[10] Their
house made of split logs and dried moss was as dilapidated as the family

Peter and Paul fortress.

fortunes with Nikolai unable to work, dying and depressed at the failure of his revolutionary endeavours. Not much is known about his involvement with People's Will but some of its members' desperate intrigues are documented by Richard Pipes in *The Degaev Affair*.[11] His brother Vasily was heavily involved.

This convoluted Dostoyevskian saga concerned the murder of the Secret Police hounder of People's Will, George Sudeykin. Sergei Degaev, one of the group's leaders, had double-crossed scores of its members, including Peter Yakubovich and Vera Figner, one of Alexander's regicide plotters. At a St Petersburg meeting of the People's Will Executive Committee in October 1883, Degaev was given the grisly task of killing Sudeykin as part expiation for his calumny. One of the committee's members, a Pole, Stanislaw Kunicki, was soon to figure in Ethel's future husband's life.

The murder took place on 16 December 1883 at Degaev's St Petersburg apartment where Sudeykin and another police officer were lured. Sudeykin was shot twice and the other's skull was smashed in with a crowbar. Kunicki escorted Degaev on his escape by train to a Baltic port with orders to shoot him if he caused any trouble and then commit suicide himself. Finally in Paris, a trio of People's Will members, including Vasily Karaulov, granted Degaev a stay of execution if he emigrated permanently. After several years hiding in London he, amazingly, lived out his life until 1921 as a maths professor in South Dakota, US.

Shortly after, in February 1884, Vasily was ordered to Kiev to start a printing press but was arrested. At his trial, afraid for his future life with Praskovya, he revealed details of the Sudeykin affair but resisted a large bribe to help locate Degaev. Schlisselburg jail was the result. Did Praskovya know the whole story about her husband's not entirely virtuous involvement; if she did, did she explain it to Ethel? Regardless, it must have been a sombre household in such wild country alleviated only perhaps by the need to minister to Nikolai, his brother and the village sick. Praskovya, who had nearly qualified medically found herself in demand there as a worker ministering to the victims of the endemic TB and VD. Ethel accompanied her by helping out, giving injections, boiling instruments, bandaging and ministering generally. This Nightingale role did not soften her stiff view of humanity's lot.

Back in St Petersburg in the autumn she was able yet again to help with Praskovya's troubles. Vasily had been transferred to a detention centre in the city, awaiting transfer to Siberia. Ethel had secured permission to deliver food to him. In the prison building not far away on Shpalernaya Street, she endured the seedy underworld of criminality – one old woman offered to recruit her to a brothel. The capricious whims of the officers would keep her waiting deliberately. Surprisingly, she sometimes took Sergei, Preskovya's son aged only six, with her to this nightmarish environment. On one occasion returning on a horse-drawn bus he was asked by a passenger whether Ethel was his mother. He replied that she wasn't and when asked about his father volunteered that he was 'in a cage'.[12] Soon after, the dates of the exile were confirmed; Vasily was sentenced to four years' hard labour in Siberia. Praskovya resolved to accompany him there with Sergei. Ethel saw them off on the long journey by train.

Praskovya's story after she left St Petersburg for Siberia is unclear, although she was yet to play a key role in Ethel's life shortly afterwards.

Maria Tsebrikova (1835–1917), writer of the appeal to Tsar Nicholas I.

She died in 1902.[13] Not all of Ethel's recorded St Petersburg encounters were with desperadoes. Maria Tsebrikova was a well-educated, middle-aged public figure and writer of socially critical stories and articles. She rather naïvely hoped that an appeal directly to the Tsar by a reasoned letter might draw his attention to Russia's ills. Maria wanted to enlist Ethel's help to smuggle a copy out when she left and have the document mailed to the European press.

Meanwhile the winter of 1888–89 had to be lived through. Owing to Ethel's teaching, the household was able to survive, unlike the many poverty-stricken, hypothermic residents of the city who suffered in the beautiful surroundings of snow-clad parks, gilded domes and the broad frozen reaches of the Neva, solid enough to take the weight of a carriage. It was a setting Ethel was to remember twenty years later in her novel *Olive Latham*, peopled with the originals of which she was meeting at the Karaulovs. Ethel clearly occupied a precarious position during her stay. As an Englishwoman the worst that could happen would be her arrest and deportation. After two years she had sampled the range of Russian life: the degradation of the peasantry, the low-life of the proletariat, the underground lives of the politically active and the inside of a prison – but also the opulence of the aristocracy and bourgeoisie whose children she had taught. One cannot but help admire Ethel's strength of character and purpose. When a date for the Karaulovs' exile had been set, she decided to return to England. Just before she left, an event occurred that must have been emotionally powerful.

On the night before her twenty-fifth birthday, 10 May, the news of the death of the writer Shchedrin/Saltykov had broken, a figure much liked and respected by all from progressive circles. A current of sentimentality was expected at his funeral, too big to be thwarted by banning it. Ethel joined the cortege through the police-lined streets to the Volkov cemetery where Shchedrin was to be buried next to Turgenev. Somehow she found herself near the grave site. Speeches were declaimed and wreaths laid; one she recalled was inscribed, 'to the unmasker of obscurantism, to the champion of truth'.[14]

One of the many problems that faced the Karaulovs was what to do with Praskovya's sister, Alexandra, who didn't want to return to their country home where Nikolai was dying, or stay on her own in St Petersburg. The solution offered by Ethel was to lend her the money to come to England and live with her other sister, Fanny, Stepniak's wife. In early June 1889 the pair left on the train via Paris. Among her luggage was concealed Maria Tsebrikova's missive to the Tsar. We can allow Taratuta some purple prose for once. 'Lily was anxious to be with her mother and sisters, to rest after her terrible experiences, to see Stepniak and meet his friends. She felt from now on her fate was inevitably linked with Russia.'[15] It was, in fact, although she never returned.

On Tuesday, 18 January 1983 I left on the long train journey from the Yaroslavl station in Moscow to Peking. Mark Reitman had given me an article on mathematical symmetries in the Cold War. I agreed to take it east with me on the Trans-Siberian instead of west, like Ethel, worrying that baggage would be searched at the border.

The diary account of my journey on the Trans-Siberian as a Cold War courier I later turned into a more polished version.

### 18 JANUARY 1983

It was the lemons that saved the day. A whole consignment of them had suddenly arrived in Moscow shops to add an exotic touch to the windows mostly displaying no more than packets of sugar and bottles of gherkins. I had this image in my head of me, sipping G and T's as the miles of the Trans-Siberian Railway slipped by. There wouldn't be gin, I was told, but vodka and T would suffice, and both tipples needed a lemon. Perhaps I'd seen too many Orient Express-type films and fancied myself as a suave and rakish spy but I was quietly relishing my role taking Reitman's clumsily typed copy of his peace document with me.

The first thing I did after the train crept out of the station was to unpack my Swiss Army knife, unscrew part of the heating duct in the toilet, and there secrete my precious contraband. This Le Carré-like ingenuity provided me with an inner warm glow for the first day of the journey. Subsequent days were more routine. Myself and the handful of other, mostly Scandinavian, western travellers occupied one carriage in splendid isolation from the packed and noisy, vodka-swilling Russian part of the train.

We plod on relentlessly at about 50 mph. Most of the landscape to Irkutsk is flat or gently undulating with endless snow-clad birch and pine forests. The settlements are either pretty clusters of wooden shacks or unpretty new towns with apartment blocks. No doubt the inhabitants prefer the latter. The train stops every four or five hundred miles. Dressed like an Eskimo, I would descend the train and run up and down the platform to get my insides moving. Occasionally I'd perform an arms-clapping-in-the-air number I remembered from school PE lessons, much to the amusement, and on one occasion applause, of Soviet spectators. At one station a glorious clanking steam engine appeared pulling an ancient steam crane. Soviet engines are usually very tall yet squat and don't look very friendly. Otherwise, apart from reading and exchanging overhearty ritual gabs of conversation with the Polish guy in my compartment in the universal language of football – 'Bobby Charlton … Preston North End' – food and drink were the only other diversions.

We're on the edge of Lake Baikal now, the sun shining on the frozen lake. On it are what look like stranded seals, but are in fact fishermen lying out flat with their rods through the ice. Inside it's a bit like being in an old people's home, cramped and warm. The bloody intercom, which I can't switch right off, is playing 'The Banks of Loch Lomond' having warbled its way through 'Camptown Races', 'Calinka' and 'Return to Sorrento' (2nd Class). There is food to hand. Breakfast consists of the unvaried serving of three oily fried eggs floating in a silver dish, plus rye bread. At one sitting, to my paranoid horror, in a completely empty restaurant car, a stolid, silent heavyweight sat himself at my table. Uncomfortable, I made a mental note to heed Mark Reitman's warning and be vigilant.

This became increasingly difficult as any semblance of reality waned; partly as a result of large doses of self-provisioned vodka and lemon (no tonic unfortunately), but more importantly, the frequent changes of time zones and altering of watches that confused my entire metabolism. Nevertheless, in the small hours, at the Soviet and Mongolian border, I roused myself to counter any KGB-like investigation. To my great relief, and demoting Mark's cautions, everything went without incident and I slept peacefully, to awake at the dazzlingly sunlit, but bitterly freezing, station of Ulan Bator, the capital of Mongolia. A strange city, high on a plateau with modern tower blocks next to compounds of yurts. I bumped into the British Ambassador by the train and exchanged a few sentences; here not to meet me but presumably to send off some dignitaries. A little later, at lunch, as the train trundled ponderously across the Gobi Desert, as we rattled round bends the plates and glasses attempted to slide off the tables. As in some Marx Brothers film we each tried to cover a couple of tables to stop gravity taking its toll. From the window we could see occasional two-humped camels, yaks and military aircraft sites.

One morning as I sat down to yet another fried-egg feast, I encountered the mirage of two very stiff-lipped Brits tucking into toast and marmalade. Almost the first thing I heard one of them say was, 'I'm told their conserves are awfully good.' These were Queen's

The 1983 west European travellers aboard the Trans-Siberian. (The Queen's Messengers are seated behind the partition on the left.)

Messengers, Phillip and James by name, far-flung emissaries of Her Majesty, carrying diplomatic bags between Mongolia and Hong Kong.

We were in the same business in fact, albeit serving different masters – theirs a regal mistress, mine, according to the absurd western press, in the pay of Moscow. A polite compatriot exchange led to the invitation to rendezvous chez Phillip and James in the evening. I didn't dress for the occasion; no need, for there, among the homely spread of peanuts, digestives, beer and gin, we commenced a jolly exchange of news. Lubricated by the booze, however, we quickly locked into our very own Cold War as we rancoured over such issues as public schools, immigration and the Falklands conflict. A sullen truce eventually crept over us as embarrassing miles carried us nearer to the Chinese border.

At one point James incidentally bemoaned the pity of having a decent bottle of Booth's and some tonic, but 'no bally lemons' for a proper G and T. Like an arriving mountain rescue team, my offer to fetch a couple of them left in my larder, two coaches up the train, proved the most effective peace offering one could imagine. Later they politely enquired about my travels and I felt it was safe enough to let them in on my mission and its now successful accomplishment. This was greeted by an almost derisory admonition; surely I must realise that the real exit from the USSR was between Mongolia and China, the former being no more than a Soviet satellite. I explained, my innocence somewhat dented, that, never mind, the article was well hidden in the toilet. They guffawed some more; surely I must realise that the customs guards frequently tear the train apart, starting as like as not in the toilet, that most obvious of places. We were at this point slowing down as the train approached the border. If I can get the document to them, Phillip and James magnanimously suggest, they'll take it through in a diplomatic bag. I race back to my carriage and, hands trembling, with difficulty retrieve the package and return to my compartment.

The train has now stopped, doors are slamming. Nervously I set off back towards the Queen's Messengers. Immediately on hitting the corridor I am confronted by two stony-faced soldiers in Russian uniform, replete with rifles and fixed bayonets. I stagger back inside and sit on my precious asset to await my fate. They eye me coldly. 'Passport,' says one. I rummage in my bag and offer it up. They turn smartly about and carry it away. Following them gingerly, as in some silent movie, I retrace my steps to where, to my great relief, Phillip and James obligingly unlock their little black bag and deposit the document.

Big smiles all round. Once safely on Chinese territory, having experienced, thankfully, no demolition of carriages and the polite return of my passport, I reclaimed the document. It was surely the only time that a couple of little citric offerings had persuaded HM Government to succour a subversive like me.

Mark Reitman's contribution to world peace stayed with me as I journeyed on eastwards through Japan, across the Pacific, across the US by bus and back to Blighty. In April 1984 I returned to Moscow and met Mark Reitman again. He gave me yet another mystifying document to take back home. My fortieth birthday was celebrated on 1 May close to Red Square watching the parade, standing, weirdly, next to Larry Hagman from *Dallas*. He, like our party, must have looked on with some disdain at the festive, popular display that had 'organised' written all over it. The balloons all had the same colour and size, mostly red – clearly Soviet progress had moved things on from the days of Kropotkin's escape from prison. Too much cheap alcohol on the train back through Hungary got me into trouble.

At the border with the USSR and a check of luggage, Mark Reitman's document was discovered. I was taken to a quiet part of the train where I was asked to sign an untranslated document declaring me, no doubt, to be an enemy of the Soviet state. Mark later joined other members of the Trust Group who were given exit visas to the US. I met him there a few years afterwards in Boston. His prophecy, however, that my goose would one day be cooked did turn out to be true – I'd clearly plucked myself.

That same year in October, I enrolled for an MA at the Centre for Russian and East European Studies at Birmingham University. My idea was to acquire background knowledge in order to become a journalist. Unlike Ethel, I found Russian hard going. Meanwhile our peace group back in Bishop's Castle, Shropshire had launched its very own international initiative on a grand scale. Never mind a twinning, this was to be a citizens' diplomacy tripling – linking our small town with a Russian and American equivalent. Exploratory contacts had been made with somewhere called Hartland in Vermont, a frontier's version of our own wild Welsh border. On my 1983 global circumnavigation with Mark Reitman's document, I eventually ended up there, as we shall see, and met another Ethel, the group's representative.

Pursuing the project further, I went with a delegation from Shropshire to Leningrad in January 1985 to meet their Peace Committee again. I was surprised to be allowed back in after the Hungarian border dispute. At our peace meeting Madame Tcherekova, vice-president of the Leningrad Peace Committee, suggested that Petrodvorets would be a suitable place with which we could make contact. This did not quite fit with our criteria. Not only was it almost a suburb of the city but had a population of 80,000 compared with our 1,500. It is also where Peter the Great squandered many tons of silver building a vast palace that set out to be the 'Russian Versailles' with its elaborate gardens and fountains.

We declined a meeting there with townspeople. Bishop's Castle, Shropshire does have a Norman church, a beautiful, diminutive town hall and a drinking fountain in the recreation ground, but we did not feel like a well-matched twin. Their suggestion was, of course, due to the fact that they wouldn't want any westerner to see any equivalent town in the real countryside, vastly improved from Ethel's time, but still poor and backward. The tripling was unhitched. So was I from my Russian flirtation – almost. The Soviet bureaucracy finally sprang into action and declared me *persona non grata*. The 1987 END delegation had given me a last reprise.

The unravelling of the Eastern bloc legitimised by the fall of the Berlin Wall in November 1989 effectively ended my involvement in 'détente from below' world affairs. The 1980s had been a very heady decade. Ethel Boole's return to London at the end of the 1880s marked for her only the beginning of her involvement with émigré Russian dissidents hoping to challenge the tsarist regime and educate progressive circles about its horrors. My journey on the Trans-Siberian to Beijing took me, unknowingly at the time, to the second location that featured in the history of the Boole/Hinton dynasties.

# ROUND AND ROUND
# THE GARDEN

My failure as an alternative diplomat had ended in statelessness as far as the Soviet Union was concerned. Any sort of career as a Russo-journalist had evaporated as a result of the embargo on my return. I did write a 'Letter to Brezhnev', without any response. A century before, Charles Howard Hinton, James Hinton's son, had, somewhat like me, effectively banished himself from his own career in England. Ethel's eldest sister, Mary Ellen and her four young sons had emigrated with him in 1887 to Japan, a completely foreign land. The exile was not a result of state decree but the pressure of public opinion. Both father and son had become enmeshed in scandal.

Ten years after James Hinton's death in 1875, his highly unorthodox views on sex, prostitution and polygamy still bubbled to the surface like gas escaping through mud in volcanic terrain. Despite the last-gasp disavowal of his views to Charles Howard, his social notoriety had merely become dormant. As ever, sexuality was an irrepressible force.

Ellice Hopkins set out in *Life and Letters of James Hinton* (1878) to revise public opinion about him. Her glowing account of his teachings contained hardly a reference to his liberal views about sex and marriage. Not for her the idea of sex being a pleasure to be indulged fully as long as it was in a spirit of 'service'. She had met him in 1872 in Brighton to hear of his campaign against prostitution. For her it meant campaigning against sexual 'beastliness'. Hinton did not share the idea that there was a beast to be repudiated; it was rather to be tethered and enjoyed altruistically. Hopkin's life's work was to tour the country preaching against the evils of vice and in 1883 to found the White Cross Movement under the banner of which men could sign a pledge to protect women, 'To keep thyself pure and put down indecent language and coarse jests'.[1]

Of the other three main evangelists of James Hinton after his death, Havelock Ellis was the only male. While living in Australia, looking for a direction in life he came across *Life in Nature* and experienced a revelation that dispelled his disillusion with both orthodox Christianity and modern science's lack of concern for the individual. Hinton stressed the unity of both. 'I trod on air; I moved in light,'[2] Ellis wrote in My Life, his autobiography, sufficient to persuade him to move back to England and to take up medicine like his mentor. Pursuing Hinton's ideas in 1880 he visited his wife, Margaret and Caroline Haddon, her younger sister.

The picture of Hinton's life they portrayed carried promiscuous overtones but none of this dented Havelock Ellis's faith in Hinton. His reification of women, analysis of prostitution and his frankness regarding sexuality were considered positive aspects, although he found his Christianity less appealing. Ellis wrote several articles about Hinton, helped edit unpublished papers and wrote the preface to a collection entitled The Law Breaker. He had good reason to continue to be on good terms with Caroline as she had lent him £200 to pursue his medical training. As we have seen in Chapter 7 the 1880s were a whirlwind formative period in the evolution of English radical views. In this light James Hinton's were ahead of his time. Intellectual circles overlapped and the dramatis personae intermixed; it was inevitable that Hintonism, as it came to be called, would be raised as an issue eventually after his death.

In October 1883 both Margaret Hinton and Caroline Haddon, plus Agnes Jones, three of the alleged Hinton seraglio, were present at the inaugural meeting of the Fellowship of the New Life, co-founded by Havelock Ellis. Its dedication to a simpler life made little mention of the sexually sensual. They were present again a few months later in January 1884 when the Fabian Society was formed to pursue more practical social issues. James Hinton's life and theories, championed by the trio, became a subject for intense debate in radical circles at that time and a personal issue between Havelock Ellis and the South African, Olive Schreiner, whose novel Life on an African Farm had become feted in 1883. The pair acted as a conduit for allegations about Hinton that were flying around.

Ellis and Olive had met in 1884 at yet another radical club, the Progressive Association. He was taken by her 'short sturdy vigorous body in loose shapeless clothes, sitting on the couch with hands pressed on thighs, and above the beautiful head with the large dark eyes, at once expressive and observant'.[3] She was somewhat disappointed with him; although handsome, his voice was unmanly, as

Olive Schreiner (1855–1920), South African novelist.

were his rather *gauche* manners. They soon began a strange liaison of copious letter-writing and soul-searching which was frank about Hinton's own special subject: sexuality. Ellis's libido was mainly centred on himself; he writes in one letter how there are fewer 'spots' in his journal that denote masturbation. His growing interest in sexual 'anomalies', however, allowed her to be frank about her own masochistic disposition which fuelled an ambiguity towards men and marriage. Their meetings involved innocent sexual encounters but no consummation.

Regarding Hinton's writings Schreiner seems initially, having only read Ellice Hopkins, to have responded to his generalised emphasis on the power of love and his empathy for women. She writes, however, to Ellis on 2 May 1884, perceptively, 'I have a feeling that Hinton is not quite showing the real man.'[4] She is looking forward, however, to meeting Caroline Haddon. Ellis in return, in a long explication of Hinton's thoughts, extols the freedom that 'any relation for good might be entered into, in obedience to the highest moral law of service and the encouraging of the free showing of what is beautiful in the human body'.[5]

In a foretaste of events yet to come, Schreiner writes to him anxiously about a friend of hers, Mrs Walters, who had passed on certain letters. In one a widow is mentioned who Hinton had made advances to. Mrs Walters reports of the time she had met him, mentioning 'his genius perhaps and a very sympathetic nature', but did not take to his voice, 'soft, insinuating and unmanly. I remember how he took my hand and held it with the softest tenderest pressure and raised it to his lips as if he would kiss it.'[6]

Such indiscretions seem now like the most minor tittle-tattle. In the context of Victorian society, however, they were to be taken seriously. Ellis was inclined to forgive Hinton his foibles; Schreiner declares in July of that year, 'I love Hinton because he had a free-loving soul.'[7] Yet another list of accusations, however, surfaced soon afterwards, told to Olive by a Mrs Farcourt Barnes and recounted in Ellis's journal, probably written in July 1884. For the Boole/Hinton dynasties in particular it is highly revelatory and fleshes out Mark Blacklock's scenario of a group of women infatuated by Hinton. The details were also brought to the attention of the chief prosecutor of the sins of James Hinton, Karl Pearson, a professor at University College, London.

Firstly, Ellis reports that Farcourt Barnes had told Schreiner that Hinton's daughter, Daisy (Ada), aged about fourteen circa 1869, 'was seen running about naked in the room when Hinton and his son were'. Secondly, she says, 'Caroline Haddon [then about thirty-two] as you could tell from her face as she sat on his knee was the woman who more than any other felt intense passion for Hinton.' Mrs Barnes does not wish anything she said to be used against him but considers that he had 'every vicious instinct'. Ellis writes, 'and furthermore, according to her [Farcourt Barnes] Mrs Boole really was his mistress …'.[8]

The other member of the Hinton entourage, Agnes Jones, is referred to in the Schreiner/Ellis letters over a period of months. Reading between the lines it seems that she had also developed a passion for Ellis and assumed that if Olive and he were in a relationship she could be a 'spiritual wife' to him – probably meaning anything but. Olive is sorry for her, alone and weak in body and suggests Ellis may have encouraged her. By December, however, he'd had enough of the devil in Miss Jones. 'I shall come and kill her,' he writes.[9]

Although this does not completely bear out Blacklock's idea that Jones was living in sin with Hinton, it seems apparent that there was sexual closeness. Other indicative letter details had emerged: how Hinton had encouraged Agnes to write 'an intimate account' of her feelings for him and had then showed it to someone else. She declares that if she'd merely heard James Hinton's words there would have been a complete unity between them, she'd known it 'from the first moment we met'. She lauds his voice as if it was her own and how intellectually and emotionally their consciousnesses developed, 'till they proved our union complete'.[10]

Surprisingly, given all the titbits that were being revealed about the scurrilous Mr Hinton, Olive continued to respect the man. She writes to Ellis in July 1885 of his value being one of stimulation: '… he makes all thoughts live and throb which is the work of true genius', and adds, charmingly, in the next thought, 'I got such nice apricot jam, you must have a little when you come.'[11] Her friendship with Caroline Haddon also seems to have continued unabated, despite what had been alleged about the latter's passion for James.

Caroline is a most interesting character. A photograph of her shows the small shoulders of a probably diminutive figure; her middle-parted, tightly-bunned hair over her forehead echoing precisely the slant of her heavily-lidded eyes. Her thin lips betray the wisp of a smile. Underneath all this, however, is a character of great strength. In 1885, in her late forties, she was running a liberally-inclined girls' boarding school in Dover that had started back in the 1850s as a joint project with her sister Margaret.

Although Caroline may have been Hinton's disciple, her independent

Caroline Haddon (1837–1905), sister of Margaret Hinton.

mind shone through. In her book *The Larger Life* (1886) she assesses his thoughts and concludes with a series of interrogative letters that had passed between them. Hinton is fond but patronising: 'Carrie you are a splendid chameleon ... so fine a specimen that you puzzle even your owner (that is me).'[12] She, however, puts the obvious questions to Hinton's dictum, 'desire the good and then do as you like'. What is the good, how will we recognise it in ourselves and others and act out of service unless there is a just moral code to guide us? The problem is evident if we take one of Hinton's own examples: if God gave us the means to perpetuate life, what right would we have in ending it? Could someone assisting suicide be said to be acting out of service? How would we ascertain their motives? The law has stepped in to judge individual motives.

The basis of morality is a question aptly put to those of an anarchist persuasion. How do we live without a bible of some sort or the legal codes of a shepherding elite? Hinton had a simple faith, as did George Boole and Mary Everest, in the bounty of a loving God enabling harmony if we use his gift of reason. An anarchist might perhaps agree with a secularised utopian version: of the potential, self-motivated goodness of the individual co-operating within society. Should however, the institutional sacraments and law of the status quo become oppressive, the justified challenge arises to bring change by robust means. In Haddon's chapter in *The Lawbreaker* she draws up two ranks akin to the perspectives dividing the Fabians and the anarchists:

The side of order ... the just, the temperate, the law-abiding, the conscientious; but also the timid, the conventional, the self-interested, the pharisaical, the unmerciful. On the side of rebellion there are the licentious, the unscrupulous, the violent, the self-indulgent but also the generous, the enthusiastic, the self-sacrificing. The 'heroes' are to be found on the side of revolt – amongst Nihilists, Communists, Carbonari; not amongst the aristocrats and bourgeoisie and the partisans of authority in the Church and State.[13]

This is strong stuff somewhat reminiscent of Mary Everest and colours Ethel Boole's novels yet to come. Not for Caroline Haddon is change to come from the childlike single genius or by Hinton viewing himself as a saviour. Haddon places individuals in a collective context of action. Hinton is a romantic dreamer, although not without a social conscience.

In a curiously entitled tract of 1886, *Where does your interest come from: a word to Lady investors*, Haddon embraces a socialist-tinged equality, berating women who live well in a family based on the long hours' labour of other women, 'whilst your daughter … sits reading under the trees in your garden'.[14] Women must act against hypocrisy and be pioneers. She asks, 'who shall deny … their right of leadership?'[15] This is a call to positive action – not Hinton's sacrifice. She ventures that women have a right to an independent career and even lists some co-operatives she knows that act equitably. It was another tract, however, that got her into real hot water.

*The Future of Marriage, an Eirenikon, By a Respectable Woman* was published in 1885 but had circulated privately before. On the copy I found in the London School of Economics, someone (a Victorian maybe) had added '?!!!' presumably referring to 'respectable'. The graffitied exclamations are indeed justified. She reiterates Hinton's argument that conventional marriage exists on the basis of the wifely puritanism in the home that necessitates the debasement of the prostitutes on the street. 'Men only drink of those poisoned fountains because the purer streams are barred from them.'[16] Law itself cannot deal with this evil; what is needed is a liberation of the 'power for joy and service', in other words more fulfilled consensual sex within marriage.

Haddon gives what must have been seen as some outrageous examples. There are women, she asserts, 'whose normal health and full mental efficiency depend not only upon a large amount of sexual intercourse, but even upon its variety …'.[17] More importantly, of course, polygamy would eliminate sexual secrecy and hypocrisy and draw in single women otherwise reduced to 'honourable' celibacy. It would enable familial burdens to be shared. It might lead to servicing others in a literal sense: the childless wife might ask a 'sister' to bear her husband's child, or the fertile wife might choose to honour her husband's friend about to risk his life in war by reproducing a copy of him. Haddon, despite her more sociological thinking, does not, however, even begin to consider the numerous downsides of her ideas. In the 1880s Haddon's brave off-the-wall assertions inevitably resurrected the dormant teachings and, especially, the philanderings of James Hinton.

Caroline had sent a copy of *The Future of Marriage* to Schreiner and Ellis. The two women had been meeting socially during late 1884 and all three seem to have kept up contact with Mrs Hinton and James' daughter, Daisy. Olive, like Edith Ellis, questioned whether polygamy should only favour men but both approved of the Hintonian notion of fluidity in personal relationships. She was too mixed-up sexually and emotionally delicate to commit to one person. At the beginning of 1885, however, her liaison with Ellis waned somewhat after she had met Karl Pearson, attracted by his good looks and intellectual focus. It was his relative straightlacedness, however, that persuaded him to become a crusader against Hinton by attacking his disciple, Caroline Haddon. The highly germane issue of the link between polygamy and prostitution was in July 1885 thrown into public debate by revelations in a popular paper, the *Pall Mall Gazette*.

Karl Pearson (1857–1936), mathematician.

W. T. Stead's pioneering 'new journalism' in a series of articles under the heading *The Maiden Tribute of Modern Babylon* had garnered first-hand evidence of the trade in young virgins for the idle rich, 'the procuring, certifying, violating, repairing and disposing of [the] ruined victims of the lust of London'. Although the Criminal Law Amendment Act of 1885 raised the age of consent to fifteen, Stead's views followed Hintonian lines, that 'sexual immorality must be dealt with not by the policeman but the teacher'.[18] The general issue of the relations between men and women against this tide led to the formation of yet another radical club prosaically entitled the Men and Women's Club. The undertow dragged in Hintonism and effectively drowned it.

The club was to be an ambitious affair of monthly meetings of twenty members and their guests for the 'discussion of all matters in any way connected with the mutual position and relation of men and women from a historical and scientific as distinguished from [a] theological standpoint'.[19] The first meeting took place on 9 July 1885 and included Pearson, Schreiner, Agnes Jones and Annie Besant. Over the four years of its existence a wide range of topics was discussed and papers written for example on the emancipation of women and marriage in Russia, (Stepniak attended several meetings). The question of prostitution arose regularly.

At the second meeting on 12 October Henrietta Muller delivered a paper on the subject declaring that men were slaves to the practice compared to women's moral superiority and even indifference to sex. At its fourth meeting in December, Haddon, who was a guest and not a member, must have lived up to her reputation by asking if a doctor had ever recommended prostitution for women. The nature of women's sexuality was to be a recurrent theme; polygamy, however, was a step too far. One member with the delightful name of Ralph Thicknesse described Hinton as either a sexual monomaniac or a hypocritical scoundrel but could see some value in polygamous experiments such as the Oneida community in the US.

Karl Pearson had by his side a stack of missiles about James Hinton's behaviour to discredit his and Haddon's theories. One involved Emma Brooke, a radical friend of Charlotte Wilson's and like her, a pioneer student of Newnham College, Cambridge and a member of the Hampstead set. In a letter to Karl Pearson of 4 December 1885, she expands fulsomely on her complaint to Ellis about the 'improper advances' of James Hinton.

Sometime in her late twenties Brooke had been staying at a friend's house where Hinton had turned up. She found herself interested in his ideas until an absolute terror came over her when the rest of the household went out walking. She was left alone with him in the drawing room, 'in a condition of intense yet silent struggle', unable to be 'secure from his touch'.[20] Her hostess had, however, reassured Emma that there was nothing to fear from the behaviour of a person of genius.

The following day after early Sunday dinner, she joined him alone on the lawn and he began to talk on the subject of self-sacrifice, exciting in her 'a glow of glory … my spirit lifted up'.[21] Hinton proposed that she should show him round the garden but led her away to a nearby wood by a pond and asked her to kiss him. She refused, to which he replied in a whisper, 'You will not … because I have told you I was only a poor office boy to begin with [referring to Whitechapel] and you are a proud girl.' His plea that he was trying 'to teach her a lesson in unselfregardingness'[22] did not impress her. She escaped back to the house after another amorous attempt only to find that the household, realising the likely turn of events, found her the subject of mirth. 'For the remainder of the visit I locked myself in my room and only came down when the gong sounded for meals.'[23] Hinton apologised, however, before he left. Brooke suggests that her friends had undertaken to convert her to 'free love as my guide in life' and that '… while seducing their minds with splendid talk and brilliant images, he helped himself liberally to such favours as he could get by the way … The influence of Hinton upon the morals of others was ruinous.'[24]

It was a heady time with gossipy letters flying backwards and forwards; even Charlotte Wilson had entered the fray. The story of Haddon sitting on Hinton's knee, for example, now involved her nakedness while he fondled her. The Hintons, fronted by lonely Caroline Haddon, were getting a bashing. Although Olive, like Ellis, had been aware of the growing dossier against Hinton and had come to detest him, she continued a loyalty to Haddon. Ellis is strangely silent on the scandal in his autobiography but protests in a letter to Olive that the 'vulgar herd' cannot distinguish between having ideas and practising them. He implies that if Hinton had done so it would have only made him more interesting. Ellis's loyalty is not that surprising considering the spiritual debt he owed to Hinton and his emerging objective interest in sexuality – not to mention the financial debt he owed Caroline. There is also a personal issue too. Olive had now fallen in love with the prudish Karl Pearson.

Haddon put up a spirited defence of Hinton in her letters to Karl Pearson of December 1885/January 1886. We do not know Pearson's exact charges but Haddon found them painful and at times beyond the limits of courtesy. They necessitated a rebuttal that her views on sexual subjects were taught at her school in Dover. His chief line of attack must have been that Hinton's views were a sham and a front for his own lasciviousness. You do not know the man, she answers; he may have made grievous mistakes, 'but he was driven by an intense burning passion for the good'. In the pursuit of it he was 'idea-led and theory driven'.[25]

Although she does not specifically allude to the notion that the 'spiritual wife' idea was designed to test the theory, this would seem to follow. He never got round to it, she avers, because he was anxious on behalf of his wife and did not advocate 'free love' as such. Haddon's defence, even to her, must have appeared that she had protested a little too much. She admits that there may have been a 'temporary aberration', which happened when his mind was unhinged in the last years before he died amidst remorseful confessions. The 'aberration' may well have been the encounter with Mary Everest Boole that had so distressed him.

The ongoing fracas left its toll on the women involved. Agnes Jones left the Club in October 1886, Caroline Haddon's health declined under the strain and she disappeared from view. Olive Schreiner retreated from everything to a convent in Harrow. In early 1887 she was having a breakdown ascribed to Hintonism. The different underlying emotional perspectives between the Men and Women's Club members led to its eventual demise in 1889. After the dust had settled the verdict regarding the Wizard was that he had failed to cast his spell beyond his immediate disciples. The timeless appetite for voyeuristic titillation is, as ever, the spume on the deep unexplored sexual ocean below. Hinton might indeed have survived historical obloquy, his minor ideas sullied but intact, if it hadn't been for the escapades of his son. In 1886 Charles Howard Hinton was found guilty of bigamy.

The facts of the case are relatively straightforward. Sometime, probably

Charles Howard Hinton (1853–1908), husband of Mary Ellen Boole, wrote on the fourth dimension.

in 1882, he had begun a liaison with one Maud Florence resulting in her pregnancy. Hinton married her in January 1883 under the name of John Weldon to give legitimacy to the twins who duly arrived in August later that year. He returned to live with Mary Ellen. The assignation continued until, probably anxious about Mary Ellen's imminent fourth child, on 14 October 1886, Howard owned up to the offence at Bow Street Police Station. On 27 October Hinton was found guilty and sentenced to three days in jail but released as he had spent that time already on remand. Hinton had obviously got off very lightly owing presumably to 'old school' connections. The national press had taken up the story.

The trial not only shamed the culprit but his father, James Hinton, whose theories and behaviour were held to be directly responsible. If *le père* advocated polygamy and couldn't keep his hands to himself it was hardly surprising that Hinton *fils* would follow suit. The London radical circles that included the Men and Women's Club were pleased to see James Hinton's works consigned to the litter basket of history.

Olive Schreiner and Havelock Ellis had become involved in the saga while it was brewing. Good-natured Olive, although being driven mad by the events, had been ministering to the guilty pair before the trial. She noted to Pearson that Maud had threatened to run away or kill herself, which might in turn have led Charles Howard to suicide because, she maintained, he actually loves her rather than his wife. Mary Ellen hardly gets a mention, only being described as clear-headed and precise. After the trial Olive writes to Ellis and delivers her verdict on the lessons to be learnt: 'Hintonism falls like a blight on everything it touches.' The social shame brought upon Charles Howard's and Mary Ellen's family must have been intolerable. Most likely as a direct result they decided to leave the country.

Havelock Ellis had made little or no effort to visit young Hinton or comment on the whole business. He did, however, attack Pearson for his fanatical anti-Hintonism, and hoped that Hinton's writings might be published to help forward the issue of sex relations. Emma Brooke, as one might expect, did not agree. Vindicated, she writes haughtily to Pearson, 'Is there not something feeble-minded about the whole proceeding ... arranged on the Hintonian principle straining after transcendental right whilst doing very common-place wrong?'[26] Caroline Haddon, of course, gets a drubbing. Charles Howard worried deeply about the besmirching of his father's teachings was still obsessively concerned with his fourth-dimensional theorising. Pearson writes to a fellow Club member that a paper of his had been generally held to be nonsense, 'part of that muddle-headedness for which the family is famous'.[27]

Mrs Weldon disappeared; we know only that Olive Schreiner was trying to help her into exile too in South Africa. Caroline Haddon continued at her school in Dover but had become incapacitated. She died in 1905. Poor Miss Jones was nudged off the stage only to reappear as a romantic go-between when Havelock Ellis and Edith Lees, the future Mrs Ellis, met at her Cornwall cottage. The other alleged member of Hinton's constellation, Mary Everest Boole, seems to have been untarnished by the shenanigans, whether victim or villain we shall never know.

Ellis himself, versed now in the eccentricity of human sexuality, continued his academic studies into its far shores. James Hinton had been a great stimulating force in his life. It was not, however, until he wrote the sixth volume of *Sex in Relation to Society* that Ellis paid homage to Hinton's belief in 'the possibility of a positive morality on the basis of nakedness, beauty and sexual influence ... regarded as dynamic forces, which when suppressed make for corruption and when wisely used serve to inspire and ennoble life'.[28]

Caroline Haddon had felt that such ideas would probably not gain acceptance until a century later. She was somewhat optimistic. Requiring a British Library copy of Ellis's volumes in *Rare Books and Music*, I discovered that it was 'Special Material'. Researching it required surrendering my Reader's Ticket as hostage and sitting with the contraband at an appointed table with the command, 'may not be left unattended at any time'. One and a half centuries later *plus ça change* ...

Olive Schreiner wrote to Havelock Ellis that Hintonism 'makes out that you can freely and recklessly play with the gratification of the [sex] instinct – it's like teaching a child you can strike matches and throw them down just where you think good'.[29]

She is possibly right: people do play erratically and flippantly with an incendiary impulse like sexuality, but in the great scheme of things the results are more limited than those devastations resulting from the urge for power or wealth.

And the consequences were ... Olive didn't know what she wanted at all; sex probably frightened her. Ellis and she never coupled, it seems – him suffering from the concept he labelled as urolagnia. He later married a lesbian after he'd possibly got over a passion for his younger sister. Schreiner's love for Pearson was never requited despite its intensity; he seemed also to be somewhat scared of sex. Another Club member, Dr Donkin, fell twice on his knees to Olive, fruitlessly. Poor Miss Jones' offer of spiritual-wifery was ignored by Ellis. Caroline Haddon never seems to have got over her crush on her sister's dead husband. Was Emma Brooke playing Hinton along in the garden? Only Charlotte Wilson, living her simple life at Wyldes, managed to stand aloof from it all – although Anne Fremantle states that she was the lover of Prince Kropotkin – a highly unlikely scenario concerning the basically asexual Ethel Voynich whose older sister had been jilted by her brother-in-law, Charles Howard Hinton, son of the Wizard.

# THE RIFF-RAFF
# OF RASCALDOM

According to Evgenia Taratuta, on 4 June 1889, Sergei Stepniak wrote in his little grey notebook in red ink, 'Sasha and Bulochka have arrived, but how our dear Bulochka has changed, how mature she has become. What a bitter university she has attended for the last two years.'[1] 'Bulochka' is, of course, Ethel Boole, just returned from her two-year stay in Russia having witnessed at close hand the huge injustices of tsarism, against which a few brave intellectuals had rebelled. 'Sasha' is Alexandra, one of Stepniak's wife's sisters who had come to join them in exile. Their other sister Praskovya would by now be in the vastly severer internal exile of Siberia with her husband Vasily Karaulov. Vasily's brother, Nikolai, in terminal ill health, had died in 1889.

Ethel too was far from well, thin, anaemic and in a state of extreme nervous exhaustion. Fortunately, a friend of the family, John Falk, came to her rescue and offered a quiet respite in a Lake District guesthouse with her sisters, Alice and Lucy, to recuperate. This would probably have worked wonders for Ethel, who seems to have revered more the beauty of nature than the waywardness of men and God. It was Stepniak's praise of her descriptions of nature that brought the first notion that she might have talents as a writer. Back in London she wasted no time renewing her commitment to the struggle against Russian autocracy. There was unfinished business to attend to.

The manuscripts of Maria Tsebrikova that Ethel had smuggled out were part of a carefully planned campaign to shock the world. Early in 1890 Tsebrikova herself had left Russia for Paris concealing in her clothes her own protest copies and the draft of a letter to Tsar Alexander III pleading for reform. These had been printed and given to friends to send simultaneously to leading newspapers such as *The Times*. Ethel and others like George Kennan in the US had also received packages. Tsebrikova then returned to St Petersburg and after distributing more copies to government and newspaper offices awaited arrest. In a letter to Ethel she explained that her desire was not for martyrdom but to, 'deliver a moral slap in the face of despotism'.[2] Alexander's reaction was at first to 'let the fool go' but later felt bound to send the sixty-five-year-old into exile in the frozen Arctic north. She died in 1917 in the warmer clime of the Crimea. Ethel described her as one of the bravest people she had ever met.

In March 1890 Ethel herself went to Paris couriering news from the Russian émigré community. Taratuta notes that her first port of call was to the Louvre to gaze once more on the iconic Franciabigio portrait of the young man in black, 'now that she knows living people full of such tragedy, force and courage'.[3] She was also in Paris to visit Peter Lavrov, exiled there for over twenty years, another veteran of Stepniak's old days, now an editor and publisher. Searching for her own perspective regarding Russia's future, it seems likely that she was influenced by his belief that revolution would come from its intellectual elite organising the masses, primarily the peasantry.

In Britain too at the time there was a general appetite for information regarding Russia's condition. The Tsebrikova affair would have resonated with the popular endemic view that Russians were a 'Slavonic menace' and the fear of the 'Bear's' designs on India. George Kennan's more recent reports of repeated brutality towards political prisoners at the Kara gold-mining camp had focused public opinion. One young woman had died as a result of a flogging, five others followed by committing suicide. Gladstone raised the matter in the House of Commons.

Tsebrikova's individual bravery, however, required a much more co-ordinated focus to expose Russian brutality. A group of concerned MPs and English radicals joined forces with Stepniak in April 1890 to launch an umbrella group, the Society of Friends of Russian Freedom (SFRF). Shortly after, its periodical, *Free Russia*, inveighed against 'the wrongs inflicted upon the millions of peasantry, the stifling of the spiritual life of our whole gifted race, the corruption of public morals, created by wanton despotism'. Tsebrikova's letter to the Tsar was printed and in 1891 the text of her letter to Ethel beginning, 'My Dear L ...', using her middle name, Lilian. After only a few issues a circulation of 5,000 was claimed with a wide outreach.

Ethel pitched in energetically with Stepniak's mission as a committee member, editor and translator for *Free Russia*. Ethel had obviously become an object of affection to Stepniak despite Taratuta's rather stern description of her as a 'very reserved, impressionable person, diligent and unaffected, painfully shy, possessed of concealed internal forces that would surface but rarely in impassioned ruthlessness'.[4] Ethel had taken up residence again in London with her mother, Mary, at 103 Seymour Place, Bryanston Square but spent much of her time at Stepniak's flat in St John's Wood. In his company over the next few years she would meet with many people brought together in opposition to the Russian regime.

The playwright and Fabian, George Bernard Shaw came often, liking Ethel and calling her 'the little nihilist'; a connotation, one assumes, not of terrorism, but in its original sense as someone outside social conventions. Ethel, it seems, did not entirely repay the friendliness to him, feeling averse to his handshake. They also disagreed regarding her rendition of one of Bach's *Preludes and Fugues*. Ethel had obviously not entirely given up on her passion for music. Oscar Wilde, Edith Nesbit, William Morris and the Garnett family were other literary visitors to Stepniak.

English radical opinion was of many shades. The writer Ford Madox Ford recollected, 'I don't know what I was … but I fancy I must have been a very typical young man of the sort who formed the glorious meetings that filled the world of the 80s and 90s …'.[5] William Morris's meetings at his Hammersmith house were legendary; discussion clubs abounded, such

William Morris (1834–96), designer, writer and socialist.

as the Men and Women's Club. Ford's precocious teenage cousins, Olivia and Helen Rossetti, had themselves begun an anarchist magazine, *The Torch*, in the basement of their home at Primrose Hill. In their own novel of 1903, *A Girl Among the Anarchists*, written under the name of Isabel Meredith, they painted a picture of Stepniak as 'Nekrovitch'. His guests are an assortment not unlike the gatherings at Wyldes: 'Russian nihilists and exiles, English Liberals … socialists and Fabians, anarchists of all nationalities'.[6] The setting is a villa-cottage in Bedford Park, Chiswick, where Stepniak moved in 1893, an early garden suburb, whose tree-lined streets and arts and crafts architecture nevertheless suited a one-time terrorist.

None of 'Meredith's' labels would have entirely suited Ethel one suspects. Literature, art and music moved her rather than any declared political viewpoint. She would have joined Madox Ford somewhat on the fence, aloof like him from the 'terrific rows between socialists and anarchists in those days … the socialists were unreasonably aggressive … holding meetings at which the subject for debate would be, "The foolishness of Anarchism"'.[7] She would be sympathetic perhaps to Ford's declaration of how he had 'early developed a hatred for tyrants and the love of lost causes and exiles'.[8] 'Meredith' separated the political theoreticians from those whose sense of justice was prompted by '… an act of personal revolt, the outcome of personal sufferings and endured by the rebel himself, by his family or his class'.[9] Ethel fits here; Taratuta, of course, Soviet apologist, had to claim Ethel firmly as a socialist.

Ethel was busy and committed to Stepniak's circle of Russian exiles, 'shaggy though not unwashed',[10] as 'Meredith' described them. Many of them had come to London after perilous adventures escaping Russia, their routes radiating from St Petersburg as from an airline's hub. They had, in fact, reconvened the Chaikovsky circle of the early 1870s. Nikolai Chaikovsky himself had left Russia for the US in 1875 before settling in London in 1880. Not so much an ideologist, his importance lay in acting as a focus for activity and propaganda.

Felix Volkhovsky, somewhat older than his compatriots, born in 1846, had already spent over two years locked up in the Peter and Paul Fortress before joining the original circle. He was tried with the 193 *narodniks* and expelled to Siberia. It was he who had met George Kennan there and helped alert him to the real horrors of prison life. Escaping east through Canada and the USA he lectured with Kennan, shackled and wearing convict clothes. He arrived in London in 1890.

Members of Stepniak's circle. Left to right back row: Volkhovsky, Chaikovsky, Stepniak in doorway. Fanny 'Stepniak' seated left, Voynich perhaps far right.

Of the immediate circle of Russian SFRF émigrés, apart from Stepniak, Ethel felt most drawn to Peter Kropotkin, the other dominant anarchist figure in England. In later life she told Taratuta that he and Enrico Malatesta 'were the only truly holy men I ever knew'.[11] His full face, St Nicholas beard and bald pate gave the impression of the prophet that others also found charismatic. Then in his fifties and in poor health because of his jailings, he still found the time and motivation to write monthly columns for *Freedom* and go on lecture tours.

157

After her return from Russia Ethel and he spent time together; she visited him with her nephew, Geoffrey Taylor, her sister Margaret's son, who recalled many years later how Kropotkin had made a ship-in-a-bottle for him. She also remembered a visit to London Zoo where, according to Fremantle, he commiserated with a Russian wolf, 'that it was a victim, in exile amongst foreigners, that no one appreciated or understood its spacious soul'.[12] The wolf apparently responded appropriately by raising its head and howling. This poetic moment mimicking the prince's own distance from his roots did not echo necessarily with his intellectual views. Given his belief in the profound sociability of mankind, the wolf was probably pining for the pack as much as the Steppes. Within the SFRF, however, he was not always in agreement with its policy. In 1891, strongly supported by Ethel, he criticised what he felt was a lack of support for the masses and attention to the land question.

The Slavic rebels had a certain romantic allure for English women. Charlotte Wilson was rumoured to be more than intellectually drawn to Stepniak. Olive Garnett, daughter of Richard Garnett, Keeper of Books at the British Library, and her sister-in-law Constance, were also attracted to him as well as to Felix Volkhovsky. Ethel one suspects stood aloof from anything less noble than pondering the world's suffering. This was all to change overnight on 5 October 1890, however, with a knock on Stepniak's front door as the company was sitting down to supper. One new visitor, both shaggy and unwashed, seeking sanctuary, had arrived circuitously from Russia. The newcomer was to prove of great attraction to Ethel and alter the course of her life yet again. Taratuta outlines the tale.

Answering the door, Stepniak was confronted by a young man. 'His grimy clothes hung from him in rags, his large grey eyes revealed extreme exhaustion ... the stranger revealed that he was a Pole fleeing from Siberia and that his name was Michael Wilfrid Voynich. He spoke Russian fluently without any accent, beyond a slight lisp. Voynich told his tale jerkily, skipping from one subject to another.'[13] The full story of his Siberian travails we shall learn of later – sufficient for now to note that he had arrived from Germany at London docks with a letter given to him in Irkutsk by Fanny's sister, Praskovya Karaulov. According to Taratuta, when he was introduced to Ethel:

He stared at her as if wanting to ask her something but didn't quite dare. Had she been in Warsaw during Easter 1887? She concurred that she was on her way to St Petersburg and had visited the Citadel. 'I was imprisoned in that Citadel,' replied Voynich, 'I was looking out onto the square at the people who had freedom and I saw you that day.' Thus destiny introduced Lily Boole to the man who was to become her companion for life, and whose name was to become her name.[14]

This remarkable story is the stuff of operatic drama and movies. Such a coincidence somehow stretches credibility. The story nagged at me; later in my Boole explorations I decided to check it out somehow in person.

Wilfrid Voynich was immediately welcomed into Stepniak's inner émigré circle. Ethel recalled how the very next day after his bedraggled arrival he went out onto the streets selling anti-tsarist tracts. With her *penchant* for the revolutionary hero he must have represented a seductive figure, combining handsome good looks and tales of adventure: the romantic young man in the Louvre picture had stepped out of it into her life. The bond between her and Wilfrid that lasted their lifetimes seems indeed to have come from a deep, shared, compassionate empathy. We have no details of any wooing, (if such there was).

On 19 August 1892 Wilfrid wrote to Lazar Goldenberg, the New York representative of the SFRF: 'and now for some personal news I am married ... my wife is a committee member [of SFRF] nee Miss Boole, a dedicated person who speaks Russian well and is secretary of *Free Russia*. My warmest regards, Kelchevsky. PS I have dropped my pseudonym.'[15] He had adopted the name to avoid exposing his Polish relatives to reprisal. The marriage was by deed poll; sharing a surname made living together easier. In the Hoover *Volkhovsky Archive* collection, Wilfrid mentions that they don't have enough money to rent a flat and are apparently in early 1894 still living at Ethel's mother's, now in Notting Hill.[16] Shortly after the 1892 letter Ethel wrote to Goldenberg about Wilfrid, 'Don't you think the Tsar should thank me sincerely for taking such good care of his quarry ... It is not likely that this meat for the gallows is going to die of consumption.'[17] He nevertheless experienced bad health for the rest of his life, in part perhaps due to his Citadel incarceration.

The editorial disposition of *Free Russia* was relatively open-minded, as *Freedom* had been, apart from its unequivocal opposition to the Tsar. This may have been to engage with a wider audience, but the begged question was whether violence should be employed to overthrow the system? Kropotkin's life in his native country had shown him that against the depth of tsarist repression violent revolt might be justified. He was against, however, the systematic use of terror that had become a fact of life in

England as a result of Irish Fenian policy. Bombings were regular occurrences in London involving Scotland Yard, Victoria station, and in a reminder of more recent times, the London Underground. Kropotkin's stand was to deplore terrorist explosions but feel unable to judge those who were driven to such ends.

Violence capable of sparking a revolutionary upsurge had in fact been taking place in London. Demonstrations against unemployment in

Match girls on strike, 1888.

November 1887 at Trafalgar Square, remembered as Bloody Sunday, led to three deaths owing to police action. Morris, Stepniak, Kropotkin and Shaw attended. Morris spoke of the 'reckless brutality', akin to that used by the Tsar. Annie Besant wrote that 'the horse-police charged in squadrons at a hand-gallop, rolling men and women over like ninepins while the foot-police struck recklessly with their truncheons'.[18] The victorious 'Match-girls' strike overseen by Besant and the London dock strike a year later showed, however, that change could come more profitably from direct action and trade union organisation.

Stepniak's own views seemed to have been tempered by his stay in England as demonstrated by his 1891 tract *What We Need*, translated by Ethel. In it Stepniak lauds the second nature of civil liberty to all English people; in Russia it necessitates the 'temporary' action of terrorist insurrection. Only then would it be possible to establish a constitution that guaranteed freedom and the right to move towards socialism. The role of intellectuals like him to lead and unite would be paramount. In Russia, however, actual violence had led nowhere except to further suppression and Siberian camps. It was as well that Stepniak did not advocate violence for his host country's political dilemmas: western Europe was in the throes of a panic about anarchist bomb plotters enacting their 'propaganda by deeds'.

In the 1890s things had got serious. The 'dynamitards' were linked to a mysterious 'Black International'. In March 1892, four anarchists were sentenced heavily for conspiring to make bombs in Walsall. The accused, it later emerged, had been stitched up by the police using *agent provocateurs*. The Special Branch, begun in 1883 to monitor the Irish, turned its attention now to screening activists and infiltrating meetings. There was no grand anarchist conspiracy, however; like-minds linked up and fantasised as much as acted. As now, the popular press exploited public fears to increase circulation and sales. The authors of popular novels mused that if the vague factions could combine the destructive power of dynamite with a new way of delivering bombs, society would be at their mercy.

Often serialised in magazines, 'anarchist' conspirators employed bomber dirigibles as fast and versatile as the aeroplanes that later became reality. If unavailable, there were still germs or chemicals that could be used to ransom and bring down society. The imagined anarchist became a figure of mass destruction, not just mere assassination. E. Douglas Fawcett in *Hartmann the Anarchist; or, the Doom of the Great City* in 1892 even expounded a realistic anarchistic credo: '… a return to a simpler life … men will effect all by voluntary association and abjure the foulness of modern wage-slavery and its city mechanisms'.[19] Their huge airship *Attila* eventually toppled the House of Commons. Tsarist oppression sometimes even figured in the doom scenarios. In George Griffith's *Angel of the Revolution* (1893), beautiful Russian nihilist Natasha is rescued from the Kara mines by co-conspirators from the Brotherhood in charge of a three-masted, six-propeller, cigar-shaped airship.

The apocalyptic sensationalism of the popular genre was also reflected in more serious novels attracted to the dramatic potential of extremism. Dostoyevsky had already explored its dark world; Henry James followed in 1886 with *Princess Casamassima*. Set mostly in England, its male hero, Hyacinth Robinson, is obliged to turn his gun on himself when caught between the reprisals of the state or the revolutionary circle to which he had been committed as an assassin. Oscar Wilde wrote a play, *Vera – or the Nihilists* in 1880 loosely based on the life of Vera Zasulich, the assassin of Trepov. The play bombed in New York in 1882.

Conrad's *Secret Agent*, published in 1907, but set in 1896, was based on an actual event of 1894. Young French anarchist, Martial Bourdin, had accidentally blown himself up carrying a bomb in Greenwich Park. At his funeral a large crowd of anarchist sympathisers rioted with the police. Conrad's anarchists are of the nihilist persuasion, prepared to act amorally either to further materialist ends or to bring down civilisation. The 'Professor', who carries a bomb he is ready to detonate at will, is a precursor of the suicidal jihadist. Behind the plot lurks a foreign power, implied to be Russia, that seeks to discredit anarchism by financing terror, hoping that the government will thus crack down on its proponents.

Stepniak himself had written a novel, *The Career of a Nihilist*, published in 1889. Set in Geneva, St Petersburg and Dubrovnik at the end of the 1870s, it mirrors his own experiences. In his Preface, however, he wisely notes that the revolutionary violence portrayed is for an 'artistic purpose', neither 'to extol terrorism as to decry it'.[20] He is interested in showing 'the inmost heart and soul of these humanitarian enthusiasts, with whom devotion to a cause attains the fervour of a religion …'.[21] The basic plot describes the intrigues of young hero, Andrey Kozhukov realistically involved in a St Petersburg group 'going to the people' as educators but driven to defend themselves and take revenge against the repressive state. A failed expedition to release revolutionary friends in Dubrovnik that results in their execution leads him to an epiphany, 'more than enthusiasm, more than readiness to bear everything … a positive thirst for martyrdom … a dream of supreme happiness'.[22] This is challenged by others as 'a futile rage of self-immolation', rather than a constructive move towards reform. Undeterred, and with the help of the party organisation, Andrey attempts to remove the over-arching 'keystone' of the system, the Tsar.

The portrayal of the characters and their motives are skilfully realised, including Khozukov's love for Tanya, his wife, and her sad but inevitable acceptance of his self-sacrifice. The novel ends with a rushed and risible account of the assassination attempt: the Tsar, out on his daily walk, zig zags as he runs away from Andrey's revolver shots. Stepniak, however, provides a final rallying call declaring that the struggle 'goes forward from defeat towards final victory which in this sad world of ours cannot be obtained save by the sufferings and sacrifices of the chosen few'.[23] The story makes clear the distinction between the poles of anarchism: the desire for grassroots agitation creating social alternatives and violent apocalyptic change. The true nihilist is the everything-is-permitted character: Verkhovensky in *The Devils*, Conrad's 'Professor' and the bomb-making 'Stutterer' in Stepniak's work.

Emile Zola's powerful 1885 novel, *Germinal*, was based on the author's visits to see for himself the oppression of coal miners and their families in northern France. Driven by poverty and despair they strike and riot but are crushed by the state. The shadowy Russian, Souvarine, another true nihilist, sabotages and wrecks the mine in retaliation, before disappearing to wreak more havoc elsewhere against the system.

Despite the move away from violence and terror, awareness of its challenging presence affected any organisation tinged with its advocacy. Stepniak's SFRF and its offshoots were prime targets. Ethel herself commented in a 1931 letter that Conrad's book *The Secret Agent* gave a very fair picture of the intrigues they were fighting against. In 1902 she described a personal instance probably dated about 1893–95. Late one rainy evening, 'a tall, stalwart individual' rudely entered their house. The intruder brandished a 'short, stout, oaken club, tipped with iron', demanding the no doubt compromising papers on their desks. Wilfrid coolly took the man to a window and pointed out a gaslight opposite. With 'deadly menace and a tone of concentrated passion, he hissed, "If you are not there in two minutes I will kill you".'[24] The bully fled.

Inevitably the SFRF's activities were scrutinised by both the British and Russian establishments, the latter hoping for its downfall. Its pressure on the tsarist regime took the form now of propaganda by document rather than deed. In the summer of 1891, Stepniak had launched the Russian Free Press Fund, an organisation aimed at Russians, publishing and distributing a broad range of otherwise censored material destined for the homeland. It included writings by Marx, Herzen and Tolstoy. With her knowledge of Russian Ethel undertook translations and secretarial work.

We know very little about their lives beyond political involvement. Wilfrid would have been hampered socially by his poor English. There was a lighter side for Ethel as she developed her literary interest. As a labour of love, she edited and translated Russian short fiction, perhaps in memoriam to the two writers who had played a part in her St Petersburg stay. In *Stories from Garshin* (1893) Ethel had included in the collection 'The Scarlet Flower', the last story Garshin wrote before he committed suicide. From Taratuta's description of Ethel's personality she would have seemed an unlikely editor of *Humour from Russia* (1895), but she obviously appreciated satirical wit and language when not too frivolous. Three stories by Shchedrin were translated, two of which, however, were grisly first-hand accounts of the Russo-Turkish war of 1877 in which he was a volunteer.

Ethel and Wilfrid as a political double act had been very busy during the first half of the 1890s, in the thick of things. The couple went together on business excursions abroad. In early 1891 Wilfrid was in Paris but seems to have not gone down too well. A year later he became the actual business manager of the Press Fund, putting in fourteen hours a day but not entirely happy with his colleagues. He complained of little help from Stepniak and

Volkhovsky and that Chaikovsky was too busy with his family. In a letter to Goldenberg in June 1893, Wilfrid is worrying about Ethel's health because of overwork and a heart ailment. They set off, however, for Switzerland to negotiate the purchase of a printing press and to visit other Russian émigrés such as Vera Zasulich. Later in 1893 Ethel travelled to Lithuania and Poland staying in Warsaw and Cracow where Wilfrid's mother lived. Ethel had by now learnt Polish.

Her favoured position within Russian émigré circles was, however, becoming somewhat challenged because of Wilfrid's truculence. Egoistic and ambitious, his domination of the Fund group had become obsessive. Felix Volkhovsky wrote to fellow Fundist, Nikolai Chaikovsky, 'he is greedy and envious and unconsciously strives to fill the whole world with himself ... all conspiratorial communication and a considerable part of overt communications are in his hands. I insist that Voynich be put in his place.'[25] The 'conspiratorial communication' refers to the coding and decoding of documents that Ethel also took part in. Chaikovsky complained too of her assertiveness. Taratuta bolsters the adverse impression of Voynich, writing of him as 'quite loquacious and affected, an impetuous dreamer and phrase-monger, fond of creating an effect ... contradictory in his treatment of people, even those closest to him ... [he has an] extreme imbalance, a kind of exaggerated sickly sweet amiability ... and bursts of disorganised fantasy ...'.[26]

Another account comes from a very different hand, commenting that a mission to unite émigrés in late 1893 had failed. 'The main reason for this was Voynich who ... managed to create a very unfavourable impression because of his intolerance of other people's ideas.'[27] This assessment was made in a report of October 1895 and sent to the Russian head of the Secret Services Abroad in St Petersburg by the Okhrana agent, Peter Rachkovsky.

After the assassination of Alexander II in 1881 the tsarist regime had set up an elaborate system to infiltrate exiled revolutionaries by spying at meetings, stealing documents and filing a flow of reports to HQ. Following a successful stint in Paris, Rachkovsky set up stall in London in 1891 with six agents under him. Stepniak's Russian and English associates were naturally a main target. In 1892 one wrote to his masters that among the women conspirators was a 'Miss Bull, a good orator ...' who he had assigned the number 35. William Morris took number 7 and 'Kelchevsky' (i.e. Voynich) 13.[28]

In London Rachkovsky contrived more means of discrediting the SFRF including accusing émigrés of misappropriating funds and trying to persuade them to counterfeit money – an action that could be later, as in *The Secret Agent*, used to slur the group.

A less crude ruse created quite a scandal. An article in the *New Review* of January 1894 contained a damning tirade, 'Anarchists and their Methods of Organisation'. Its first part spoke of them as 'the very dregs of the population, the riff raff of rascaldom, professional thieves, bullies ... cut-throats ... despicable creatures ...'.[29] The second part named some of the villains attached to the SFRF, including Volkhovsky and Voynich, and alluded to *What We Need* and its call for the government of Russia to be 'annihilated by violence'. The aim of the article was to mobilise opinion against the government's open-door policy towards political immigrants. It also recounted at length, without naming names, the story of the murder of General Mezentsev in 1878. Stepniak was of course the author of the pamphlet referred to and the assassin. In reply he published the article *Nihilism As It Is*, translated by Ethel, which did indeed include terrorism among its various methods of revolution.

It is strange that Stepniak's early nihilist career had not been acknowledged for so long by his English circle of friends, even though his writing sometimes lauded violence. He claimed that he had never concealed his past; those who knew preferred to forget all about it. Ethel may have known from her close contact with him and although not an advocate of violence, was somewhat fascinated by it. She would soon display blood-thirstiness on a grand style in a novel of her own.

The episode is of interest too for the reactions of Olive Garnett, one of Stepniak's young admirers. The rumour about his past came as a dismal shock to her. She wrote in her diary, 9 December 1893: 'Selfishly, I feared that I might lose "my Stepniak" – the artist – in the S. I do not know, the N'ist, T'ist and _____.'[30] The long dash presumably meaning 'murderer', the exclamation of someone unable to reveal her true feelings even when writing to herself. Olive was dedicated to the cause of Russia but also in love with Stepniak. Later she writes, 13 January, 1894, 'I am so happy. This morning a postcard came from S asking me to come and revise an article. Then we had tea, a real English tea with pink and white sugar-cake.'[31] Later still that day, dressed in a Russian blouse to join New Year Russian celebrations in Gower Street, she observed a meeting of the SFRF: 'After a while there was voting, the result was read out, I think that Mrs Voynich headed the list and she made a speech to which everyone listened attentively ... she was the only woman who seemed to take part ...'.[32]

The small episode reveals the difference between the two women. Olive is the poetic extrovert from an established middle-class family dreaming of love and sacrifice. The other is the romantic introvert wedded actively to ideals but possibly not able to express personal love easily, if at all. Olive makes clear their difference, expressing in her diary a dislike of Ethel. What she wouldn't have appreciated probably is Ethel's business-like manner and competitive closeness to her Russian hero. Ethel in 1894 was about to court danger again in the service of revolutionary struggle.

At about the same time of Stepniak's exposure, things had been coming to a head in a growing rift between Wilfrid Voynich and the Fund; he was just not an easy person to work with. A further witness to this was Theodore Rothstein, yet another Lithuanian Pole from Kovno, who had gone into voluntary exile and ended up in London working for Stepniak's comrades. In the 1890s he was a close friend and colleague of the Voynich duo.

His low moral qualities made him impossible even for his closest friends and in the end he was forced to leave not only the RFPF but the Russian émigré colony as well. This was a great tragedy for his wife and she made great efforts to save him. But to no avail, so she loyally stayed with him in a state of ostracism.[33]

According to Taratuta the quarrel became 'catastrophic … they accused Voynich not only of thoughtlessness and unsound, hare-brained schemes but also of improper use of monies designated for revolutionary work'.[34] Perhaps unsurprisingly, supported loyally by Ethel, he broke away some time in 1894 to form a rival group, the League of Book Carriers.

It was an ambitious affair reflecting Wilfrid's inflated view of himself. In order to raise money in January 1895 he sent out an appeal. One of the wished-for donors was double agent Evalenko in New York who'd coughed up before for the cause. Little did Wilfrid know that his request would find its way to the Okhrana. The appeal was signed grandiosely by members of the Central Bureau: Ethel and Wilfrid plus a mysterious young chemist, S. Stein.[35] Continuing the ambitious programme, the triumvirate issued an open letter to the new Tsar, Nicholas II. It started boldly, 'You first began the battle, and it won't be long in coming',[36] and went on to rebuke him for his treatment of land activists. Impressively, 10,000 copies were produced by February; some found their way to St Petersburg.

Ethel, meanwhile in early 1895, armed with a letter of introduction from Stepniak, had travelled to Lemberg (now Lviv), then part of Austria, to meet Michael Pavlyk, a writer and Ukrainian political activist to establish how to send books. She seems to have become very friendly with him, returning in March exhausted. In a letter in December later that year Wilfrid asks Pavlyk, 'How have you so enchanted her that she has quickly fallen in love with you?'[37] One assumes that this is an ideological closeness – although he is only eleven years older. Any sense of intimacy may have been due to nuances both of translation and social mannerisms. The correspondence does reveal, however, that Ethel could turn on the charm when she wanted – a fact highly relevant to developments shortly in her life.

Earlier in the year Pavlyk had written to a friend, 'my only joy has been the memory ... of an Englishwoman from London ... a well-educated and human person'. Ethel had, in fact, invited Pavlyk to come to London, but he continues, 'this cannot be. I could not live with her husband yonder.'[38]

The open letter to the Tsar and the Lemberg visit may have been the only successes for the League. Money was not forthcoming, probably Wilfrid's reputation and rift with his Russian emigres frightened people away. The enterprise that Stein had been masterminding came to nothing. Wilfrid writes of a 'smash' financially and general ill health. A glum mood is found in Wilfrid's letter of 2 December to Pavlyk, 'Reaction is everywhere ... even in England,' but adds with his usual optimism, 'If we live, we live to better times.'[39] It tempted providence.

Three weeks later on Monday, 23 December 1895, Sergei Stepniak walked from his home in Woodstock Road, Chiswick to meet Volkhovsky in nearby Shepherd's Bush. The route took him across the unprotected level crossing of a railway branch line. Despite braking and blowing his whistle, the driver of the 10.20 train to South Acton knocked him over. Stepniak, aged forty-three, was killed instantly. Rumours abounded that somehow it was the work of the *Okhrana*, but the more likely truth is that he was either deep in thought or reading a book.

Five days later Chaikovsky read out an address at Stepniak's house before a procession led off with the body to Waterloo station. Here a crowd of 1,000 gathered in the pouring rain to despatch him on the special 'necropolis' train to the newly instituted crematorium at Woking in Surrey. Earlier that summer Friedrich Engels had journeyed finally on the same route. Demonstrating the polyglot nature of London's political refugees, the gathering included a large group of Russian Jewish workers from the East End carrying red and black banners. Speeches were made in Russian, German, Italian, Polish and Armenian. Marx's daughter, Eleanor, spoke on behalf of women.

Stepniak's death was a watershed for many people close to him; the list of speakers and mourners at Waterloo attested not only to his personal significance but coincided with a time when political perspectives were changing in Britain and abroad. Vera Zasulich, whose violent act had goaded Stepniak to kill Mezentsev decades before, was a terrorist no more but a Marxist hoping that revolutionary change would come through the emerging strength of the European working classes. The St Petersburg Romanov palace was indeed eventually successfully stormed in 1917, ushering in a dictatorship of the proletariat that, as Bakunin had warned, would lead to a vicious stranglehold by the state, not its evaporation. While continental anarchism of the terrorist kind continued on a path of assassination and outrage, in England at least there was to be a relaxation of the fear of violence. Support for the SFRF itself continued on a decline.

William Morris's speech at the station turned out to be his last in the open air: he died in October 1896. Like Stepniak, he had embraced a compendium of ideas. While a believer in revolution, his backwards look to a guild form of socialism was not compatible with a nascent mass-industrial society. Kropotkin, of course, also spoke but his idealistic anarchism of grassroots communality also felt increasingly out of touch. Shortly he would take on a losing fight against the ideological exclusion of the anarchists by Marxists at the Second International held in London in 1896. This ascendancy was not, however, a future indication of Marxism's adherence on the English side of the Channel. Anarchism as a positive force nevertheless found some concrete effectiveness abroad by allying with the workers' movement through trade union syndicalism. Tom Mann and John Burns, both at Waterloo, represented the move towards the alliance of trade union power and representation in parliament. This social-democratic path was one that the Fabians who had met at Charlotte Wilson's farm in the mid-1880s would have welcomed. In 1896 she decided to take a back seat from managing the anarchist platform *Freedom*.

It's highly likely that Ethel and Wilfrid, despite the rupture with Stepniak's circle, were participants in the funeral in person. Stepniak had been her Russian teacher, close friend and guide. Neither of the Garnett women attended. Olive apparently suffered for a long time from the blow of Stepniak's death and cut off all her hair. Both had followed in Ethel's footsteps to Russia, prompting Olive's book of short stories, *Petersburg Tales*. Constance undertook authoritative translations of Russian literature.

Ethel too had been working for some time on a novel, 'about Italian life', that drew its inspiration from the heroism she had witnessed in St Petersburg and Stepniak's milieu. The revolutionary circle of the 1890s would have been completely astounded to find that her writing gained an extraordinary fame lasting 100 years. It is charmingly ironic too that remnants of that formative period in British political history can be found today in well-heeled salons draping William Morris textiles and wallpaper. In more down-market haunts Charlotte Wilson's still-published *Freedom* may yet grace a humbler coffee table.

# COMMUNING WITH SPACE

*Seat 58, Rare Books and Music, British Library. 15 February 2011.*

Back here again. I shall expect a small plaque on the chairback to reward my perseverance. As with his father, James Hinton, I've read the books, consumed the articles and cogitated; but to little avail. Charles Howard Hinton's topsy-turvy world of the fourth dimension is as befuddling as another fantasy within the Boole/Hinton aegis: the *Voynich Manuscript*. His universe involves leaps into planes where one is likely to disappear up one's own whimsy.

When I opened *A New Era of Thought* (1888), the spine covering of the book was unhinged as a long flap. The quires beneath had been bound with a strip of nineteenth-century newspaper advertising some nameless pills. Worth a guinea a box, they professed relief from 'Bilious and Nervous Disorders such as Wind, Pain in the Stomach, Sick Headaches, Giddiness … Disturbed Sleep, Frightful Dreams and all Nervous Trembling Sensations'. Opening the pages of Hinton's volumes I went through the gamut of these symptoms, becoming somewhat unhinged myself.

We know a good deal about his adult life but very little about his childhood except that he was born in 1853 and brought up in a middle-class family with three other children, Margaret, William and Ada. (She was also known as 'Daisy', he as 'Howard'. I have kept the 'Charles' to distinguish him from another 'Howard' relative.) The family must have been much dominated by their headstrong father, James Hinton. His obsessive writings on sex and marriage in particular, as we have witnessed, generated considerable notoriety in conventional Victorian society. With such a pedigree it seemed not entirely surprising that his son's life would follow suit. The ensuing bigamy scandal of 1886 was seen almost as a natural consequence, leading to Charles Howard and his family's exile to Japan in 1887.

Charles Howard's education had begun at Rugby School, followed in 1871 by Balliol College, Oxford studying mathematics. In 1875 he taught at Cheltenham College, completing his BA two years later and an MA in 1886. His interest in the world of the fourth dimension may have begun when he was quite young. While at Rugby School in 1869 his father James had written to him, 'I am glad you like the idea of studying geometry as an exercise of direct perception.'[1] He adds that he does not share his enthusiasm for a formula of Professor Boole. This is probably the earliest reference to the link between the Hintons and the Booles on which our present historical saga is based. George Boole had been dead for five years by that time. Charles Howard would have met Boole's widow, Mary Everest, through James' visits to the top floor of Queen's College, Harley Street where the Boole girls lived and the pair unofficially instructed pupils. He was probably influenced by Mary's love of pure geometry. After the Booles had left their college lodgings and James had died in 1875, the connection continued. His growing interest in the fourth dimension was passed on to Mary's third daughter, Alice; his growing interest in the eldest daughter, Mary Ellen, led to their marriage in 1880. To support the family he took up a post as assistant science master at Uppingham School in the East Midlands.

The fourth dimension had been fascinating the public about that time owing to the séance antics of an American, Dr Henry Slade, magicking slate-writing and levitations. Although convicted of fraud in 1876, Slade had convinced the scientist, Johann Zollner that rings could be mysteriously encircled around an unlifted table leg and steel rings could be interlinked. The only explanation it seemed could be that of some transportation of matter into another dimension. Zollner published his views about the time that Hinton had begun experimenting with the idea of experiencing higher dimensions. What was to become his life's work was expounded in a first collection of 'scientific romances' on the subject published in 1886. The pure essential geometric principles that underlie the supposition of a dimension higher than our accustomed three, were in some ways best set out in the same year by Edwin Abbott, headmaster of a London school.

In his own romance, *Flatland*, he hypothetically conceived of a 'cartoon' world existing on a two-dimensional plane that would appear to us as no more than an outline like a map.[2] (Travelling by train to 'meet' George Boole in Chapter 3 across the polder countryside from Peterborough to Lincoln was almost like visiting it for real.) Flatland's inhabitants are geometrical figures ranked in society according to the number of their sides: triangles are workers and soldiers, squares and pentagons the

Right: *Flatland* by Edwin Abbott.

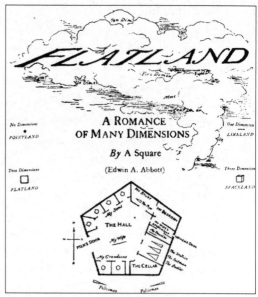

professional classes. Hexagons pose as nobles; above all are the priestly circles. Lowest of all are the women – mere, almost invisible, fragments of lines, insubstantial and given to hysteria. Unlike Hinton, Abbott is consciously using his suppositions also to parody Victorian society.

Communication between the Flatlanders is by hearing, feeling and inference. Without any upwards or downwards, movement is sideways only; its citizens cannot pass over each other. Art is limited, there being no perspective. The novel's hero, A2, known as 'A Square', lives comfortably with his wife in their pentagon house. Here he muses about the even more restricted Lineland world of one dimension, peopled by lines only varied by their length and unable to see more than a point to their left or right. Even this is more substantial than Pointland, having no dimensions at all: life is a completely solipsistic existence with nothing beyond each individual.

A2's complacency is shattered in the year 1999 just before the dawning of a Third Millennium when a circle magically appears possessing the awesome ability, unknown to Flatlanders, of being able to change its size. The visitor is a Sphere from a three-dimensional world, its size growing from a mere dot through increasing circles as it progressively appears in A2's own plane, as a fish would see a ball descending through the surface of water. The presumptuous Sphere lifts A2 up and beyond into Spaceland from where Flatland's houses and people can be viewed, even their interiors and innards. This god-like revelation inspires the now visionary A2 to preach the gospel of a third dimension to his countrymen. Unsurprisingly, he was considered heterodox and treasonable, and like other figures of his ilk locked away as mad. Before his incarceration, however, A2 had asked Sphere if there might not be another world in which Sphere's fellow solid, a Cube's inside might be similarly seen. Sphere replies that this is utterly inconceivable but A2 precociously suggests that there might be a further higher dimension if one projects by geometric analogy.

He explains that if you look at a cube you can break down its attributes into points, lines, squares and cubes. A point by definition has only itself as an entity but any one-dimensional line has two points at either end. A line given breadth as well as length forms a two-dimensional square with four points at each corner. When raised into a three-dimensional cube it possesses eight points, twelve lines and six squares (top, bottom, two sides, front and back). If we project an added fourth dimension to the cube by a continuing progression it must possess sixteen points, thirty-two lines, twenty-four squares and eight cubes. (If you are beginning to feel 'Bilious', with 'Nervous Disorders', etc, take it from a co-sufferer that this is indisputably true.)

Such a figure, unlike the friendly Cube, can only actually exist in the imaginary realm of hyperspace as a tesseract, an appellation probably given by Charles Howard Hinton himself. We can glimpse it in various ways, analogously to how with perspective distortion and artistic tricks we can 'see' a three-dimensional cube expressed in a two-dimensional drawing. Our retina, after all, is a two-dimensional plane on which we see three dimensions.

Claude Bragdon in Illustration 1, figure 1, demonstrates that if a glass cube is held in front of the eye a two-dimensional image as shown would be seen: a square within a square. In a similar way, as in figure 1, the analogous, in reality unrealisable, projection of a four-dimensional tesseract as a three-dimensional representation can be approximated on a two-dimensional sheet of paper as figure 2. The 'exterior cube' exists 'further away' in the direction of the fourth dimension. Another version can be represented if, as a three-dimensional cube can be 'unfolded' into a two-dimensional cut-out (Illustration 2), a four-dimensional cube can be unfolded into a cuboid cross complete with the requisite properties of points, lines and cubes.

Extraordinary properties would exist in a fourth-dimensional world inferred from those progressively existing in lower dimensions. Symmetrical but left-handed and right-handed objects such as gloves would be able to rotate to fit each other exactly. If Sphere were flexible enough he could turn himself inside out without tearing. Moreover, as A2 could see inside his house when Sphere raised him up into the third dimension, a fourth-dimensional being would be able to see all around an object or remove itself from a dungeon-cube like a spirit and re-appear again. In the fourth dimension Slade's tricks might be realisable.

To aid the visualisation of four-dimensional space Hinton developed

## THE REPRESENTATION AND ANALYSIS OF THE TESSERACT, OR FOUR-DIMENSIONAL CUBE BY A METHOD ANALOGOUS TO THAT EMPLOYED IN MAKING A PARALLEL PERSPECTIVE

FIG 1.

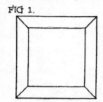

A GLASS CUBE, HELD DIRECTLY IN FRONT OF THE EYE, WILL APPEAR AS SHOWN IN THE ACCOMPANYING DRAWING. THIS–BEING A PLANE FIGURE OF TWO DIMENSIONS–MIGHT HAVE BEEN PRODUCED BY DRAWING ONE SQUARE INSIDE OF ANOTHER AND THEN CONNECTING THE CORRESPONDING CORNERS. THIS COULD BE DONE WITHOUT ANY THOUGHT OF THREE DIMENSIONS, YET ON THIS PLANE FIGURE MANY OF THE PROPERTIES OF THE CUBE CAN BE STUDIED. BY COUNTING THE FOUR-SIDED FIGURES, WHICH WE FIND TO BE SIX, WE LEARN THE NUMBER OF FACES OF THE CUBE. BY COUNTING THE NUMBER OF CORNER POINTS, WHICH ARE EIGHT, WE LEARN THE NUMBER OF THE CORNERS OF THE CUBE. BY COUNTING THE LINES, WHICH ARE TWELVE, WE LEARN THE NUMBER OF EDGES OF THE CUBE.

FIG 2.

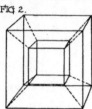

IN THE SAME WAY THAT FIGURE 1 REPRESENTS THE CUBE, FIGURE 2 REPRESENTS THE THREE-DIMENSIONAL FORM CORRESPONDING TO THE TESSERACT. JUST AS WE DREW A SMALLER SQUARE INSIDE OF A LARGER ONE, SO WE REPRESENT A SMALLER CUBE INSIDE OF A LARGER CUBE. AND JUST AS WE DREW LINES JOINING THE CORRESPONDING CORNERS OF THE SQUARES, SO WE FORM PLANES JOINING THE CORRESPONDING EDGES OF THE CUBES. TO FIND THE NUMBER OF CUBIC BOUNDARIES OF THE TESSERACT, WE COUNT THE LARGE OUTER CUBE, THE SMALL INNER CUBE, AND THE SIX SURROUNDING SOLIDS–EACH A DISTORTED CUBE–EIGHT IN ALL. A FURTHER STUDY OF THE FIGURE DISCOVERS 24 PLANE SQUARE FACES, 32 EDGES, 16 CORNER POINTS.

Illustration 1. Representation of the tesseract.

the use of small multi-coloured bricks given names and arranged in blocks so that they could be memorised to allow an envisaging of the cross-sections of four-dimensional figures as they entered into three dimensions. Fortunately for Charles Howard he had acquired the phenomenal memory that this necessitated from his father. For others less well-endowed it has been rumoured that it drove some adherents to near madness. Martin Gardner reported that one Czech inmate of a sylvan Cotswold commune had been unhinged by cube-gazing.[3] There was, however, an important adjunct for the practitioner. The mind-stretching concentration of the perceptive process, as if by auto-hypnosis, could open up an inner world of being seemingly in touch with a higher reality.

In his most seminal work, *A New Era of Thought*, Hinton borrowed

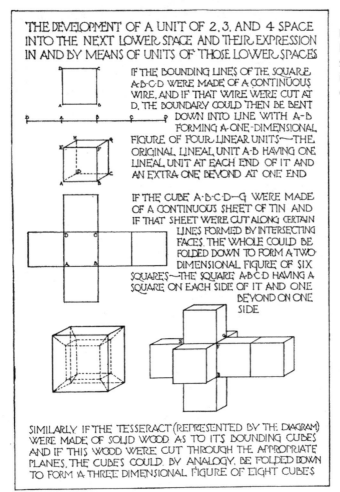

THE DEVELOPMENT OF A UNIT OF 2, 3, AND 4 SPACE INTO THE NEXT LOWER SPACE AND THEIR EXPRESSION IN AND BY MEANS OF UNITS OF THOSE LOWER SPACES

IF THE BOUNDING LINES OF THE SQUARE A-B-C-D WERE MADE OF A CONTINUOUS WIRE, AND IF THAT WIRE WERE CUT AT D, THE BOUNDARY COULD THEN BE BENT DOWN INTO LINE WITH A-B FORMING A ONE-DIMENSIONAL FIGURE OF FOUR LINEAR UNITS—THE ORIGINAL LINEAL UNIT A-B HAVING ONE LINEAL UNIT AT EACH END OF IT AND AN EXTRA ONE BEYOND AT ONE END

IF THE CUBE A-B-C-D—G WERE MADE OF A CONTINUOUS SHEET OF TIN AND IF THAT SHEET WERE CUT ALONG CERTAIN LINES FORMED BY INTERSECTING FACES, THE WHOLE COULD BE FOLDED DOWN TO FORM A TWO-DIMENSIONAL FIGURE OF SIX SQUARES—THE SQUARE A-B-C-D HAVING A SQUARE ON EACH SIDE OF IT AND ONE BEYOND ON ONE SIDE

SIMILARLY IF THE TESSERACT (REPRESENTED BY THE DIAGRAM) WERE MADE OF SOLID WOOD AS TO ITS BOUNDING CUBES AND IF THIS WOOD WERE CUT THROUGH THE APPROPRIATE PLANES, THE CUBES COULD, BY ANALOGY, BE FOLDED DOWN TO FORM A THREE DIMENSIONAL FIGURE OF EIGHT CUBES

Illustration 2. Development of a square 'unfolded' into the fourth dimension.

from his father James the idea that to perfect society individuals should 'cast out the self', allowing us to enter into a condition of 'altruism'. Charles Howard was looking for some form of certainty in moral life that went beyond the religious and social beliefs, for example, of his father and George and Mary Everest Boole, that depended on faith. He hoped that this would occur naturally through both the intellectual process of acquiring objective knowledge of four-dimensional figures but also with our thus coming in touch with a unifying existential state we can all share.

Charles Howard's proposal is to remove the individual from our constraining three-dimensional world and, like A2 being lifted by Sphere out of Flatland, enter a four-dimensional state in which 'God becomes known as an inward spirit … allowing the question of altruism against self-regard ... almost to disappear'.[4] With this awareness comes a sort of geometric socialism: 'the sense of isolation is gone. The bonds of brotherhood with our fellow men grow strong, for we know one common purpose.'[5]

Hinton here seems to be echoing William Blake, 'If the doors of perception were cleansed the whole world would appear as infinite.' Blake could see the universe in a grain of sand – Charles Howard could view it in a stack of bricks. Such a visionary state, a glimpse of selfless nirvana, would render us equal in a beautiful harmony. It will affect our view of nature. 'How set out in exquisite loveliness are all the budding trees and hedgerows on a Spring day ... to where, in the distance, they stand up delicately ... distinct in the amethyst ocean of the air.'[6] Whether this is a form of transcendental experience Hinton is unclear. Is this heightened awareness only transitory, as many witnesses report, and usually unbidden, or repeatable at will? Will the effect of a sense of the unity of all things be long-lasting? He provides us with few psychological insights. His claims are far-reaching though: Hinton hoped that his method would eventually 'bring forward a complete system of four-dimensional thought – mechanics, science and art'.[7]

Hinton even speaks of 'the relationship which we have to beings higher than ourselves',[8] who, like A2's friend the Sphere will lift us up into a new realm of consciousness. What he probably means is gained from the inference that if an n-dimensional being can perceive the one below then we as three-dimensional beings must be viewable from a dimension higher. At present Hinton asserts we aspire socially to a higher state, 'in those ways in which we tend to form organic unions ... in friendship, in voluntary associations, above all in the family we tend towards our greater life'.[9] Such sentiments regarding everyday life seem a bit hollow, however, coming from someone who had shown little respect both to a mistress and his wife.

As seems to be so often the case with the Boole/Hintons, there is almost no personal correspondence on which to assess the man apart from his public writings. One letter of February 1887 to his publisher, William Swann Sonneschein, written just after his trial but before *A New Era of Thought*, says a good deal. He is in a self-pitying mood: 'I have had to give up everything and go through disgrace such as rarely falls to anyone's lot ...'.[10] He feels there are prospects for change, however: '... the first and absolute condition of any true life as I understand it now lies in absolute openness'. Perhaps like A2 he has become sufficiently above things to be able to see more clearly his connections with others and cast out the self.

At the time he certainly saw his relationship with his life's work more outlined. He writes in the same letter: 'What I have come to see is that in the mere facts of the material world there is an evident and clear proof of a higher existence than that which we are conscious of in our ordinary bodily life.'

It is difficult to feel any great empathy with Charles Howard Hinton. The dimensions of his character are not laid bare through the stages of his life in comparison to his geometric progressions: point to line, square, cube and tesseract. Like the *Voynich Manuscript* he is an enigma, a one-off. Both James Hinton and son seem to lack a grip on the real world. Staring at a block of cubes is not really going to provide any sensible basis for grounding the morality upon which our social institutions are based.

*A New Era of Thought* was published one year after Hinton's exit to Japan in 1887. Alice Boole, hooked into the fourth dimension too by him, was responsible with family friend, John Falk, for editing and supplementing the second part's much less philosophical and more down-to-earth exercises in envisaging tesseracts. They urged readers to give objective treatment to his ideas and hoped that cube models as patterns would be supplied by the publishers. If not, coloured cubes could be made by borrowing kindergarten ones, painting them with watercolours and sizing with varnish. Hinton's 1904 *The Fourth Dimension* was accompanied by the purchase offer of a set of eighty-one differently coloured cubes for sixteen shillings. Hinton certainly expected people to take up his challenge to acquire the power of higher-dimensional thought. In *Many Dimensions* (1885) he suggested that it would be more beneficial if lads on Underground trains instead of 'bending over the scraps of badly-printed paper reading fearful tales', were, 'communing with space'.[11]

Unlike Edwin Abbott's satirical Flatland, Hinton's forays into other worlds seem mostly to be a platform for further dimensional exercises. *A Plane World* (1884) comes closest in inferring human characteristics from geometric templates. Its token two-dimensional population inhabit a flatland existing on a sphere instead of a plane and live in crevices and channels. Having no way of passing they build chambers to go below each other. Hinton describes their houses and carts for transport. They possess gender in 'a crude kind of polarity', and have sensitive and hard edges but, unable to come face-to-face do not generate friendships. His last work, *An Episode of Flatland* (1907), also anthropomorphises two-dimensional flat beings with personality and a history.

It is a very garbled *Boys' Own* affair spilling out the conflicts between the races of Scythians and Undeans against a cosmic background of earthquakes, collapsing lakes, planetary collision, and somewhat presciently, global warming. The hero of this cosmology is Hugh Farmer, who while describing himself as 'a drivelling dotard speaking in a scarce intelligible language',[12] has nevertheless mentally habituated himself and become familiar with three-dimensional shapes, professing that the 'tool of thought' is space.[13] Fortunately, amid all the chaos, children are naturally acquiring higher-dimensional consciousness, realising that they are beings on a higher plane. It is by the collective realising of its power that their planet eventually avoids collision.

Threading through these apocalyptic times weaves a feisty but tender woman, the essence of all being, all loveliness. This is Laura, a triangle, niece of Hugh, beloved by her many suitors. She professes to him that the unknown third dimension is where souls end up, much to his scorn: 'All that is a vast fiction.'[14] He has a dim view of women in general as all they think of is to be flattered. Men get carried away by them, 'a thing likely to happen to all of us'.[15] (Is this a belated reference to the bigamous events of twenty-five years previously?) In other places Hinton seems in thrall to women – like his father. In many of his writings a female model is found who somehow represents purity and innocence in the same way that a geometrical figure could be said to. The very act even of discovering higher space is like an act of worship of the female form. Space itself to Charles Howard is a 'she', 'in all her infinite determinations of form'. In *Many Dimensions* he compares our blindness to this realm as of Egyptian priests bedecking a goddess, 'until with a forward tilt of the shoulders the divinity moves and the raiment and robes fall to the ground … leaving herself revealed, but invisible; not seen but felt to be there'.[16] This is the exact theme amplified in his novella, *Stella*, published in 1895.

Another Hugh – Hugh Stedman Churton – finds himself up north in three-dimensional Yorkshire attending to the estate of a deceased recluse, containing a library and his extensive collected writings. One wing of the mansion is shut off but Hugh (bulls-eyeing a few pre-Freudian targets) discovers a ladies' boudoir and in a secret garden with a dense undergrowth of clematis hears the voice of a young girl, Stella. She has been made invisible by the dead hermit with a potion in order to cast out the egoistic self-element in women: the need to adorn and express physicality. She has not been allowed to be clothed so that, as in the Garden of Eden, she would be 'content as she was intended to be'.[17] If this seems a bit gender one-sided her mentor also wanted to devise some way of removing the self-element in men, who, like grown-up boys, care only about eating, drinking and acquiring things.

Needless to say, Hugh falls headlong in love with Stella, her voice at least, as they discuss a higher level of consciousness unclouded by considering matter as evil, but capable of being revealed by a discipline of thought practised by her master. We can guess what this might involve. Stella is later abducted by a mad professor who wants her for his own experiments, but is rescued by Hugh after searches in the US and London clubland. For some obscure reason they embark on an adventure in China, besieged by pirates and rescued by the guile of invisible Stella. She later materialises and they have a transparent son who, however, 'soon took in enough of this wicked world's nutriment to become as opaque as the rest of us'.[18]

*An Unfinished Communication*, published with *Stella*, is a more serious work, ramblingly attempting to clothe a philosophy of life. The fourth-dimensional element is less prominent. Written in the first person, the narrator, dejected in New York, comes across a notice on a door advertising 'Mr Smith, Unlearner'. The idea appeals to him of finding certainty, wiping the educational slate clean of physics, philosophy, Darwin and 'those foul ideas that pollute man's strength and women's beauty … and turn the man to brute'.[19] The narrator muses, as might Charles Howard, that although 'most reasons for wishing to forget come through women, she is also the means of forgetting … her own antidote'.[20] Mr Smith, however, is elsewhere. He catches up with him by a lonely, grey coastline. Echoing a recurrent theme, recounted also by Stella, the Unlearner teaches him that the perceived present is just one moment in an uninterrupted whole; all our moments exist in an unfolding eternity that links us to others. This is grandiose stuff. Interpreters of Hinton have dwelled on Nietzsche as an influence.

Seeking oblivion, the hero finds himself on the barren immensity of a shore – a theme with possible sexual overtones. He encounters and falls in love with a young girl, Natalia. She has warned him of quicksands (perhaps the pitfalls of amorous immersion), but he falls prey and feels the waves close over him. In a long and rambling stream-of-consciousness coda, time reverses and a series of retrospective life-glimpses and visions ensue at the moment of drowning. Hinton's imagination is here wild as in a dream sequence.

Yet there is an undeniable flow: the narrator meets his father and mother and a German woman named Gretchen who, with intimations of prostitution, he had once married to save her from life in a music hall. Most powerfully, under the 'courts of heaven', in Central Park he sees St Paul and St Simon Stylites, perhaps symbolising the intellect he is escaping and the need for self-denial. Nature in the guise of an old woman comes before them asking in another version of Hinton's motif, to be forgiven for wearing men's praise. It masks innocence yet brings the beauty of nature to the world seen in sunsets, and in 'the flash of white limbs through translucent water … vermeil lips, the glance of loving, passionate, ardent alluring eyes … the raised arms of Venus'.[21] A higher Presence bids her take off her garments of self-love, freeing herself and allowing men to enjoy nature untrammelled. No sooner, however, is this purity envisaged than the leering voice of a low-faced beast licking its chops is heard: 'my intimate friend, whom I know better than I do anyone else'.[22] This too smacks somewhat of autobiography.

The waves wash over the narrator's sodden inert mass. His meagre existence has been recalled and reviewed but is experienced as empty and worthless. Yet he realises that his memories are really actual presences, nothing is lost, as with nature, 'everything is in a state of development. Similarly in our many lives the will is free to alter its path.'[23] In some future life he expects to re-meet Natalia: 'I know she awaits me in that most holy union of soul and soul.'[24]

Hinton seems, in both *Stella* and *An Unfinished Communication*, to be putting clothes on the starkness of his four-dimensional geometric exercises. The analogies are still there but, embellished with his experience of life and symbolised in human terms, less abstractly. The sea, for example, is also a powerful metaphor of our collective being; individual consciousness is found on the crests of waves. The plane between water and air is often used by Hinton to demonstrate four-dimensional concepts. Here it possibly represents the divide between death and life. Below lies all that has happened, the world's soul; above, the shapes of things to come as yet unknowable as a four-dimensional object would be to our limited three-dimensional awareness, only able to see part in the present. All is stretching into Eternity – the geometric figure to suggest this is one from the Boole/Hinton store: the spiral.

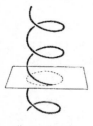

Illustration 3, Path of a spiral.

Using a plane of water as a paradigm, if time were imagined as a stick progressing vertically through its surface it would appear as a point, but if the stick was a spiral it would describe an O as it passed through: a moving recurrence (Illustration 3). Stella tells Hugh: if you feel eternity you will know you are never separated from anyone whom you have ever been. This is the greater interconnectedness Hinton felt in *New Era* that he has now tried to express in a literary form.

Most important, it seems, is the place reserved for one's soulmate, attainable for Hinton in woman and her beauty stripped of the self-elements of vanity and temptation; separated from his own beast within. His allusion to idealised women is repetitive through its various guises. The fourth dimension is woman bared: Laura the triangle is an object of beauty, Stella though coquettish is wiser than Hugh but he needs her like Natalia is needed. 'I felt I could bear anything if this time with her were always with me, never past and gone – I in reality always with her whatever scenes I might be in my time consciousness.'[25] Hinton has unfolded like his geometric figures but more poetic and feminised.

*Stella* and *Unfinished Communication* were presumably written in Japan, far from his own convoluted past and the scandalous affair. We shall never know if Maud Weldon was an inspiration. We have no information either on how the Boole family reacted to those events but it can't have been uninvolved. The most affected, of course, would have been the Booles' eldest daughter, Mary Ellen. She had to suffer the indignation of being rejected. We know very little about her compared to the other four sisters. In family photos she does not appear as a woman of great beauty, in fact rather manly in outline. By the time of their exile in 1887 she had borne four boys: George, Eric, Billy and Sebastian (known as 'Ted'). The few of her letters extant written to her sister Lucy and a small diary spanning September 1888 to 1890 reveal something of the life she and Howard led in Japan and the influence of eastern life on both their outlooks.[26]

In 1889 their address was 179 The Bluff, an area reserved for foreigners, overlooking Yokohama harbour. A little further down the coast at Tokyo Bay the US Navy under Commander Perry had forced the Meiji empire to open up for trade in 1853. Yokohama developed rapidly from a small fishing village to a trading centre. Western-style mansions had been built for the new occidental influx. Mary writes that she is disappointed with their house because it is only half-Japanese, 'shabby and dilapidated', but by 1890, with a Boolean love of flora, she seems to have put it in horticultural order. She lists magnolias, daphne, wild camellias, viola aurea and English narcissus. She admires Japanese interiors: the metal charcoal-burning fire bowls, convenient instead of back home the 'endless black-leading of stoves and friction with servants about them'. She thinks, however, it must be difficult to keep warm given the deep snow for many

Charles Howard Hinton and Mary Ellen with their four boys in Japan.

months of the year. The simplicity of the *tatami* mats in the rooms appeals to her and bedding that can be 'cleared away in the daytime so that every room can be used. No dirt or dust because no one wears boots or shoes there.'

She has been putting together a trunk of objects to send back home, chief of which is a shrine complete with brass vases, oil vessels and incense. It has cupboards for wooden tablets to remember one's ancestors. She is worried it might be a white elephant in Lucy and her mother Mary's small house in Notting Hill but hopes that they will share it without quarrelling. For other family members there are bows and arrows for Ethel, a small white box for Alice, two kimonos lined with silk wadding for Maggy's son, Geoffrey (Taylor), a larger white box for Mrs Hinton (James Hinton's wife, Margaret), and an inlaid wooden box for her daughter Daisy (Ada). With a Hintonian Chinese hint of habits to come she states that 'We are thinking of hiring a cow.' Charles Howard her husband is not described in any way compared to her fulsome praise of the boys. In a letter of November 1891 she writes:

Georgie [is] beautiful but suffering. Everything that requires effort [is] a pain to him because he has not the strength. Eric is the exact contrary. He has a thoroughly good, sound brain … he is very quick at calculation. It is perfectly possible that he may be a very clever man some day. Willy is a sturdy boy, like Eric in face, not so good and loving … usually cheerful and healthily contented.

Sebastian is clearly her favourite with bright red hair like Ethel's. 'I could make him a big overgrown brain but am not going to do so. He keeps us all in a state of interest for he is always talking brightly and he is like a piece of sunshine in the house. An artist is painting his picture. I know I should envy anyone whoever else had it.'

In the school holidays the family took off on visits around Honshui, the main island. In September they are at Kaneda Bay, a fishing village, some fifteen miles further south on the east side of the Miura peninsula. They are living in two small rooms in a temple having made friends with its priests, 'such gentle and refined simple people'. They witnessed a long service on the beach invoking protection against spirits, but somewhat bored moved off only to be followed by half the congregation staring at these European strangers. In the evening they joined fishermen and priests on the shore from where little rafts made of crossed lathes holding lit candles were launched as charms against danger. 'Soon there were hundreds … stretching away in a broad path over the sea.'

I too have a clear image of this setting. In February 1983, on the third stage of my global 'Peace' journey I was staying a further ten miles down the Miura peninsula on its west side with a good friend from Leeds University days who had been living in Japan for ten years. Her house was situated fifty yards from the sea with Mount Fuji across the bay. I wrote in my diary on Sunday, 20 February 1983:

> I walked round the bay to the neat fishing village, pretty with people at work, washing radishes, mending nets, pounding the sand out of seaweed. I found two wooden holy relics washed in by the sea. I have taken on the role of supplying driftwood for the fire that I cut up with a power-saw. Today, mysteriously, a heavy scaffolding-board leaning upright against the house hit me thwack on the head making me dizzy. Later I'm told that my relics have been thrown into the sea intentionally and are not meant to be recovered. Clearly the gods are paying me back for my interference. Cursing at my appeasing stupidity I take the offending items down to the beach and launch them to the waves saying 'Sorry mate I didn't mean it.' [I actually kept one of the *sotoba* relics; it still gathers dust in my attic – might this explain some of the ups and downs that have befallen me since?]

On another vacation in 1889 the Hinton family ventured north to Gunma Prefecture, sixty miles from Tokyo. The precariousness of the country emerging frantically from feudalism is illustrated by the 'Gunma Incident' five years before, when the Japanese army, equipped with rifles, gunned down protesting farmers. Foreigners like the Hintons must have felt somewhat uneasy. Here the volcanic mountains bubble up with *onshen* – hot springs thought to have healing powers. At Ikao, another spa site, she writes to Lucy how delightful it is 'to sit up to your neck in very hot water with the sky overhead with rain pouring down'. It was January and bitterly cold with stalagmites emerging strangely from the frostless ground 'like spars or splinters. Howard can't imagine why they form. The mountain earth is coloured red, orange or clear white and supports prairie and bamboo grasses.' It is the first time she has seen begonia growing wild.

Mary has become fascinated with Buddhism and what she believes is its beneficial effects on people. 'Here you may walk through a crowd with your purse in your gown pocket and forget its existence.' Christianity, she admits, can influence special people, 'but it has not the power to take hold of a nation and hold it'. This she relates to the personal inclusiveness of Buddhism and its faith in the Amida, the Japanese version of the wider religion of China, Vietnam and Korea. In a passage highly reminiscent of her sister Ethel she writes, 'I remember as a child being troubled by doubts as to whether Providence was paying attention.'

She has been having instruction from a Buddhist on Sunday afternoons and now seriously questions the Christian notion of the Atonement and Christ's suffering for our sins. This is unsurprising: no member of the Boole/Hinton dynasties was conventionally religious. She objects to the selection of only a few souls for salvation. Buddhism's various levels of existence allowing the finding of one's own level through *kharma* and stepping off the wheels of rebirth and death, 'that is the one thing that is most attractive to us now'. The 'us' here is of significance. One can imagine Charles Howard's attraction to the eastern concepts of planes of consciousness, particularly the Amida, Pure Land, available to all, existing infinite and outside ordinary space and time that yet can be embraced by prayer, contemplation and visualisation.

In July 1890 they had ventured north of Tokyo to the famous temples at Nikko set on the side of a mountain with gigantic pine trees all around. She mentions 'The lanterns of stone and the dragons and the great sandalwood porches. How can I describe the temples to you ... I would want the eloquence of Demosthenes ...'. The colours particularly strike her:

the underside of the roof is brilliant yet sombre red and gold, the upper side black, the shutters are a soft green ... produced by their being made of a fine kind of copper. Inside the ceilings are gorgeously painted, blue being the prevailing colour ... you could spend the whole day in one room.

She describes the brass banners and bells and lotus flowers in gold, each many feet in breadth:

in the middle of the largest ... the image of Amida ... that incarnation of Buddha which saves women sinners and the weak (in fact he is a Jesus Christ). The only unsaved people are those who don't call on him. Outside there is a stone walk winding up among the trees to a tomb. Howard went up. It was three hundred steps.

I remember it well.

NIKKO, 23 FEBRUARY 1983

Set off to local shrines past a long line of Buddhas some with their heads missing and round pieces of rock perched instead. They look great calmly sitting there overgrown with moss but uncomplaining. One is

Nikko shrine, possibly where both Howard and the author communed.

Nikko shrine, with offerings and *sotoba* memorials.

185

festooned with garlands of paper birds and surrounded by offerings of empty Coke tins and milk jars. I suppose it's the thought that counts. The Japanese seem to have a very manipulative view of the world. Seeing as it's full of demons and bad luck, e.g. famine, earthquake, flood – it pays to propitiate with charms, messages, incense, prayers, bells, gongs and gifts. Walk to new batch of shrines, one is approached up steep steps and has stunning workmanship. Ascend further to a dilapidated hut shrine. I think I prefer it. Sit outside in the warm sun listening to the birds and the snow melting and crashing down through the leaves of the enormous pine trees.

Mary Ellen is very intrigued by a mandala, a sacred picture of scenes of the Buddhist paradise.

The 'paint' was an embroidery made with human hair … given by 4,000 women who are supposed to have contributed each one. In the middle sits Amida. The mandala is being taken on a tour around Japan, the priest who has care is glad to help us to understand it. Think of us this afternoon. The old priest, a young priest, a merchant who is a devout Buddhist … a philosophical gentleman … acting as interpreter. Howard with pen and ink rapidly writing down all their explanations and I making coffee and drinking it all in … of course I meant metaphorically.

On 28 October 1891 a major earthquake over eight in magnitude destroyed towns on the southern coast of Honshui such as Nagoya, 150 miles west of Tokyo. Over 7,000 died.

We were startled early in the morning by a shaking, got up, rushed to the children and took them down in to the garden where we stood night-gowned and bare-footed. The time of an earthquake is a moment of pure terror. It is strange to think how near we have been to serious disaster … There was a post at Nagoya at the time that Howard was offered this, he would have accepted it … after October 28th none of us would have lived to tell the tale. Shocks continue daily. It is enough to make one think seriously is it not? Only think till you have no more feeling left – and you arrive at no result. It is aimless groping. If these were the last words I should ever write I couldn't add to them.

We cannot be sure why Charles Howard Hinton sailed with his family from Yokohama on 16 August 1893 to Tacoma, Washington state, USA. Despite their roots in England, there would have been probably no way of staunching the disgrace of his bigamy and finding work and respectability back home. Apart from the massive earthquake and fear of another, he may well have wanted to engage in 'western' university life again.

He took up the position of Instructor of Mathematics at Princeton University, New Jersey, with a not very rewarding, untenured salary. Hinton was in contact from 1896 with William James, the brother of novelist Henry. William's best-known work is *The Varieties of Religious Experience* (1902). After their first meeting James had obviously taken an interest in Charles Howard's career. He wrote to him that Japan had represented 'extinction for so much'. James lent him money and suggested that he should give a series of lectures on mathematical geometry at Harvard. Charles Howard declined possibly because he was anxious at being on such a public stage but may already have become disenchanted with American academic life.

In a newspaper article in 1902 entitled *The Oxford Spirit* he contrasted its rarefied tenor with the 'business man in control of the halls of learning of US university life'. 'Bull' as he was popularly known to his students, owing to his physique, was probably not going to fit in anywhere given his Boole/Hinton background of rebellion. Idiosyncrasy, another familial trait, was probably the cause of his undoing at Princeton.

A far cry from the demure cricket pitches of Oxbridge, Hinton had taken a great liking to baseball and invented a pitching gun. It was a kind of shoulder-mounted cannon about two and a half feet long which fired balls at speeds up to 70 mph. Their trajectory could be altered by guides within the barrel. It must have presented a fearful sight. *The Boston Daily Globe*, December 1896 reported that the puff of smoke it emitted made many batters dive out of the way. To compound the idea's zaniness and no doubt further irritate the Princeton hierarchy, he used the initial lecture about the gun to a college crowd of 300 to perpetrate an inventive hoax.

Early in his address a special-delivery postman interrupted him with a letter that he at first declined to accept. The students, expecting that someone was playing a trick on him, were in turn 'horsed' by the contents being a report of a basketball match played in 1950. Whether this saga was directly responsible we'll never know, but he either resigned or was himself fired in 1897. What it does evidence is a hint of a quality of the

Boole/Hintons not always in great supply – humour. Charles Howard clearly possessed an oddball proportion. A reviewer of *Stella*, for example, noted, 'One is often at a loss as to whether to take the author seriously or to recognise that he is in a chaffing mood.'[27]

The Hintons consequently moved to Minnesota in the Midwest where he took up a position as Assistant Professor of Maths at its State University. He became a US citizen there on 9 October 1900. No four-dimensional writings emanated during his stay there for four years. This seems to be a recurring interval in his life as he moves restlessly, disappointed and unable to settle, hoping for recognition of his ideas.

Baseball seems to have been an addition to his dimensional mono-mania, having given another lecture about his gun at the YMCA in Minneapolis in December 1897 and writing an article *The Motion of a Baseball* in 1898. In October 1901 the family trekked back eastwards again to Washington where he took up a post in the Naval Observatory. His attachment to his *oeuvre* continued. He read a paper, *The Recognition of the Fourth Dimension*, to the local Philosophical Society in 1902. *The Fourth Dimension*, a collection of essays published in 1904, mostly recapitulates his theories of the previous twenty years including a simplified version of tesseract-spotting with his system of cubes. It received mixed reviews: favourable by *The New York Times*, poor in *Nature* and a whimsical appreciation by Bertrand Russell in *Mind* that probably has echoed many other minds since. While praising the author's 'most pleasing enthusiasm', Russell wonders whether his elaborate models 'require as much thought for their use as would suffice for the understanding of the fourth dimension without their aid'. His speculations, however, 'stimulate the imagination, and free it from the shackles of the intellectual'.[28] In 1903 Hinton went to work in the US Patent Office as an examiner.

His Japanese years must have made him more open to the possibility of transcendental experience. In the US he gave several talks on Japan. Even before he left England, in *Many Dimensions* (1885) he writes of the western habit of 'seeking formal causes always', compared to the eastern world's 'living apprehension of the proximate'. Such 'an inward communion' provides a 'delight whose presence in my mind for one half-hour is worth more than all the cosmogonies that I have ever read of.'[29]

Correspondence with a close friend, Gelett Burgess, about this time is illuminating. Burgess (1866–1951) was a free-thinking poet, novelist and writer of satire who bequeathed the world the word *blurb*. He was drawn no doubt to the craziness and humour of Hinton, although

Hinton's mood is often self-doubting and despondent. In a letter to Gelett he laments that his own writings are 'on one note of life altogether. You are in life.'[30] Always looking for converts, however, he had sent Gelett a set of his cubes and instructions but bemoans 'the ridicule I create in the professional mathematicians ... maybe only poets can see'.[31] Alice Boole, who he is still in contact with in England, thinks it 'simply a crime to say in words anything at all'. On 22 August 1906 he wrote to Sonnenschein about 'flying my great and solitary kite ...'.[32] But his imagination is still at work: he has new ideas of birotation, a novel kind of algebra, and is writing short parables and a paper on *Electricity, a Service of Mathematics*. One unpublished story in the Burgess archive entitled posthumously *Travels of an Idea* is a highly revealing pilgrim's progress with oriental overtones.

Sloman Selby, its hero, leaves London in 1890 to bring western science to the Buddhists of Japan but becomes instead drawn to the sights and sounds of its temple life and the monks' meditative peace. He learns to appreciate that 'an instrument can be formed within the mind itself and of the eternal being that sweeps by outward sense in raiment of fleeting things'.[33] True to erotic form he meets the pliant, erect and slender Tessara. She is an Idea, 'demanding an empire and giving the joy of a smile which lighting on things lives in them forever. I am launched, says she, into the career which shall bring all men's thoughts to me.'[34] Travelling in America, with Sloman, however, she finds little success with theosophists and professors. At Niagara Falls he dreams of 'the sweet, drawing life of England's countryside for ever the same'.[35] Ultimately, homesick and defeated, he admits to Tessara the hope that 'when he has gone, in the future ages some great walls of dark-lined trees and deep alcoves of solemn temple roofs, like we saw in Japan, will rise to your honour'.[36]

Charles Howard's health had begun to fail about this time although doctors seemed unable to locate the cause. He continued to espouse his ideas, giving a talk before the Washington Society of Philanthropic Enquiry on 30 April 1907 on the somewhat weary title *Psychological Entrance into the Fourth Dimension or Heaven or any Other Place*. At the ensuing dinner on the sixth floor of the YMCA building where he apparently responded warmly to the toast to 'Female Philosophers', on entering the lift he suffered the same demise as his father from a cerebral haemorrhage. One cannot help thinking that it was an apposite space to have exited into some other dimension.

Only two weeks before he had written to his eldest son, George, 'lots of people are coming round to my way of thinking and perhaps I'll get a chance after all'.[37] It was not to be. Perhaps he was just simply wrung out after years of proselytising his views. The *Washington Post* reported the day after that his wife Mary Ellen was 'prostrated at the news and was placed under the care of a physician'.[38] Sebastian ('Ted'), his youngest son, wrote to a friend about his father's 'jolly laugh quieted into a warm kindly smile which seemed to be the stamp of the clean manly soul of his – brimming once with love for everything'.

Hinton was obviously likeable but occupied a strange intellectual niche: a kind of combination of the materialist and idealist that must have bemused people. While on the one hand he grounded himself with masculine baseball guns, jokery and mathematical principles evoked with the mundanity of children's building blocks, he also wished to take flight like A2 into a state of consciousness suffused with romantic female images and a fuzzy but embracing morality. The leap from the kitchen table stack of bricks to the mystic 'All is One' is highly obscure; his works never clearly established the connection. The strangeness that surrounded his *idée fixe* produced an inability to find common ground with the institutional world: entirely in keeping with the Boole/Hinton dynasties.

Although Charles Howard never lived to see the mainstream accept-ance of his four-dimensional theories, they had a lasting effect on ideas on both sides of the Atlantic.

His scientific romances were cosmological stretchings of the mind – mostly, I found, verging on the incomprehensible. One typical model was inspired by the then recent drawing-room invention of the phonograph and the flat disc. Hinton used it to posit our relation to the ether.

Subsequently made redundant by Einstein, it was assumed then to be the luminiferous medium, a transparent space-filling substance surrounding us, through which light is propagated, as sound travels through air. Conveniently, conceptually it answered the question: why if there exists a higher dimension more powerful and omnipresent than the other three is it so elusive to our senses? He surmised it must be because if a two-dimensional world must have some detectable three-dimensional thickness, a fourth-dimensional one must also, although it would be infinitely minute or we would be aware of it. This ethereal hyper-thickness cannot be something that we are in but on, 'about the thickness of an atom along which we slide'.[39] Like the recurring images of boundaries between dimensions such as the surface of water or a state between which gases turn to fluids, the ether is a medium between dimensions, 'more solid than the vastest mountain chains, yet thinner than a leaf; undestroyed by the fiercest heat of any surface ... bearing all the heavenly bodies on it and conveying their influence to all regions of what we call space'.[40] If this uber-Wagnerian notion doesn't leave you with Seat 58 'nervous trembling sensations' you may begin to suspect you are of fourth-dimension origins yourself. There is more.

The ether it seems has grooves in it like those in a record that control the behaviour of the substance of the world which slides temporally over its definite marks and furrows. Once in a cosmic while it all begins again and therefore appears cyclical – although a record groove is actually a spiral with a beginning and an end. To avoid life being merely a repetition of the same old tune replaying our pasts and futures, during each performance we manage to jog and etch the indentations with our free will. The Grand Entity that winds up the phonograph Hinton never tried to envisage, hinting only that just as A2's two-dimensional plane was made visible as he was dragged upwards by Sphere, viewed from the fourth dimension, the physical and moral interiors of our three-dimensional world would be made apparent. Our openness to view and collective existence as we track our ethereal corrugations lift us from being a mere herd of followers to beings sharing and realising the unity of all things.

Hinton's popularisation of the idea of a fourth dimension was taken up by others similarly on the fringe of orthodoxy. They used the concept to explain otherwise inexplicable psychic phenomena and provide keys to the meaning of life and death. Theosophy was based on the personal spiritual welfare to be gained from revelations supplied often by eastern mysticism and symbols: immortality could be verified empirically by higher life forms reached through séance and the evidence of telepathy, mesmerism and automatic writing from the beyond – the etheric layer of existence, construable as the fourth dimension. If not in heaven, where more likely was the spirit world to be found than a realm only a hyper-thickness away? Theosophy gained many adherents including Annie Besant, who had now abandoned socialism. There were, however, more down-to-earth and broader links being embraced on both sides of the Atlantic.

Drawing heavily on Charles Howard Hinton's explorations in the fourth dimension, Claude Bragdon (1866–1946) was both a theosophist and architect. If a theosophical fascination with universal symbols included the properties and attributes of the pan-cultural 'magic square', even more compelling was the hypercube, the square writ cosmically large.

Despite its traces being unrealisable concretely, it could be nevertheless approximated in architecture. The public buildings he designed in his native town, Rochester, New York – the railroad station, Chamber of Commerce and Universalist Church in particular – show definite cubist elements.

Trying even further to universalise life aesthetically, he advocated in his designs what he described as 'projective ornament', the use of geometric patterns derived from nature. During the First World War he staged huge ornamental choral festivals to try to bring citizens together in harmony. Bragdon's theosophical vision, true to its own cultural setting, bore the imprint of democratic America and less a reverence for the seer and master, compared to European leanings. What he was attempting, no less, was to implement in his own way Hinton's 'complete system' of four-dimensional thought that had been projected in the *New Era* of 1888.

Perhaps nowhere in pre-war Europe was more open to all-embracing new ideas than Russia. As Ethel Voynich had discovered in her stay there back in the late 1880s, the suppressive tsarist order was ripe for change. If the dangerous espousal of political revolution did not appeal, art could express new waves of thought. Influenced by theosophy both the composer Alexander Scriabin and poet Andrei Belyi looked to a spiritual revolution in consciousness. Wassily Kandinsky sought for the secret laws of the artist through 'an aesthetics of the future'. It was a mathematician, Peter Ouspensky, who brought Hinton's challenging ideas to the Russian *avant-garde*, publishing a book on the fourth dimension in 1909 and translating Bragdon's book *Man the Square* in 1913. Bragdon returned the favour by publishing Ouspensky's major work *Tertium Organum* in the US in the early 1920s. Ouspensky is, in relation to Hinton, a very illuminating character: he was responsible for projecting his influence historically in a direction Hinton could never have envisaged.

Born in Moscow in 1878, a generation later than Hinton, by 1905 he was living in St Petersburg as a freelance journalist and soaking up the heightened atmosphere of expected dramatic change following both the revolutionary political turmoil of 1905 and the theosophists' perennial hope for a new era. Like Hinton, Ouspensky had little interest in the former, but he possessed a panoptic curiosity in what comparative religious culture and knowledge of heightened states of awareness could add to a scientific view of the world.

Ouspensky based himself at the Stray Dog café off Nevsky Prospekt where drugs, music, poetry and the search for the miraculous in life prevailed. In *Tertium Organum* Ouspensky examines Hinton's geometric analogies and the insights of higher space but feels he is merely walking alongside a wall. Like Hinton, he believed that if we can know of the possibility of the fourth dimension we must in some sense be part of it – our lives are mere shadows of a higher perceivable reality. Even though our glimpses of it are puzzling and fragmented there must be something, a consciousness more complete, astonishing and meaningful, more available than just a realm of spirits.

Ouspensky called this higher dimension, cosmic consciousness: a variety of religious experience that William James had induced with nitrous oxide. Occurring spontaneously, it has also been called an 'oceanic' experience. This is exactly what Ouspensky glimpsed on board ship in the Sea of Marmara in 1908 as he watched waves break upon its bows: 'The white crests were running towards us from afar ... drawing my soul to themselves. It was only a moment ... maybe less than a moment. I became All.'[41] Trying to recapture the elusive sensation, he too experimented with drugs but recognised that certain mind-states were conducive: yoga, meditation and the study of ancient religion.

Ordinary, everyday objects might provide a trigger for elevation but symbols such as a lotus mandala that the Hintons had encountered in Japan might provide a focus. Christianity, compared to eastern religions, has mostly been wary of such ecstasies as they bypass institutional authority and teaching, appearing idolatrous and pagan. For Ouspensky, the natural world too is highly conducive: 'The yellow leaves of autumn with their smell and the memories they bring ... the first snow dusting the fields ... thunderstorm, sunrise, the sea, forest, rocks. The sensation of being completely at one ... with nature ....' [42] 'The feeling of sex places man in the most personal relationship with nature ... indeed it is the same feeling ... only in this case more vivid.'[43] This perception of unity, he maintains, when accepted by others leads to a higher morality of tolerance and love.

It is fairly obvious that all these elements were also part of Charles Howard Hinton's world view. Hinton is explicit about definite results. 'The true good comes to us through those ... desiring to apprehend spirit ... willing to manipulate matter.'[44] The hyper-geometry of the tesseract, Ouspensky would maintain, is in itself a mystical experience, the mathematics of an infinite and higher logic. As a form of consciousness it exists in an evolved fourth dimension, 'timeless and miraculous', far removed from the third-dimensional logical world of codified morality which imposes subjugation of the individual to society and law. Way below all, suggests Ouspensky, lies the two-dimensional world of animals (and our old friend A2 who briefly managed to escape it) and further below still a one-dimensional world where 'everything takes place as it were, on one line', approximating to 'the state of the cell'.[45]

Ouspensky, like Charles Howard, felt that a new system of thought would usher in a new form of society, although for the former, the introspection of a stack of tiny cubes was not going to be the means. Amazingly, however, Ouspensky was instrumental, through his work, in providing a direct channel to the memorialisation of Charles Howard's worship of the cube.

In a 1913 series of lectures in Moscow and St Petersburg, Ouspensky expounded his ideas to popular audiences. It is highly likely they included the painter, Kazimir Malevich. Although Ouspensky decried the idea of pictorial representation of the fourth dimension – like trying, he suggested, to make a sculpture of a sunset – in December 1915 at an exhibition entitled *0.10 The Last Futurist Exhibition*, four of Malevich's paintings employed the term 'fourth dimension'. The collection included coloured geometric figures emerging or dissolving in differing planes from a white void into our world.

In one of Malevich's most seminal works, *The Black Square*, he painted a large black square almost dominating a white background in such a way that it could be interpreted as emerging from another world or disappearing into one – the ambiguous plane between dimensions. Bragdon's figure 1 in Illustration 1 (page 174) is a template. In his futurist opera *Victory Over the Sun* the costumes are composed of cones and rhomboids almost as realisations of Hinton's geometricised characters. A2 had come a long way from humble Flatland beginnings, now part of 'suprematist' art, as Malevich dubbed the new order.

The final reification of the cube came in 1924. After the death of Lenin on 21 January 1924 a hurried commission appointed Malevich's disciple, A.V. Shchusev to create the (initially wooden), structure of cubes in Red Square in which the embalmed body was placed. As theosophists looked to earthly inspired leaders to place a construction on immortality, so the atheist Soviet state needed to overcome death with a symbolic edifice to His immortal presence. Nothing less than a cube representing a higher dimension where death did not exist could enshrine the great Bolshevik leader. Tiny cubic icons were also authorised by the Party to be set up in homes, offices and factories. Would Charles Howard Hinton, one wonders, have laughed himself silly or torn out his hair at this manifestation of his 'complete programme'?

It was not only in a disintegrating Russia that new art movements were founded; revolutionary ways of experiencing the world were also being mapped in western Europe. Stravinsky's 1912 *Rite of Spring*, fragmented and angular like Malevich, burst into the world of classical music almost from another dimension. Cubism was essentially a way of looking at reality from a multitude of angles. Picasso and Braque's representations offered a wrap-around tour of an object as a four-dimensional being would be able to view all sides at once. Marcel Duchamp maintained a lifelong interest in the concept. His monumental *Bride Stripped Bare by her Bachelors, Even* 1915–23 contains allusions to higher spaces. Salvador Dali joined the surrealists in the 1920s, more reliant on free-ranging release of dream-like images from the unconscious than mathematical intuition. Much later Dali made explicit use of the fourth dimension in his 1954 painting *Corpus Hypercubus*.

It depicts a Christ figure crucified against an unfolded octahedral hypercube as in Illustration 2 (page 175) while below the Virgin looks up standing on a chequerboard of black and white squares. The reference is explicitly modern, alluding to what he termed 'nuclear mysticism'. As for many others, Hiroshima marked the shift into a new world of science and politics in which faith would need to reassert itself. Christ explodes into it sacrificially. Another painting, *Landscape of Butterflies* (1952) alludes to the projection of shadows: two beautiful butterflies cast unreal huge shadows onto a fiery red, ruptured cliff face. It was at Hiroshima that other innocents were instantaneously extinguished, leaving behind only their two-dimensional shadows projected onto the backgrounds before which they had stood.

Although Hinton was instrumental in generating an interest in the fourth dimension with his two volumes of 'scientific romances', it was H. G. Wells who in a number of more accessible stories excited public imagination. It is more than likely that he read Hinton and Edwin Abbott but his interest lay less in the possibility of four-dimensional space, rather more in the scope it gave for theorising about social realities and creating gripping stories through fantasy. *The Remarkable Case of Davidson's Eyes* rests on the idea that a 'kink in space' could, like a two-dimensional sheet of paper, be circled on itself to put one point in contact with another. *Plattner's Story* took up the idea that an object's ability to turn through itself in the fourth dimension would reconfigure a teacher blown apart in a chemical explosion leaving a heart on his right-hand side. *The Invisible Man* of 1897 makes a number of references: the use of the idea of a formula to create invisibility has a clear echo of Hinton's 1895 novella *Stella*.

Other writers played with the idea of higher dimensions: Conrad, Ford Madox Ford's *The Inheritors* (1901), Aleister Crowley's *Moonchild*, for example, but it was Wells' versatility of ideas that led to the wider readership in Russia and France influencing Jarry and Appollinaire. In the US popular magazines became infatuated with the subject. *The Scientific American* in 1909 ran a competition for the best exposition.

Interest declined subsequently after Einstein's theories became established but the concept continued to fascinate and provide a vehicle for relaxing the shackles of the mind, as Bertrand Russell put it. Jorge Luis Borges included Hinton and Wells as authors among the thirty books he most treasured in his Library of Babel. He refers to Hinton's 'alterable past' in the short story 'The Secret Miracle'. The tesseract is the hero of Robert Heinlein's *And He Built a Crooked House*. The house itself is in the shape of one: walking four rooms in a straight line brings you back to where you started. Unfortunately an earthquake unfolds it into an actual tesseract. More recently Ian McEwan's short story 'Solid Geometry' harks back to the nineteenth century in which the modern-day narrator learns, from an old diary associated with the Hintons, the secret of making things disappear into another dimension. His annoying wife becomes the victim.

Although the concept prompts many directions in literature, they do not lead far. Several films have featured a tesseract (*Flatland: The Movie*, *Cube 2: Hypercube* and *S. Darko*). The advent of sophisticated computer graphics has enabled the depiction of hypersolids convincingly on screen. This has led to a renewed interest in the work of Hinton. We can assume that if he had available the means to convey what we can on screen now, he would have indulged it to the full. My hunch is nevertheless that he would most likely still be found in some quiet room completely absorbed, scrutinising a stack of multi-coloured cubes.

# 12

# THE CITADEL

Three years after the 2004 publication of my collaborative volume, *The Voynich Manuscript*, the mysterious tome still lurked at the back of my mind. Our review of its history and possible theories had drawn some good reviews in the press and been published widely. It had not, as we hoped, outshone the *Da Vinci Code* as a subject to excite the general public. Several other books had addressed the substance of the manuscript without 'solving' anything; the internet discussion by the posse of 'Voynicheros' had not abated. They greeted our contribution more or less with silence.

It was probably my dusty hoax theory that had stuck in their throats. Nobody likes to find that an object of fond attraction is in fact worthless, manufactured by a twentieth-century maverick like Wilfrid Voynich. The dating of the manuscript to the early fifteenth century had made the suggestion unlikely but not completely impossible. Those aspects of the manuscript's history that looked suspicious still seemed so to me: its centuries-long disappearance, the long secrecy over its place of discovery in 1912, the high coincidence of the de Tepenecz signature and the sheer impenetrability of exposition by gigawatts of computers and battalions of linguists and historians. For me, lingeringly, Voynich's opportunistic drive to succeed seemed to waggle still a finger of suspicion in his direction.

More than anything I felt a continuing interest in Wilfrid and Ethel as individuals and as a pair during that inspiring part of their lives when they were both involved with Russian anti-tsarist activities. It was their connection with Poland and Wilfrid's romantic tale of the glimpse at the Citadel in 1887, prior to his arrival in London from Siberia, that fascinated me. Wasn't it just too good to be true? Was it an example of Taratuta's assertion of his 'bursts of disorganised fantasy'? I needed to make a visit there to put some flesh on the bare bones of the distant story. Warsaw had a definite appeal.

I knew the fact of the city's destruction in the Second World War by the Nazis and its rebuilding under a communist regime from which it struggled towards self-determination in 1989 and the growth of a capitalist society. I didn't expect at Warsaw airport's tourist desk in November 2007 that this last development would put even cheap hotels beyond my reach. Instead I took a bus to a hostel address in the Old Town and was agreeably surprised to find a warm room and a bunk space in an ancient, clean house of great character for 45 zloty, about £10 per night. It is always very unsettling arriving late in a foreign town without a guaranteed place to stay, so after decanting my rucksack I celebrated by promising myself a good meal.

The cobbled streets around the Old Square were almost deserted, but after peering through various steamy windows I found a family-feeling restaurant. I ordered borsch (when in Rome) from a blousy, blonde waitress and some fried carp because I needed to know more about the 'fish topic' of the day back home. Urban legend had it that this aquatic delicacy disappears fast from English ponds during autumn heading for a Polish Christmas table. My carp didn't seem anything to become a petty criminal for; it mostly tasted of fish. The borsch was ribena-clear like the wine I ordered, and delicious, as were the boiled spuds and plate of grated crudités.

The following morning I set out in tourist mode to explore the modern city centre. It looked somehow like I expected: broad, grid-patterned boulevards, once all the better to move tanks down, now gridlocked with private cars. In the middle of one chequer-square towers the massive Stalin-gothic edifice of the *Palac Kultury i Nauk*, the Palace of Culture and Science, a gift to the Polish people from Uncle Joe Stalin that they wish had never been made. Not having experienced thirty-five years of totalitarian oppression I could appreciate its almost fairy-castle quality compared to the sleek but cheap corporate towers that dot the skyline. It now serves as offices, shops, a multiplex cinema and exhibition hall. I couldn't resist going up the magisterial steps and through the vast doors that were clearly intended to dwarf any notion of individuality. At a vast round console in the foyer sat a very young student apparently in control of the whole shebang who helpfully translated current film titles for me. The café nearby reminded me exactly of being back in the USSR – stock snacks grumpily served and munched at ungainly plastic tables and chairs.

Adjacent to the *Palac* and from a very different planet stands an intergalactic, geodesic-roofed shopping vehicle studded with escalators crisscrossing skywards. It's inhabited by the new middle-class species of consumer, for whom Paradise has landed. It entails learning an alien tongue. Alongside the colonising Next, M & S and Body Shops are local attempts to catch the tricky, cosmically meaningful but vacuous titles we're used to in our western orbit: Apart Exclusive, Beyond the Fashion, Time Trend, Your Real World Skin and, I assumed, the unintentional pun of Chopin Luxury. It must be really difficult to grasp some merchandising concepts we never think twice about. Take Pizza Hut, where I took one lunch having tired of carp.

'What is Hut?'

'Well it's a sort of shed'.

'Like where you keep lawn-mower?'

'Well, yes'.

'And you eat pizza there – maybe only in summer?'

Worlds away, outside the bubble, roam drabber proletarians. They continue to shop at little stores window-dressed with pre-1989 piles of detergent or dummies wrapped in synthetics, but set against Pinocchio-fashioned, wooden, rustic props. (The Geppetto-style moustache lives on too, I noticed: one Lech Walesa lookalike sported what could have served decently as a wallpaper hanger's brush.) The subterranean arcades of little shops at metro stations and road intersections selling trash food, keys and plastic ware reminded me exactly of the troglodyte malls of Moscow twenty years before. These mycelia seemed to feed the vast state fruiting bodies above, like the *Palac*, reaching to a triumphant workers' heaven. Giant billboards no longer show smiling families acknowledging the leading role of the Party but their more stylish descendants enchanted by a multimixer.

It is, of course, somewhat superficial to judge a society by its shops, but it's certainly meaningful to note that that universal, capitalist bestseller, sex, kept under protective wrapping previously, has arrived in town. Little 'private' shops are pocked around even in the suburbs and in the space mall there were displays of G-stringed mannequins that made even me blush. I was in Warsaw, however, primarily to look backwards into its grimmer past; not for me the glories of its royal palaces and parks. I was here to learn about the activities of a small group of socialist revolutionaries who 120 years before had rebelled against tsarist autocracy.

Poland, occupying a fertile plain north of a barrier of high mountains, has always been jostled throughout its tortuous history by invaders east and west: Russians, Lithuanians and Prussians. It briefly emerged as a unified state in the sixteenth century only to be dismembered in the eighteenth by Prussia, Austria and its (forever overbearing) neighbour to the east, Russia. At the Treaty of Vienna in 1815 which re-drew mainland

Europe's boundaries after the defeat of Napoleon, Poland emerged truncated as the Duchy of Warsaw under firm Russian control, reinforced after insurrections in 1830 and 1863. Like Russia, while remaining basically a peasant society, it began to industrialise in the latter part of the nineteenth century producing new tensions along class lines as well as nationalist.

*Proletariat* became the first avowedly socialist party in Poland, founded in Warsaw in 1882. Wilfrid Voynich was an associate. Its never-sizeable ranks were drawn mainly from intellectual circles and espoused a clear Marxist line proposing that the newly emerging means of production should be claimed by the working class, by force if necessary; the peasantry were of secondary consideration. Despite its call for the liberal freedoms of parliamentary democracy, running counter to traditional Polish radicalism, it did not propose the setting up of democratic institutions. Its strong connection with Russia's *Narodnya Volya* included the acceptance of terror as a weapon, but as Lucjan Blit points out in his study of *Proletariat*, it was never much more than 'a fighting squad of eight men, who did not know how to produce a bomb, and for whom a revolver or stiletto were the highest form of personal ammunition'.[1]

This pan-slavic affiliation meant that, despite Marx's own urging, independence from Russia was not its foremost political aim even though Polish culture was suppressed and public life russified. Its chief activities lay in creating links with worker cells, organising strikes and printing its own newspaper. The actual practice of terror consisted mainly in reprisals against the property of oppressive capitalists and victimising state officials. In this it was largely ineffective, dogged by traitors from within and crackdowns by Russian authority. The party's general naïveté was shown by its assumption that a worldwide revolution was in the offing. By 1885 it had had three changes of leader as each in succession had ended up in prison.

To get to grips with all this I decided my first port of call in Warsaw should be the *Niepodleglsci Museum of Independence* on Solidarnosc Street. Here I thought I'd pick up some basic archival information. This turned out to me being as naïve as any *Proletariat* member. A curator kindly showed me around several *grandes salles* within the old neo-classical building housing paintings and photos of freedom fighters – but contained no references to *Proletariat*. Puzzled, I asked her about this. Rather affronted, she explained that as the group never specifically fought for independence it did not qualify for inclusion here. Strictly speaking I could

The bastion, Warsaw Citadel.

see that this was true, but weren't they patriots in fighting against tsarist oppression? Apparently not. This hair-splitting masked the real conceptual problem: by working with Russians in *Narodnya Volya* they had collaborated with Poland's traditional enemies and by espousing Marxist doctrine they linked themselves contemporarily with the recently vanquished bane of Soviet communism. They were damned on two accounts.

The Citadel prison just north of the Old Town is a huge, bastioned, brick pentagon on raised ramparts alongside the Vistula river. Built by the Russians at huge expense after the 1830 Polish uprising, it was paid for by the Poles. It housed 16,000 soldiers during the 1863 troubles. Here in 1885 in the Tenth Pavilion block, awaiting trial and possible execution, were incarcerated the leaders of *Proletariat*: Warinski, Bardovski, the twenty-year-old Maria Bohuziewicz and Stanislaw Kunicki – the member of People's Will we met in Chapter 8 involved in the Degaev affair.

Enter Wilfrid Voynich centre stage, a member of *Narodnya Volya* at that time. He was called in as a Russian-speaking volunteer unknown to the local authorities in order to assist with a daring plot worthy of Baroness Orczy. Ethel recounted it in a 1947 letter to the London School of Economics.

Given a fund of money, Wilfrid's task was to spend it acquainting himself with Lieutenant Colonel Bielanowski, the head of the gendarmerie, 'a renegade Pole of bad reputation, socially a leper'. By deliberately losing at cards Voynich ingratiated himself with him to gain passwords and layout plans of the Pavilion. The actual escape involved a tunnel, rope ladder and a boat. True to form everything went wrong due to a *Proletariat* member ratting on the plan. Wilfrid had played his part well, organising the project, but when it was aborted he fled back to Kovno where he was arrested. He returned a prisoner himself to spend eighteen months in the Citadel without trial followed by permanent deportation to Siberia in June 1887.

A far worse fate awaited Bardovski and Kunicki who along with others were executed in 1886. Warinski was sent to the dreaded Schlisselburg fortress and died three years later of TB. To rub salt into the wounds and avenge his pride Bielanowski forced Voynich to watch the hangings and told his mother that he had been shot. To lighten this tale of pitiable woe there exists the supplementary episode of the Citadel glanced liaison that united Ethel and Wilfrid. There was a very long and tortuous build-up to it.

After months of enduring the appalling eastward journey across the vastness of Russia, Wilfrid found himself in 1887 exiled with many other Polish political prisoners 150 miles west of the city of Irkutsk, close to Lake Baikal, in the Tunka valley penal colony. Peter Kropotkin, subsequently a political exile himself in London (where he would meet Voynich), had visited the area as a geographer in the 1860s. He described the mixed, indigenous, north Mongolian Buryat and Russian Cossack population as engaged in cattle herding and small-scale agriculture in the mountain valleys. Because of the forbidding territory, surveillance of the prisoners, in contrast to those consigned to the mines, was limited to signing-in once a week and the guarding of roads east and west. North would have led through endless forest to the Arctic, south through hostile Mongolia. None of this, however, put off intrepid Wilfrid, keen to escape. The story of his exploits was told by Ethel to Jean, her American Hinton grand-niece.[2]

In one scheme Voynich apparently enlisted a Buddhist monk to help him escape via the Mongolian route by disguising himself as a horse trader. The *lama* warned that if they encountered aggression there would inevitably be a battle to avoid being shot or taken hostage for a Russian police bounty. Voynich aborted the plan and spent the next two years bored alongside his sequestered compatriots but making the most of captivity by studying the local native culture. Gifted with a strong memory, he acquired a working knowledge of some of the local languages. This led to the later legend, no doubt fostered by himself, that he was a formidable linguist. Once asked if it was true that he'd written books in twenty languages, he replied, 'How you English exaggerate. I have but written books in seventeen.' (He also once wrote to his SFRF colleague Chaikovsky that he had no aptitude for languages.) Interested in the Buddhist customs of the Buryats, on one occasion when 200 monks came in their yellow robes to a village to pray, he collected together a number of cast-off relics which he packaged and sent off to ethnographers in Paris.

Anecdotes like these reveal important sides of his character. On the one hand there is Wilfrid the adventurer and risk-taker but on another the focused and enterprising collector, not yet connected to financial gain. Linking both, the tales of his linguistic prowess suggest the life-loving, jovial egoist always ready to foster a grand impression. His eventual escape from Siberia involved breathtaking twists and turns.

He needed first to get back to relative civilisation. Some versions recount his attempted or feigned suicide; Ethel's speaks of a 'chest wound'. Either way, in June 1890 he ended up in hospital in Irkutsk. Here he made the connection that would change the whole course of his life. As we have seen in Chapter 8 during Ethel's stay in St Petersburg, she had lived with Stepniak's wife's sister, Praskovya Karaulov and son Sergei. In 1889 they had accompanied her husband Vasily to Siberia. Praskovya, as a doctor, had ended up working in the hospital in Irkutsk that ministered to Wilfrid. By giving him a letter of introduction to Stepniak he was most likely influenced to flee west to England rather than east towards the US. Retaining some freedom, it seems, to roam outside the hospital, Wilfrid made a visit to the local museum. Here on display he found his collection of Buddhist artefacts intended for Paris but labelled as being presented by the Mayor of Irkutsk. Voynich took him to task for his insolence.

Still under arrest, Wilfrid was sent to another penal colony in the Balagansk area 130 miles further north on the Angara river. (Thirteen years later it housed Stalin for two months until his own decampment.) It was during Wilfrid's transfer from Irkutsk to Balagansk that he made his bid for freedom. According to Ethel, a confusion of identity with another fugitive had led to a report that he had been killed while trying to escape. A clandestine paper printed an obituary, leading the Russian police to believe he was dead. It was at this time that Wilfrid adopted the name Ivan Kelchevsky. Without a passport he would still need to have been very careful on his journey. Ethel acknowledged that it involved 'five months of adventures and hairsbreadth escapes'.[3] Stretching credulity, at one stage he became disguised as a gold-courier, helpfully offering him fast horses at posting stations but also the risk of being robbed.

The secret of avoiding disaster, he maintained, was, 'however frightened you are you must never show it', a maxim, it seems, he clearly put to the test, by purposely sitting next to railway officials. On another leg of his trek he disguised himself as a worker on a Volga barge and was forced to rush offboard to avoid an imminent police check-up. His flight was apparently aided by the gendarmerie captain escorting some schoolgirls on a trip from Siberia, being too busy to notice. Many years later in Switzerland recounting this particular exploit to an audience, Wilfrid stated that the captain didn't realise he was on the run. In the audience was one of the very same girls who declared that this was not so; the captain had later raised a toast 'To the health of a brave man'. Noting that this escapade detail came from a reliable source – Ethel – we may still wonder how gullible was she to Wilfrid's storytelling?

Taratuta sketched the remaining story of Wilfrid's journey westwards across Europe. Managing eventually to cross Russia and Germany, at Hamburg, '… hiding in docks and stacks of timber, he at last managed to persuade the captain of a small fruit-boat to take him to England. In order to pay for his passage he had to sell all his possessions even his glasses and waistcoat. All he could buy for the journey was some salt herrings and a little bread.'[4] Unbelievably, the boat was wrecked on the Scandinavian coast and the cargo lost – yet somehow, he reached the docks of east London. From here he was escorted to Stepniak's house.

This at least is one story; there is another. In 1931, James Westfall Thompson, a professor at the University of Chicago, wrote an *in memoriam* to Wilfrid. 'His intimate friends who have heard from his lips the story of his Siberian experiences will never forget the impression. In Mongolia he joined a caravan bound for Peking and after more than a year of wandering Mr Voynich entered China through 'the back door'. From Peking he made his way to England and became a British subject.' The southern route as we have seen was virtually impossible. Ethel, however, and Taratuta explicitly support the western path. Thompson, like others, must have fallen for Wilfrid's *braggadocio*. He wrote, 'The range of his knowledge and interests was enormous, and his conversation threw a spell upon his auditor.'[5] Millicent Sowerby, one of his London book-selling employees, reported falling under the same magic as he demonstrated to her in his pidgin English how he had acquired some scars: 'Here I have sword, here I have bullet.'[6] Wilfrid, it seems, knowingly embroidered his own undoubtedly heroic tales.

In London, safe at last, the final climax was still to come – that during Easter 1887 Wilfrid had espied Ethel in the square from his Citadel cell. This is Fate on acid; it parallels Ethel's own novels in derring-do and romance and has been repeated as history in most accounts of his life. The Citadel story just seemed too good to be true. Ethel herself never seems to have retold the tale, maybe finding it too gushing for her taste. According to Taratuta, Wilfrid during his first meeting with Stepniak had mentioned that, 'Praskovya Vasilyovna has told me so much about an English woman from London'.[7] Given that this probably included the details of her distinctive black garb and golden hair, the quick-thinking Voynich was perhaps able to fashion an immediate positive impression on the gathering. Was it just one more of Taratuta's Voynich character examples: that he was 'fond of creating an effect'?

A freezing, overcast morning was perhaps the most fitting time to visit the Warsaw Citadel. I walked there from the hostel fortified by coffee and a *pain au raisin*. From a narrow strip of parkland, next to a busy main road hugging the grey Vistula, a long, ascending lane takes you past an ominous double row of nameless, concrete crosses and through the Gate of Execution that pierces the outer wall. Close to the Gate were three marble plaques inscribed, I was pleased to see, with the *Proletariat* names of Warinski, Bardowski and Kunicki. Now level, the lane leads on to the *1984*-sounding Tenth Pavilion, the only prison block to survive. It did not live up to my expected image of a Dartmoor, barred barracks. Instead it was an almost pleasant-looking, stuccoed, grey, U-shaped building, two storeys high with proper windows. The grassed interior courtyard of the 'U' was divided by stockade fences topped with offensive metal hooks.

Now a museum, the entrance and reception was dingy and unkempt, perhaps itself a 'heritage' device in keeping. The cells either side of a central corridor were surprisingly large, not the cramped space that Voynich insisted had resulted in one of his shoulders being higher than the other. Each contained a large ceramic tiled stove and a rickety bed.

Model of the Tenth Pavilion.

Once again, not the damp pit that was supposed to have blighted his health ever after. The atmosphere was grim but not entirely different from my hostel room.

I contacted the curator's office and was told that room 10 had incarcerated Voynich: this one was out of sight at the rear. It would have been almost impossible for him to have seen anyone beyond the courtyard. How could Ethel even have got this far beyond the outer bastion gates? Wilfrid's espying her seemed completely fanciful.

Taking a number 18 tram back to the centre I marvelled infantly at the way it flung its carriages around tight bends and whisked along the central tracks allowed by the width of the streets. The older carriages have a strange shape rather like an elongated coffin and are painted in a livery of red and orange; the newer 'capitalist' ones are sleeker and quieter and act as travelling hoardings. Sitting up front behind the driver felt like being inside your very own train set. What fun it must be to drive one.

To cheer myself up after the Citadel's misery I indulged myself in a 'meal' at McDonald's. Although the chain's chief advantage to a traveller is that you know exactly what you will get, I felt experimental enough to try the global food empire's attempt to please Poles with its aping of a *pirogi*. These elongated stuffed dumplings are very popular, with a variety of fillings. My example, filled with an unascertainable, grass-like substance, was, I'm reliably informed, a case of what is known sociologically as *glocalisation*. I predict that this little *pirogi* will not please its market. More a case of *cloacalisation*.

One of the many pleasing links researching the Boole family that seem to crop up out of nowhere is the fact that Ethel, obviously a fan, undertook to edit the definitive collected letters of Chopin, published in 1931. I felt I needed to know more about him plus some cultural uplift, so I decided to visit the Chopin Museum. Here in a baroque-style castle with frescoes, polished floors and salons is all that a Chopin lover could want: photos, family artefacts, original music manuscripts, a cast of his fine hand, the famous painting of him by Delacroix and the last Pleyel grand piano he owned. It was accompanied over the intercom by a selection of his music. When the 'Minute Waltz' tinkled forth I automatically heard Nicholas Parsons intoning his panel game opening on BBC Radio 4, 'Welcome again to *Just a Minute* …'.

Born in Poland in 1810, Chopin's French father had married into a poor but aristocratic family that provided the connection enabling him to teach in Warsaw and raise his pianistic genius of a son. In 1830 Frederic left the capital to tour in western Europe, turning his back on the revolt against Russian occupation that followed. He never returned. His travels took him to France, Spain, Germany and Britain, his ever-failing health giving out finally in Paris in 1849. He was buried in the Père Lachaise cemetery there but his heart at least was returned to Warsaw and resides in a finely-plaqued pillar at the Holy Cross church near the museum. The composer of the tempestuous *Revolutionary Etude in C minor* of 1831, other heroic pieces and wonderfully lyrical ballades and nocturnes is deeply synonymous with Polish nationalism.

Despite the soothing grandeur of the museum, my thoughts strayed back to the morning's episode. Chopin, distant from his motherland, had lived a longer, more luxurious and feted life than his poor, undervalued *Proletariat* compatriots who died on the gallows or en route to Siberia. They just didn't tick the historically correct boxes. Like everyone, of course, I love his music. The whole world does, because, as his fellow Pole, Arthur Rubinstein, declared, 'It speaks directly to the hearts of people.' In that sense it is internationalist – the very quality decried in his *Proletariat* compatriots.

In her preface to *Chopin's Letters* Ethel notes 'his love for his native land and the sincerity of his sympathy with its desperate struggle against alien oppression'. This she moderates with the assertion, recalling a Boolean motto, that 'he took no part in the struggle need not surprise us;

the wind bloweth where it listeth, and a creative artist ... must live under the compulsion of his art.' Despite her admiration for Chopin, ever-truthful Ethel found it 'a little startling that he treasured a diamond ring from the Tsar and ... favours from Grand Duke Constantine', the ruthless Russian ruler of Poland before 1831. What is impressive about Ethel is that despite her later-in-life desperate need to write music, ironically, she surrendered her youth exposing the evils of the same regime that had oppressed Chopin's countrymen. When it comes to distinguishing between principles, however, she rather tiredly notes, 'The human mind is a queer jumble.'[8] Outside the museum I stumbled into the first real snow I'd encountered. Walking back to the hostel, past Warsaw University where Chopin's father had lived, I acquired a sheet of snow clinging to my front as if I were a hedge. One stamp inside and it fell off as a sheet.

A jumble in my own mind and my files on Ethel had provided me with another reason for being in Warsaw. Having glanced at one particular, stapled document over several years, I'd somehow failed to read fully a brief 1997 handwritten letter attached to the front sheet. It led me to a Polish scientist, Richard Herczynski, who had worked with Sir Geoffrey Taylor, Ethel's grandson, at Cambridge on fluid dynamics in the 1960s. Taylor was perhaps the foremost researcher in their field (of whom more in Chapter 18). I managed to gain Herczynski's contact details in Warsaw and arranged to meet him.

Senatorska Street winds from the statue of Sigismund the Third on top of his column waving a cross and sword past the Grand Theatre to its end at Andersa Street. Round the back of a post office in a post-war block lived Richard and his wife, Grazyna. Their apartment on the third floor is reached by the smallest lift I've ever claustrophobed in, as narrow as a filing cabinet. Richard, small and gnomish, greeted me alongside his younger, crop-haired wife. Pleasantries over, in his small book-lined living room we discussed our mutual interest in the Voyniches. Strangely he knew little about Wilfrid's famous manuscript, but he had thoughtfully photocopied a collection of Polish documents about him that I'd not come across before relating to his experiences with *Proletariat*. Here I got into trouble.

Grazyna was in fact engaged in some research into the group for a book. I recounted my miffed experience at the Independence Museum and its unkind view of the group as unworthy of respect. My hosts agreed in strong terms that not being nationalists they certainly didn't deserve any recognition, they were something of a historical curiosity. I should have anticipated this response as I knew that Richard himself had fallen foul of the communist system.

In 1981 he was accused by the communist government of 'handing out materials harmful to the interests of the Polish People's Republic', in fact no more than *Solidarity* bulletins. He was sentenced to fourteen months in prison, in his words, 'to show scientists should be quiet and not speak too much'. He insists he was not a hero; in those days he had 'one leg in science, one in politics', but now 'the second leg has become shorter and disappeared'. Meeting him had personalised what was becoming more apparent from my brief stay in Warsaw: that I knew very little about Poland's history, one episode in particular.

In 1944 Soviet troops across the Vistula had stood by as the Old Town was razed to the ground by the Nazis in response to the nationalist insurrection of the Warsaw Uprising. The Allies too because of agreements with Stalin also stayed aloof. In the end the loss of life was worse than that inflicted on Hiroshima a year later. After years of Soviet-style communism the inextinguishable desire for national liberation surfaced once again in the Gdansk dockyards. In this context any group like *Proletariat*, communist-endorsed, would inevitably be consigned to today's dustbin of history. Yet in their clumsy and naïve way the first Polish socialist party was also a premature gesture by bravehearts, hiding their printing presses, plotting and living on tea and bread in the attics of the Old Town. Unaided, they too were snuffed out by overwhelming forces; as underdogs go they earned my sympathy. They also reminded me powerfully of Ethel's acquaintances in St Petersburg and from my own experience in 1983 with the Trust Group in Moscow. Weren't they also a premature, idealistic bunch of heroes?

One final whiff of the old Eastern bloc came at the airport. My ticket didn't specify a terminal but I knew there were two of them. The bus from the city centre stopped at terminal 1 so I got off with everybody else. Inside, the destination board stated that the Luton flight would be checked in at desk *E-1 Estudio*. I asked at information where this might be and now, confused, found myself at terminal 2. Here I asked again and was told that *Estudio* was a third small terminal on its own about 300 yards away. I stomped off irritated, heavily be-rucksacked, and found what looked like a large cow shed hemmed in by barbed wire.

Here Soviet-style surliness reigned; what had Easyjet done to deserve this, I wondered? The security staff were dressed in fatigues and looked suitably exhausted and bored. Why is it not possible, I found myself wondering, for there to be a middle way between capitalist 'have a nice day' fawning and the socialistic fobbing 'don't care if it's your last'? Still disgruntled, I checked in and delved for my tissue-wrapped slices of Pizza Hut shed pepperoni from the day before. I munched surreptitiously, enjoying feeding crumbs of it to a stateless sparrow trapped here in no-man's-land. I don't suppose it felt fawned upon or fobbed off. I found someone to ask what *Estudio* meant and was told it was 'something to do with music'. The zloty dropped: it was, of course, part of Warsaw's Frederic Chopin airport. For a lot more than one minute I'd had to endure 'deviation, repetition and hesitation' in finding it. In this morgue-like byre, I ruminated, they should have been playing over the intercom Chopin's death march from the *Piano Sonata No. 2 in B-flat minor*, as they are clearly suffering from some kind of terminal disease.

Back in London I found someone to translate the Polish documents that Richard Herczynski had kindly copied for me. One interesting snippet emerged: during the initial hearing in October 1885, Voynich asserted, in line with his fantasy-world, that he was a member of the Central Committee of *Proletariat*, which is patently untrue. Against this bombast, however, we may note that the prosecution praised him for his courage. One other fact was important.

The *Proletariat* group were normally kept together in the Pavilion and had managed, with the help of a bribed guard, to get letters regularly exchanged between them for a year or so. Eventually the letters were discovered, one of which, by brave Wilfrid, encouraged his fellow inmates to demonstrate against their imprisonment. For this offence he was put in solitary confinement. As this happened a few months before their group was sent away to Siberia in June, he could have been residing in a casemate cell in the outer Citadel walls during Easter 1887. He *might* there have seen the conspicuous figure of Ethel below in the square outside the main entrance. I learnt a valuable biographer's lesson: not to try to fit facts to wishful thinking.

Some Russian documents I also came across added a nice twist to the tale. Joseph Pilsudski (1867–1935) is still celebrated as a founding father of Poland and the fight against Russian occupation, becoming its President after the First World War. Born in Lithuania, like Voynich, at about the same time, he attended Kharkov University where he and his brother Bronislav joined the *Terrorist Faction* of the People's Will, of which Lenin's executed elder brother had been a member. It was also the same broad socialist/anarchist movement from which Wilfrid was recruited to release members of *Proletariat*. Contrary to what I'd been told at the Warsaw Independence Museum, there were at least early links between nationalists and the first socialist party that they rather unkindly, I felt, rejected. The link went even further.

Lydia Loiko, a doctor who was arrested in Kharkov for her political activities printing seditious, anti-Tsar pamphlets, wrote about her banishment in her 1928 Soviet memoirs *From Land and Freedom to the All-Union Communist Party*. From Moscow in July 1887 her group of twenty dissidents were taken by train to Novgorod and then by a combination of barges and railway to Tomsk. Here they prepared themselves for the long walk to Krasnoyarsk and on to the Siberian penal colony. The exiles' belongings were carried by horse and cart as they trudged beside for two days in the autumn rain and snow, resting on the third, each prisoner guarded by a soldier. Sometimes friendly peasants offered them food such as eggs, bread and milk, but at night-time their stays were mostly in squalid rooms bringing the typhus that would take its toll on the weaker members.

Finally reaching Krasnoyarsk after a gruelling six weeks, they were placed in a crowded jail. Although under guard, they could at least visit shops to buy necessary products. In December they moved on to Irkutsk, 1,000 *versts* away (about 725 miles) in the now freezing temperature, so cold that at least typhus was avoided. From here Lydia was sent on to the Tunka valley – welcomed, despite its privations, after the journey's hardships. She writes: 'The place helped them to bear their fate as prisoners and made them stronger to confront their future battles – which was not the aim of the autocratic government they'd fought.'[9]

The colony practised a meagre but healthy, self-sufficient lifestyle; they were able to garden, gather wood and hunt goats and rabbits. Lydia earned a little extra from sewing. Even books found their way to the inmates. She decided to escape, however, and acquired a passport for ID from a soldier's fifty-year-old recently deceased wife. As Lydia was only thirty-five, the date of birth had to be changed. A qualified chemist offered to help. Guess who? None other than our old friend Wilfrid Voynich. Unfortunately in the process he ruined it: a lapse he seems to have repeated later with the de Tepenecz signature of the *Voynich Manuscript*. Nevertheless, he merits the accolade from her as the 'good and dedicated Voinich'.[10] Other members of the colony also impressed her, namely Bronislav and Joseph Pilsudski.

Both brothers had been arrested because of their *Narodnya Volya* activities centring on a plot to assassinate Alexander III and were sentenced to exile, ending up at Tunka. Although she befriended them, Loiko, writing much later, was critical of Joseph and his subsequent politics. Exactly opposite to Richard Herczynski and the curator at the Independence Museum, she disapproved of his putting Polish nationalism before socialism leading to his class betrayal and pursuit of a personal power as President. She likened him to Mussolini. But all this lay a long time ahead. In 1892 Pilsudski was released and in 1893 joined the Polish Socialist Party, editing their journal *Robotnik*. For this he was jailed in 1900 and ended up in cell number 25 in Warsaw Citadel's Tenth Pavilion – where we started. The room is well kept and hallowed as befits the father of Polish nationalism. History as ever is written by the winners.

# DEAD AS MUTTON?

In a letter to Wilfrid Voynich in December 1907, George Bernard Shaw avers, 'Most books are as dead as mutton eighteen months after they are born ...'.[1] Wilfrid Voynich had asked him to recommend a literary agent for his wife, Ethel. Shaw had obviously kept up a connection with the Voyniches; he had taken a particular liking to Ethel within their shared political circles in the early 1890s. The link had continued when GBS had written a dramatisation of her first novel, *The Gadfly* in 1898 to secure its copyright for her. The fascinating fact about the book is that although it may now lie as dead as mutton on the slab of contemporary literature, its shelf life has lasted sixty times what Shaw predicted. It created quite a stir not only before the First World War but later became one of the most revered and widely read stories in eastern Europe and China in the twentieth century. In those countries Ethel Voynich was rated as one of the greatest fiction writers in English. It still ranks high in certain quarters.

Strangely enough, two of the figures who inspired it are historically old bones. Little recognition is now paid to Charlotte Wilson, who, although not a major political thinker like Marx or her fellow anarchist, Kropotkin, was involved in most of the emerging strands of English political thought in the 1880s. She introduced Ethel Boole (as she was) to Stepniak, who motivated her to visit Russia in 1887 and to work with him in the affairs of the Society of Friends of Russian Freedom. Stepniak too has almost disappeared from history despite his colourful life and presence, not helped by its early drab ending, hit by the buffers of a suburban train.

A novelist himself, he'd encouraged Ethel in a letter of 22 August 1889, 'You must try your talents as a writer', and suggested she should be 'capable of capturing ... the nature of man and phenomena of life, if one only observes them sufficiently long and attentively'.[2] Her life in St Petersburg, living dangerously among the anarchist fraternity there, had given her first-hand insights into the nature of political commitment. Later that year, apparently in Charlotte Wilson's garden at Wyldes farm, Ethel began to

Giuseppe Mazzini, Italian revolutionary 1805–1872.

write *The Gadfly*, the heroine of which was based loosely on Charlotte. Its setting, however, was not Russia but the nationalist struggles of Italy for unification some fifty years before.

It has been suggested that a reason for this lay in an attempt to outwit censorship of the novel in Russia itself, but there were certainly a number of possible inspirations suggesting Italy. One was the family tale, as mentioned in Chapter 5, of Ethel's eldest sister, Mary Ellen, concerning the two shipwrecked Italian patriots who stayed with the Boole family in Cork. Another was Emilie Venturi's *Joseph Mazzini: a memoir*. Although she never specifically acknowledged it, Stepniak's own life fomenting revolution in Italy in the 1870s must have been persuasive.

The underlying basic plot of *The Gadfly* reads essentially like a historical romance. Barbara Garlick has pointed out that it is 'operatic in both its setting and its passion with echoes of *Tosca*, *Pagliacci* and *Rigoletto*'.[3]

The book is written in three parts – convenient for an opera. The curtain rises on Act One at a garden scene in Pisa where virtuous and idealistic, eighteen-year-old, Arthur Burton, member of a rich English shipping firm based in Leghorn, sits under a flowering magnolia with his mentor/confessor, Padre Montanelli. The two are very devoted to each other. Arthur is training to be a priest but is struggling to reconcile his faith with the need to help expel the Austrian occupiers to create a free republic. Montanelli warns him to think carefully about committing himself to dangerous opposition. The padre is about to be elevated to a bishopric in the Apennines. Enter Arthur's younger long-time 'playmate' Gemma, aged seventeen, with whom he is in love. She is known as 'Jim', also from an English family and is actively allied to the nationalist group Young Italy. Outspoken and revolutionary, for her the Church is an obstruction, although Arthur argues that Christ was a revolutionary too.

As a result of a treacherous priest leaking information about an opposition group, told innocently to him by Arthur in confession, he is imprisoned but released. Gemma, thinking that Arthur has implicated another revolutionary rival to leave the way for her affection, strikes him on the cheek. To add to his humiliation his unsympathetic stepmother spitefully reveals that he is actually the bastard son of Montanelli and his now dead mother. Thrice played false – by Gemma, his duplicitous priestly father and the Church – he feigns suicide by drowning and stows aboard a boat for distant shores, caring no more for all 'the broken and dishonoured idols that yesterday had been the gods of his adoration'.[4] End of Act One.

Act Two. Thirteen years later. A group of nationalists including Gemma, since married and widowed, discuss how to gain further civil liberties in the more relaxed political climate with a new Pope. They opt to try to undermine social order by satirical pamphleteering and engage the maverick, Felice Rivarez, also known as the Gadfly, to sting authority. He has proven his worth both with his clever writing and attempted insurrection. The audacious Pimpernel Gadfly carries the police description, 'one foot lame, left arm twisted; two fingers missing on left hand, recent sabre-cuts across the face; stammers, very expert shot'.[5] It is, of course, Arthur returned from South America, ill and hardened after many an adventure working on sugar plantations, down silver mines and being degraded as a clownish hunchback in a travelling circus. Despite his scars he is replete with an amour, gypsy girl Zita. Gemma and he meet but she fails, somewhat absurdly, to recognise him. Montanelli, now a cardinal, has meanwhile gained popular respect for his honesty and sense of justice. Underneath, however, he grieves for the son whose suicide he believes himself responsible for.

The Gadfly is given the job of smuggling inflammatory literature across the mountains. In a touching scene he reveals his latent pity, both for a hunchback seen in a passing circus procession and a boy beaten by his uncle. Gemma and he draw platonically nearer to each other at the expense of another suitor. She is prepared to help him smuggle guns. In a scene that prepares for the great confrontation to come, the Gadfly in disguise confronts the saintly Montanelli outside his cathedral and tauntingly expounds a parable of him as a father who has killed his son and cannot find absolution. God is merciful, replies Montanelli, and will forgive. Zita deserts her lover to rejoin Romany life where she can find physical passion.

Act Three. The arms have been transported and need to be distributed in the mountainous, police spy-ridden territory. Brave Gemma decides to accompany the Gadfly there, tacitly recognising at last who he really is and accepting his expectation of martyrdom. In the market place of Brisighella the Gadfly and his accomplices find themselves trapped by spies and cavalry. Montanelli intercedes in the confrontation and is only saved from death by the Gadfly's action, resulting in his own imprisonment. Gemma and friends decide to rescue him. The military governor is anxious to court martial the Gadfly as he can't trust the guards from falling under his charismatic influence. Montanelli, capable of wielding authority in civic matters, intervenes and interviews the prisoner. An escape plan is hatched by the revolutionaries via the proverbial smuggled file and secret tunnel. The weakened Gadfly, however, is overcome and caught. Fearing another escape attempt or a riot by the peasants on his behalf, the governor insists on a military tribunal that will inevitably find him guilty.

Montanelli confronts his unacknowledged son, unable himself to escape from the dilemma of agreeing to a trial or licensing unrest. 'If I consent I kill you; if I refuse I run the risk of killing innocent persons.'[6] Asking the prisoner what he would do, the Gadfly pours out all his scorn at the cowardice of the question. 'Go back to your Jesus … After all you'll only be killing an atheist.'[7] Arthur reveals his true identity, 'I have been crucified … I too have risen from the dead.'[8] Montanelli hides his face on the breast of his resurrected son and praises God for his return, but Arthur is relentless. 'If you love me, take that cross off your neck … come away with us … this dead world of priests and idols … full of the dust of

Brisighella cathedral and square.

Cardinal Montanelli and the Gadfly.

bygone ages.'[9] Montanelli cannot and sacrifices his son to Arthur's plaint, 'Your god is hungry and must be fed.'[10] The Gadfly, unroped and with unbandaged eyes, is eventually despatched by reluctant *carabinieri* riflemen as the victim pushes away the offered blood-stained cross. As if this were not enough high drama there is a powerful postscript.

During the pomp and celebration of the Blood of the Sacrifice in the cathedral, Montanelli breaks down and cries to the congregation that no one remembers the Passion of the Father who has sacrificed his only begotten son for their earthly salvation. He flings the host to the floor and dies crying to God, 'I have given him up for you.'[11] Gemma receives a last letter from Arthur who recalls their childhood and pronounces, 'Finita la Commedia' – the show is over. He signs off with the William Blake verse, 'Then am I, A happy fly, If I live, Or if I die'.[12]

There are very few plots that will withstand the denuding of a précis. By modern standards *The Gadfly* tempts farce. Nevertheless for a young woman of her time it is an extraordinary achievement. The climactic scene in the cathedral is stirring and powerful. Despite its melodramatic and morbid tendencies the novel moves at a fast and gripping pace establishing a lot of credible detail regarding the historical setting. The dialogue is realistic; the Gadfly's sarcasm is telling. Ethel had spent several months in Florence in 1895 at libraries doing research and had even met a peasant who had been involved in smuggling arms in the 1840s. Tuscany had gained a liberal constitution after agitation; there was indeed gun-running across the Apennines.

Much more specific colouring comes from Ethel's real-life involvement in a revolutionary setting in St Petersburg forty years later. She, like Gemma, had regular contact with political pamphlets, coded documents, chemical inks and the dangers of smuggling literature, if not arms. The harrowing and explicitly bloody prison scenes in the fortress at Brisighella could have come straight from the testimonies of her Russian friends, and even closer from her husband's dramatic experiences in the Warsaw Citadel. Like Arthur he had suffered exile and physical degradation; both had one shoulder higher than the other. Ethel conveys with conviction the intensity of group meetings and their disagreements.

The politics of the novel are not central but are a strong sub-theme that must have borne out her own thoughts in the 1880s as a result of the conflicts that centred on Stepniak. Compared with the multi-variable London debates on politics, the struggles for constitutional reform and national liberation in Italy were clearer cut. Ethel could allow herself to respond to a general principle of justice. The Act One debate between Arthur and Gemma about the compatibility of Christianity and revolution had been resolved by the Church's betrayal following his confession. A political question does arise when Arthur, morphed into the bitter Gadfly, asks her about the value of a new press law. 'Half a loaf is better than no bread,'[13] she replies. He dismisses this and probably Ethel's own adherence to any Fabian gradualism.

Gemma espouses Ethel's own likely anarchistic conviction that no man 'should hold over another the power to bind and loose'.[14] The Gadfly wants to enlist her for insurrectionary action, chiding her that her talents are wasted as a factotum to a radical group of men she is far superior to. (Maybe a dig here at her work at the SFRF.) He reveals that he is a member of a group called the Red Girdles, not averse to violence, but who back up their action with education. Most of all he believes in propaganda by deed. There is a very strong resonance here with Russian *narodnik* philosophy with which Ethel was well acquainted.

Gemma agrees that a violent act may be necessary, 'to eliminate for the moment the practical difficulty … of a clever spy or objectionable official …' but she counters, 'if it becomes habitual it invites vicious retaliation, for there is nothing else which governments so dread but the secret sect and the knife'.[15] More importantly, if it is an end in itself it becomes a mere ritual when 'what we really need to reform is the relation between man and man ... the sacredness of human personality'.[16] This dialogue she must have regularly encountered among anarchists in the 1890s. Ethel and Gemma, like Kropotkin, have no truck with dehumanising violence motivated by hatred, but are not opposed to the isolated necessary brutal act.

The Gadfly's quest for social justice in the present, prompted by the rejections of Gemma and Montanelli and his sufferings at the hands of others in South America, leads to his answering pronouncement, 'War is war.' He sees his opponents as vermin and suspects Gemma's equivocation about violence. He is on the other hand aware that the non-religious can create fetishes too: 'It makes little difference whether the something is Jesus or Buddha or a tum tum tree.'[17] What is commonly ritualised by the secular and theological is martyrdom.

As has been pointed out, the character of Gemma is partly based on that of Charlotte Wilson – Ethel's own friend and, like Stepniak, a political mentor. Her role in the novel is congruent with Wilson too; middle class and dedicated, she is a good comrade willing to organise and argue her case as she did as the mainstay of *Freedom*. Unlike Gemma, however, Charlotte, living the good life at Wyldes on Hampstead Heath, did not organise and participate in insurrection. 'Gemma would fight at the barricades ... she would be the perfect comrade, the maiden undefiled and unafraid of whom so many poets had dreamed.'[18] Gemma is both a 'new woman' of the English 1880s and the brave female anarchist/nihilist that Stepniak had lauded in *Underground Russia*.

There are other underlying human and existential layers that say a good deal about Ethel's own character. Despite the 'romance' of the novel as a very readable swashbuckling adventure story, there is little trace of personal romance between the two main characters, Gemma and Arthur. What matters is their faith in revolution, not personal emotional or physical attachment. Arthur is in Act One dressed in the manfully concealing black of the novitiate. Gemma, although christened by her family as Jennifer, becomes counter-gendered as 'Jim'. Voynich does not allow them a glimmer of a relationship as man and woman – even though, as we have seen, some of Ethel's male Russian exiles exuded a Heathcliffian erotic pull. In Act Two Gemma has been married and widowed, thus dismissing any girlish unfulfilled passion as a prompt for allegiance to Arthur as a man. Gemma's slow recognition of the innocent reborn Arthur as a Clark

Gable, Rivarez/Gadfly precludes anything touching in every sense of the word. The only sop to sensuality in the book is the gypsy, Zita, whose role in the plot makes Gemma seem more high-minded.

In *The Gadfly* the greater love for the struggle is the most important. By adumbrating any love story there is a heightened tension that emphasises Gemma and Arthur's sacrifice to more eternal values. One suspects that this may also be the tie that bound Ethel and Wilfrid initially together, developing lifelong into solidarity rather than emotional intensity. It also allowed an equality between them that, given the former's independent nature, obviated any surface concern with feminism.

The personal emotions that do get explored are suffering and reconciliation, not just in a social context but existentially. Ethel and her mouthpiece the Gadfly seem to revel in the harrowing tales of his adventures in South America. The meanness and absurd randomness of life haunts Ethel Voynich. The theme of forgiveness is central to the novel. Voynich allows no heart-clutching *rapprochement* between Arthur and Gemma, their love exists unspoken. Both are innocent anyway: Gemma was justifiably mistaken in judging Arthur for what seemed to be his betrayal of comrades. Arthur's vengeance is reserved for his betrayal in the confessional and by his father, Montanelli.

The climactic part of the novel centres on what seems to be an oedipal conflict but, like Samson overturning the Temple, Ethel, garbed as Arthur, is attempting to completely overturn religion. Arthur is a Christ-like figure; loving and pure, he suffers being scourged in South America. Resurrected as the Gadfly, he still has compassion and a belief in change but has come back with a sword to smite the unrighteous and offer himself as a sacrifice. When he sets off to smuggle arms he tells Zita with his stigmata stammer, 'I'm going s-s-straight to the infernal regions.'[19]

In captivity he finds himself at the mercy of the father he still loves, who is loved in turn by his flock. As a Pontius Pilate figure with a civic responsibility, Montanelli is mocked by his son, caught between duty to prevent a riot, or to help him escape and join the revolution. The son rebukes the father and suffers his fate because the father out of love for the world has to sacrifice his only son. So far so biblical – but in the awful choice it is the father, God and religion itself that is destroyed as well. The splendour and idolatry of the cathedral cannot compensate for the destruction of Arthur who literally shared his own blood. Montanelli curses his congregation, 'It is better that you should all rot in your vices, in the bottomless filth of damnation, and that he should live. What is the worth of your plague-spotted souls, that such a price should be paid for them?'[20] This is shocking stuff in terms of Christian mythology; the son cannot forgive the father and the father cannot forgive the masses for whom he has been sacrificed. The only redeeming salvation is the assumption that the revolutionary fight will continue.

What is the source of Ethel Voynich's wrath against the Almighty? We presumably don't have to look very far. Perceptively, W. L. Courtney in his 1904 review of *The Gadfly* wonders if the artist has caught some taint of almost malevolent spite. Childhood hurts no doubt cut deeply and remain as scars. Ethel's cruel treatment at the hands of Uncle Charles must have remained with her. There is indeed a specific reference in *The Gadfly* to a child being ill-treated by an uncle. Later in life Ethel told Anne Fremantle, 'I've forgiven him now because I know he was a musician who could not express himself.'[21] But at the time, aged ten or so, it must have been traumatic for a young girl to have left a cosmopolitan London home, however poor, for a grim colliery town like Rainford, isolated from her four sisters. The despising of a God who couldn't construct an octave properly must have been another nail in His coffin. Or maybe it was a futile, powerless frustration at the fact that her own godhead, the genius and apparent saint, George Boole, had died before she could have any memory of him?

The book was published first in the US by Henry Holt in 1897 and three months later in Britain by William Heinemann. The public took to it fulsomely: it was published eight times in four years and by 1920 had reached eighteen editions. Having assumed it was written by a man, the *New York Review of Books* in June 1897 praised the author's 'wonderfully strong hand and descriptive powers'. Indeed, a reviewer in *The Critic* in 1899 commented that, 'a more masculine book has not been written for years'. D. H. Lawrence found it 'awfully good', Bertrand Russell considered it 'the most exciting novel I have read in the English language', and Joseph Conrad wrote, 'I don't ever remember reading a book I disliked so much.'[22]

There were three more novels to come before Ethel left for the US in 1921. *The Gadfly* marked a distinct watershed in her life, although it was, strangely, to play a major role much later on. Now in her early thirties, with her association with revolution and Russian politics over-shadowed by Stepniak's death, she concentrated on her literary career. Probably exhausted by work and worried about Wilfrid's health, the couple went abroad to the south of France and Florence in spring 1898. The following summer Ethel crossed the Atlantic for the first time to supervise a theatrical production of *The Gadfly* in New York. She was horrified to find that, long before Hollywood, a political drama could be reduced to a confection. She was asked to 'introduce ginger' into the 'love' scenes. Her protest was printed in a letter to *The New York Times* in September. She refused to accept any royalties.

Stepniak's death was probably a milestone in Wilfrid's life too. Ethel's and his financial crash in 1894 with the unknown Mr Stein and the animosity between Wilfrid and his Russian colleagues required another direction and source of income. R. S. Garnett in Voynich's *Times* obituary gave the anecdote that his father, Richard Garnett, had met impoverished Wilfrid at the British Museum Library and suggested to him that, 'You only have to travel and pick up incunabula and rare books and sell them in London. There is success in front of you.'[23] (Incunabula are printed books dated before 1501.) Millicent Sowerby told the story that it was a borrowed half-crown from Stepniak that had started him off in the trade. This must be taken with a large block of salt, alongside many of the tales handed down to his admirers.

More reliable is Arnold Hunt writing in 2006 that it seems for someone 'starting out in the trade with little previous experience ... positively miraculous'[24] that Voynich could have issued, in July 1898, the sophisticated first catalogue of rare fifteenth- and sixteenth-century books. Where he found the money to finance and distribute it plus invest in his stock is a mystery. The catalogue mentions a Charles Edgell, a Cambridge graduate from a well-to-do family in Shropshire, who may have provided help. Andrew Cook has suggested that Voynich may have capitalised from the raising and laundering of revolutionary funds. More prosaically, Ethel may have invested money from royalties from *The Gadfly*. Taratuta states that in Russia it provided quite substantial sums. Their 1901 census address at Great Russell Mansions, Bloomsbury indicates that they had been somewhat upwardly mobile.

Whatever the source, it was Voynich's own drive and exuberance, the qualities that so irritated his Russian associates, that must have enabled his success. Another book collector exhibited the same required qualities: 'knowledgeable and enthusiastic, self-assured and tough'. This was C. J. S. Thompson, who about the same time began acquiring books and artefacts about medicine for the pharmacy millionaire, Henry Wellcome. Although with the benefit of a native tongue and someone else's purse he also quickly transformed himself into a business man. According to Frances Larson, biographer of Wellcome, in 1904 Thompson, as 'a skilled tactician ...

began combing the continent looking for treasures ... bidding at sales-rooms ... contacting traders, private collectors and institutions like hospitals and surgical instrument makers ... he had a knack for rooting out interesting objects from little-known sources'. Like Voynich, undoubtedly, he quickly learnt the tricks of the trade: buying job lots, outwitting auctioneers and undervaluing items he wanted to buy.[25]

By 1902 Wilfrid had published eight widely-mailed lists. They numbered thirty-three by 1914; their wealth of detail and use of others' scholarly knowledge made enterprising marketing. Some useful information about his milieu and dealings comes from the diaries of Robert Proctor (1868–1903), an expert in incunabula and a buyer at the British Museum. At the very beginning of his entries, 1 January 1899, he refers to Wilfrid bringing books to buy. Later in July that year he reports, 'Voynich came with glowing accounts of his acquisitions in Italy.'[26] He is clearly a regular visitor offering books for sale and has struck up social connections too. Ethel was introduced to Proctor in January 1900 and on 26 March with another dozen diners Proctor attended chez Voynich 'a most amusing scrambly supper in a room calculated to hold four at most'.[27] Rather snobbishly, Proctor refers to Alfred W. Pollard, who was at the time Assistant in the Museum's Department of Printed Books, and his wife, as 'the only respectable people'. The others mentioned included an actress and a typographer, book binder and calligrapher – the last three eminent in fields of great interest to Wilfrid.

Voynich opened his first shop around the turn of the century at 1 Soho Square and others followed in Florence, Warsaw, Paris and Vienna. In 1904 he moved again to 68 Shaftesbury Avenue. Like Thompson, he too was evidently a quick learner of the trade, able in 1902 to give a lecture to students of the London School of Economics on early printed books. In the same year he sold a collection of 'unknown and lost books' to the British Museum, catalogued now in the British Library with the rare personalised collection title of 'Voynich'. An interview in *Tatler* in May 1904 mentions his stock worth £25,000 acquired only from 'government collections or monastic institutions in out-of-the-way places'. With Voynich, however, there were always persistent undertones of over-archness in his dealings.

His knuckles were rapped, for example, when he went behind the back of the British Museum to find private sponsors for his sale of the 'lost book' collection. Giuseppe Orioli gleefully reported Voynich telling him how he'd 'exchanged a cartload of modern trash to a convent in Italy for a mine of early printed books … and illuminated manuscripts'.[28] We have already noted the 1908 purchase of the Libreria Franceschini in Florence. Rumours of riskier pornographic material being on sale in his London shop and of his 'trying it on' as a salesman were not infrequent. It all added up to give the once-shabby and penniless dockland refugee of 1890 a decent income and status, membership of the respectable Savage Club and no trouble in securing British citizenship in 1904, two years after he and Ethel had officially married.

Millicent Sowerby adored Wilfrid. 'He was a powerful man of medium build … but with the head of a great Norwegian god …'.[29] Sometimes the bad temper that had rubbed up his Stepniak colleagues the wrong way was revealed, but often his mercurial charm would come to the rescue. A victim might be taken off to lunch at the Casino Royale. He saw himself perhaps as a rightful member of the minor Polish aristocracy his family had once belonged to. Sowerby maintained that he often used 'de' or 'von' before his surname; Dr and Count also appeared. His headed notepaper and later catalogues sported a copy of a medieval emblem featuring a cat with a mouse in its mouth. Wilfrid transformed his erstwhile tribulations as prey to his advantage, embellishing his history for effect.

Despite the withdrawal from open politics in 1895 both the Voyniches had kept political links alive, especially with the Polish cause. According to Sowerby, a stream of Polish refugees came to the bookshop and were generously treated by Wilfrid. Ethel, still crusading, had learnt Polish and undertaken to translate documents to be forwarded to the British press.[30] Ethel's name appears alongside those of John Burns, MP, G. K. Chesterton and Richard Garnett as members of the Polish Relief Fund. Her support for Poland did not go unnoticed: in 1909 'Lilya Woyniczowa' is listed as Honorary President of the Polish Circle, London. Wilfrid was no doubt still being watched by the *Okhrana* but gave shelter to Polish refugees using his bookshop as a reception point.

At the time her husband was selling ancient books, Ethel was writing new ones. While strong and distinctive, none of them compare with the flair of *The Gadfly*. At the beginning of that novel she placed the epigraph, 'What have we to do with thee, thou Jesus of Nazareth?' referring to the Gospel demons expelled by him. She may have expunged some of her own in its writing but not quite. *Jack Raymond* (1901) breathes the same air but in less exotic climes. Like for many another, Cornwall's treeless moors and Atlantic-lashed shores have provided writers with inspiration.

Jack, the hero, is aged fourteen when we meet him, the orphaned nephew of the vicar of Porthcarrick near St Ives. He and his sister Molly had been taken in by him. Jack is the rough-and-tumble leader of the gang, into scrapes but only too aware of the injustice of the world and 'the big anthropomorphic Thing which he had been taught to worship'.[31] He finds solace with the 'tender mother' of the natural world in rock pools and 'skylarks singing in far blue heights and sunbeams flaming on the yellow gorse'.[32] Voynich, like Hardy, can be ecstatic in her descriptions of nature – but also, like him, there is often a countervailing cruelty. Jack's rapture is dispelled when he finds his uncle thrashing his favourite, blind old dog. Things plummet even further: Jack is falsely accused of stealing a knife to buy a dodgy set of coloured photographs, for which he is whipped and made to suffer the vicar's 'cannibal craving … the maddening lust to see something struggle'.[33]

Jack is sent away on his own to a school in London, having drowned the blind dog rather than leave it to his uncle's lashes. Here he befriends another underdog, Theo, weak but musically gifted. His mother Helen is the Polish widow of a dissident who has died in exile in Siberia. She takes Jack under her wing knowing only too well Russian oppression, and 'the naked sores of the world owing to the years spent among a monstrous population of degenerates in a land which has been for years a sink without a drain …'.[34] She talks of adopting the boy.

Years later Jack is studying medicine; Theo is mastering the violin in Paris and has taken Europe by storm. Helen after much suffering from cancer has experienced a visionary moment. Lying in Kew Gardens, she thinks of the crocuses there as an army 'that reeks not of victory or defeat yet when each dies a new soldier fills his vacant place never turning to look where the dead comrade lies'. Even when time and weeds cover the throng, spring returns and they 'rise up from the dead and stand in armoured ranks for battle'. It is the only allusion to revolutionary sentiment in the novel – but it is soon quashed. Voynich always pulls back from overt violence: 'Yet what use, when the seed is so bitter, and all the harvest is death?'[35]

Two years later, Jack meets his unhappy sister visiting London from Cornwall. They re-awaken old kinship but it is disturbed by the arrival of Theo who, with a revealing echo of Ethel's own passion, speaks of an 'eternally unsatisfied' need for music. Years further on still, Jack is a doctor but living a lonely life until Molly turns up drenched with rain, pregnant and turned out of home by her uncle. She refuses to reveal who the father is but asks Jack to adopt the child should anything happen to her. Four years later the pair are living with Johnny, her child, who dies in a diphtheria epidemic. Molly reveals at last that Theo was the father, destined to continue his wanton life.

The novel is not a success; the plot creaks somewhat and, despite echoes of worlds Voynich knows first-hand, there is much less contextual flavour compared to the romance of Italy. There is no historical theme to drive it forward, although the Church and its hypocrisy gets a good kicking. The almost random injustice and cruelty of life predominates without the great mythical climax of *The Gadfly*. In some ways it is more unrelenting than the armed clashes in the Apennines; disease is an enemy and the uncle's cruelty visceral rather than linked to a cause. Although the life stories are more individual, their fates are more impersonal and unresolved. Molly may forgive Theo but Jack does not forgive his uncle.

I came across a rather obscure but apt review of the book by 'G. M. R.' in the *Nursing Record and Hospital World* of 15 June 1901, fresh from the novel's release. 'The author could not have written for the pure joy of telling such a story, she must have had some object in the background. What the object was the present reviewer has failed to discover, unless the wicked uncle is a real person and Voynich owes him a grudge.'

Bravo, 'G. M. R.' With knowledge of her biography we can pick out other clues. Compared with *The Gadfly*, sexuality in *Jack Raymond* is not ignored but frowned on in the reference – rather boldly – to pornographic photos. Molly's baby is almost the result of a 'transaction'. 'He has had his joy and I pay the cost of it. It was a free gift.'[36] Love has no place in this panorama. In its publication year Ethel was thirty-seven years old and childless. Interestingly, in the light of her own later life, there are three references in the novel to surrogacy: the uncle had taken in Jack and Molly, Helen came to call Jack her son and Molly raises the issue of Jack adopting her bastard child.[37]

Another novel, *Olive Latham*, was published in 1904: a tale of anti-tsarist oppression that reprised her St Petersburg stay in the late 1880s. We shall come to it fully in Chapter 17 when retracing her steps, and my own, in Russia.

# THE ENGLISH AUNTS

The five Boole sisters c1897. Left to right standing: Margaret, Ethel, Alice, Lucy and Mary Ellen. Sitting: Julian Taylor, Geoffrey Taylor, Mary Stott on Mary Everest's knee, George Hinton sitting behind Leonard Stott.

The five Boole sisters posed together, probably for the last time, in 1897 with their mother, Mary Everest, and five grandchildren. Mary Ellen, far right, looking stern, has her hand on the shoulder of her eldest boy, George Hinton, appearing young for fifteen. On her left is Lucy, somewhat thin and rather lost. Next left is Alice, then Ethel. Margaret, far left, appears amiable as in real life, her hand on her son, Julian, aged eight. His brother, Geoffrey, aged eleven, is easily identifiable as is his grandmother, Mary Everest. On her knee is probably Alice's daughter, Mary, born about 1894, aged three or so here. The other grandson to her right is Leonard,

227

aged five. In 1924 Geoffrey Taylor wrote to his cousin, George, in Mexico recalling his memory of this meeting. One assumes that on this occasion Mary Ellen had visited from the US where she had been living since 1895. Strangely, her other three boys, Eric, Billy and Sebastian (Ted), were not present.

Lucy Boole had only another seven or so years to live owing to her tuberculosis. Why she contracted the illness and not her sisters we don't know. Ethel maintained that part of Lucy's inheritance was spent on a trip to Switzerland to ameliorate it. She is somewhat lost to any biographer as all we have are the bare facts of her accomplishments, which are striking but unamplified. Born in 1862, Lucy Boole was registered for some sort of education at Queen's College only from 1866 to 1869. What happened to her during the last difficult years at Queen's College we cannot be sure; maybe she was sent away to relatives in Cork. Ethel maintained that Lucy was also trained as a nurse. Like Margaret and Alice, this might too have been via the connection with Henry MacNaughton Jones, founder of the Cork, Eye, Ear & Throat Hospital. Her aptitude for science most likely came from her mother's tutelage and aspiration.

Lucy enrolled as a student at the School of the Pharmaceutical Society from 1883 to 1888, becoming only the second woman to pass their examinations. This led subsequently to a research position at its laboratory: a remarkable achievement for a woman at that time, especially one without the benefit of a typical middle-class background. Educational opportunities then were emerging from the Stone Age; rearing a family was considered to be the highest calling for a woman. The social theories of the day found justification for the inferior place of women on the evolutionary ladder in the writings of both Herbert Spencer and Darwin. Rather more parochial, in 1868, one Sarah Sewell had commented, 'Women who have stored their minds with Latin and Greek seldom have much knowledge of pies and puddings.' A woman away from her mother in order to be educated ran the risk of being prey to all kinds of dangers. Unlike Lucy's sisters, who didn't waste much time flying the coop, she remained living with her mother, Mary.

If immorality wasn't the problem, conversely, independence might lead to an opting out of motherhood and marriage altogether. This seemed to have some justification: of the first generation of Girton girls only sixteen of thirty-five married and only seven had children. Independence and achievement had, however, begun to be an option for women by the end of the century. Regarding chemistry for women as a vocation, it has been remarked 'what comes across most strongly is their dedication … they had to succeed for the sake of the young women who followed them … it was indeed the centre of their lives'.[1] This was clearly true of Lucy Boole; she never married.

In the 1890s Lucy became well-recognised in her field. Assistant to Professor Wyndham Rowland Dunstan at the School of Pharmacy in London, she produced a paper with him, *An Enquiry into the Vessicating Constituent of Croton Oil*. This may have been a slightly risky project as the oil from the Indian croton tree is highly poisonous. In 1891 Lucy became a demonstrator at the London School of Medicine for Women and later a lecturer there. The school merged with the Royal Free Hospital in 1898, Lucy becoming the first woman Professor of Chemistry. The Royal Institute of Chemistry elected her as their first Fellow in 1893. Gender discrimination had not been debagged, however, at the rival Chemical Society. A petition was got up to rectify this; Lucy was one of the nineteen signatories. Its Council adopted a rectifying proposal but only 45 men out of 2,700 turned up at a meeting to approve it. Twenty-three voted against. In 1919 women were finally admitted after the war that had done so much to shake Victorian precepts. Sadly Lucy never saw the belated victory; she died in 1904.

There's no knowing how well she and her mother got on at 16 Ladbroke Road, Notting Hill, during their long cohabitation there. Mary admired her daughter enough at least to list her among the figures, alongside James Hinton and her husband, who had helped accumulate information for her book *Forging Passion into Power*. Ethel mentions her fondly as 'Lulu'. She too suffered from the 'Matthew' effect that afflicted so many women, her mother included. Acknowledging Matthew 13:12, 'whosoever hath not, from him shall be taken away even that he hath', the sociologist R. K. Merton coined the term referring to the way in which already underdog contributors to knowledge are denied recognition. Margaret Rossiter re-gendered this as the 'Matilda' effect after the late nineteenth-century feminist, Matilda Gage. In such a vein Lucy was remembered by a student at the School for Medicine: 'We much revered Miss Boole on account of her clever mother … who had proved the divinity of Christ by mathematics, and of her sister who had written 'The Gadfly' and then married a nihilist.' Despite her obvious successes Lucy seems to have waltzed into history as another 'Matilda'.

The eldest of the Boole daughters, Mary Ellen's life had been somewhat adumbrated by her husband Charles Howard and his almost mono-maniac

Mary Ellen Boole (1856–1908), the eldest Boole daughter.

interest in the fourth dimension. She had stuck to him and their four sons, George, Eric, Billy and Sebastian ('Ted'), through bigamy and re-settlement in Japan and the US until his unexpected demise in 1907. In November that year she had written to Gelett Burgess, Charles Howard's friend, that six months to the day after 'my beloved husband's' death, she had felt comforted by visiting the chapel where his ashes were kept. The idea of promoting his unpublished material concerning the fourth dimension had roused her from a deep apathy: 'Plans seem to be settling on my mind.' Awaiting the public were *Travels of an Idea* and a number of short *Parables*. 'He was only sixteen when I read the first two of them,' she writes, hinting at her infatuation with him when they first met in 1869 at her mother's. Mary's new enthusiasm for Charles Howard's work was, however, not shared by the book trade.

On 26 April 1908 her youngest son 'Ted' wrote to a friend, 'Mamma is well and flourishing – at least physically,' but adds, 'she is beginning to be rather despondent.' A month later, Mary Ellen Boole Hinton killed herself. *The Washington Times* of 28 May wrote that she was, 'found dead in bed with a rubber tube connected with the [gas] chandelier between her lips'. Only recently she had written, 'Life is something we have the privilege of ending when we choose. When I think it is time to die I shall end it all.'

Charles Howard Hinton's sudden death must have hit Mary Ellen like a hammer blow. The Japanese letters reveal a period of consolidation in their lives, attracted by sharing its novel and exotic culture. Her Boole creativity found a spur there to poetry, some of which was published in *Nature's Notes* (1901), a book of fifty poems dedicated to Charles Howard. Some of them appeared in magazines in the US including *Harper's*. They tend to the mournful.

Perhaps the tenor is unsurprising; as the eldest of the Boole daughters she would have felt most keenly the equally premature death of their father, George. The girls' subsequent early upbringing in poverty at Queen's College, London and the unpredictability of their mother, Mary Everest, must have all added to a lifelong despondency. The other creative salve seems to be her love of music, although even here there is a sense of disappointment and being cheated. Like her youngest sister Ethel's petulant complaint about the tempered scale she asks, 'out of an infinity of possible notes why has God limited them to twelve?' Mary may well have been frustrated at not pursuing the musical talent that her Queen's College reports recognised. In *Credo* she, however, affirms her devotion: 'Thought-wise my soul's agnostic to the end / But music-wise, is mystic to the end'.

Her poem entitled 'Nature's Notes' finds her singing praises to the lilt of falling water, the trill of bird musicians, 'when my soul goes exploring / All the heights and deeps of tone'. Had both of her artistic talents been more recognised perhaps her life would have been more fulfilled. Without Charles Howard, cut off from her sisters and her sons at that time dispersing from the family home, she must have felt that she had nothing more to contribute. Not even her 'piece of sunshine', her youngest son, 'Ted', seemed to sustain her will to live. 'A Last Confession' runs presciently, 'but now / My husband's spirit with my God's is blent / And that is well – death dews creep o'er my brow / my length of days is spent'.

Alice Boole had moved to Liverpool sometime in the 1880s to work as a secretary to the merchant, John Falk, with whom she collaborated on Hinton's *New Era*. In Lancashire she met and married Walter Stott, an actuary. They had two children, Mary and Leonard, 1892.[2] Despite his apparent insistence that she remained a housewife, during the 1880s and 1890s, working entirely as an amateur on her own, she, amazingly,

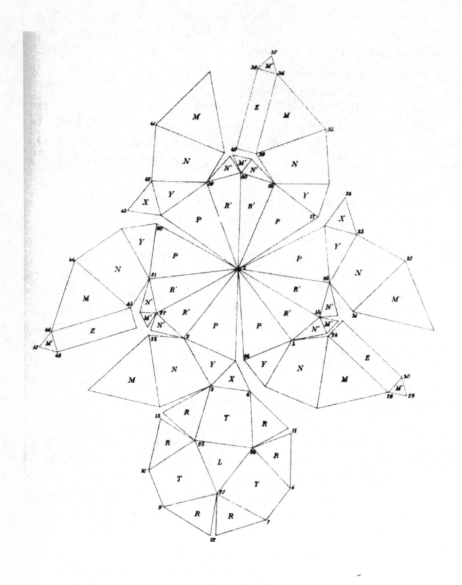

A cut-out model section of a 600-cell polytope by Alice Boole.

calculated and built beautifully constructed models of what sections of four-dimensional figures would look like as they pass through three-dimensional space. For someone lacking an intensive education Alice too exhibited impressive talents of concentration and visualisation; another case of the combination of theorising and the ability to implement the results that typifies the Boole/Hintons.

Here we need a short diversion into geometry. The two-dimensional shapes with which we are all familiar from school are called polygons. Examples are triangles and squares. These shapes have three-dimensional analogues, called polyhedra. The three-dimensional analogue of the square is a cube and of a triangle a pyramid (or tetrahedron). If you cut a section through a cube you obtain a square and through a tetrahedron a triangle.

Extrapolating from these examples, as a generalisation, an entity of x dimensions can be sliced to produce sections of x - 1 dimensions. Thus a cut cube (three dimensions) produces squares (two dimensions) and a cut tetrahedron (three dimensions), triangles (two dimensions). Reversing the logic, it should be possible to create a section of a four-dimensional entity in three dimensions. Alice grasped this notion and was able to construct three-dimensional cardboard models of sections of the six regular four-dimensional entities, known as polytopes.

Walter must have been impressed by his wife's acumen, for somehow in the mid-1890s he came across the geometrical expertise of a Dutch mathematician, Pieter Schoute, a professor at Groningen University in Holland. Photos and models of her work were sent to him and the two began to collaborate. H. S. M. Coxeter, the famous geometrician, remarked that 'her power of geometric visualisation supplemented his more orthodox methods of using co-ordinates'.[3] Schoute came to England on a number of occasions. The professor died in 1913 but her link with his university continued when she was awarded an honorary doctorate in 1914. Some of her models were presented to Groningen and are still on display; others at Cambridge University. She did not go to collect her honour but after it had arrived in a tall cylinder she is said to have used it as a store for sticks of macaroni. Alice, like her sisters, was obviously a very modest woman. 'I am such a duffer at analytical work,' she wrote, despite publishing two papers on polytopes in 1900 and 1910. In a letter to her nephew, Geoffrey Taylor in 1911 she apologetically mentions her recent life: 'I have not done anything more interesting than staining very shabby floors and suchlike household things for some time …'.[4] This was not to last; like her sister Ethel, there was to be a resurrection.

Left to right:
Margaret Taylor,
Mary Stott,
Professor Schoute,
Geoffrey Taylor and
Alice Boole (Stott).

After Mary Ellen Hinton's death it would have been unsurprising if the Boole/Hinton connections across the Atlantic had faded away. This was not the case: the links were maintained especially by Carmelita Hinton, Mary's daughter-in-law following her marriage to the youngest Hinton son, 'Ted'. On a trip to England with her children Jean, Bill and Joan in 1924, Carmelita visited her English Boole aunts, Alice and Maggy. Her diary gives a glimpse of Alice's life in Liscard across the River Mersey from Liverpool.

This aunt lived in a high, narrow, quaint brick house, dark stairways and odd steps up and down leading into various rooms. The living room was a comfortable place, charming in colourful quintz and a few beautiful heirlooms. The kitchen was particularly fascinating with a cheery open fire for boiling the tea water and out behind a little garden where flowers grew that some people could never make grow at all. [Upstairs], was the home of scholars. Books lined the walls … lay on the tables and spilled over onto the floor. This room was the private domain of the uncle (Walter). But the crowning event was the morning I was taken into the dining room and shown by my aunt her geometric models made of different bits of bright enamelled cardboard, fitted and glued together into a hollow ball, yet not a ball because of the half-inch facets on its surface.[5]

In 1930 her nephew, Geoffrey Taylor, at Trinity College, Cambridge, knowing of her interest in geometry, introduced her to H. S. M. Coxeter. He was much taken by her interest and knowledge and asked 'Aunt

Alice', as he called her, to participate in seminars; she was aged seventy, Coxeter, twenty-three. She brought along her models. He described her as 'Quite amazing. She had such a feeling for four-dimensional geometry. It was almost as if she could work in that world and see what was happening. She was always very excited when I had things to tell her … and she helped me in what I did.'[6] This late flowering of her quiet passion must have been very satisfying for her.

In the sisters' league table of the Boole International Hall of Fame, Margaret, the second-born in 1858, would probably rank herself as last. She apparently regarded her life as something of a failure in relation to her siblings. We don't really know much about her; she seems to have led a family life of stability that eschewed public notice. The Taylor home in St John's Wood was described as a friendly, happy and interesting environment. Both Margaret and Alice, as 'housewives', in different degrees shared mental skills of intuition and concentration that probably relate directly to their mother, Mary. Both lived long and relatively quiet lives, 'Maggy' until 1935, Alice until 1940.

### OPHTHALMOSCOPES, etc.

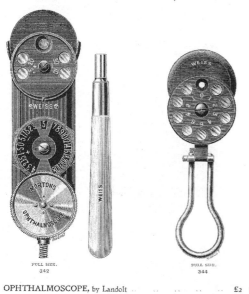

FULL SIZE.
342

FULL SIZE.
344

| | | | £ | s | d |
|---|---|---|---|---|---|
| 340 | OPHTHALMOSCOPE, by Landolt | | 2 | 2 | 0 |
| 341 | Ditto by Loring | | 2 | 10 | 0 |
| *342 | Ditto with double revolving mirror, 29 lenses and extra disc containing 4 lenses, by Morton | | 3 | 10 | 0 |
| 343 | Ditto ditto ditto with three mirrors, by Morton | | 3 | 12 | 6 |
| *344 | Ditto by Oldham | | 1 | 5 | 0 |

*The objects illustrated have an \* prefixed to their number.*

During Margaret's employment as a clinical assistant for three years until 1878 at the Eye, Ear and Throat Hospital in Cork she developed a highly useful skill. Analysis of disease and conditions of the ear and eye depended on accurate depictions of the state of the organs. George Boole's friend, Charles Babbage, invented an opthalmoscope in 1847 with which the retina could be seen clearly. It consisted of a magnifying glass held close to the eye illuminated by a light source refracted from the side. The pupil of the eye was dilated by administering a mydriatic drug. By this means the retina could be examined for signs of

An opthalmoscope.

Below: Watercolour drawings of ear and eye pathologies by Margaret Boole.

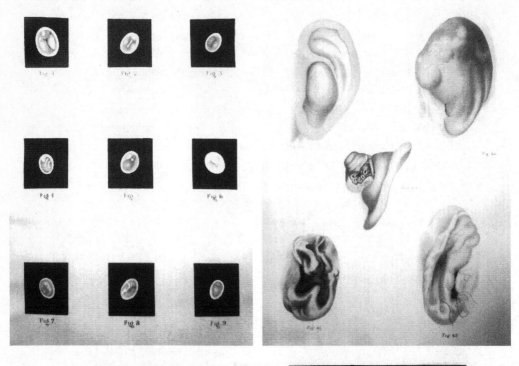

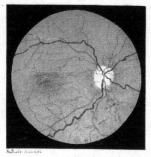

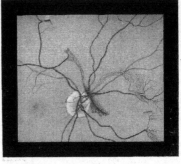

New vessels in the vitreous humour in a case of Syphilitic Retinitis. (2nd drawing – March 1884) The vitreous has cleared

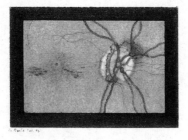

New vessels projecting from optic disc into vitreous humour in a case of Syphilitic Retinitis (1st drawing, Sept. 1883) with haemorrhages, & vitreous opacities.

disease – the only problem being that what was observed could not be recorded by photography at that time. Specialised artists who could retain an image in their mind and reproduce it in watercolours for examination were scarce. Ear specialists even more so needed great expertise. In both cases this could take many hours of discomfort for the patient. Fortunately Margaret Boole was exceptionally talented and able to create a good image relatively quickly.

Henry MacNaughton Jones published in 1878 his *Atlas of Diseases of the Membranum Tympanum*, illustrated by Margaret Boole. In the preface he acknowledged her for the 'labour, care and patience which she bestowed … I am indebted for these most truthful representations …'.[7] There are fifty-four coloured drawings laid out like precious stones on black squares in a jeweller's display. Each one is described. We find, for example 'recurrent *furunculus* in the meatus, origin of mischief, sea-bathing', 'polypus (mulberry) in cavity of tympanum of insane patient, producing vertigo'.

Margaret had stayed on in Cork after her training as a nurse using her artistic talent at the Art College, gaining a prize in 1879. On returning to London in the early 1880s she took up her skills relating to the eye. The *Transactions of the Opthalmic Society of the UK* at Moorfields Eye Hospital Library show a collection of drawings made by her between 1883 and 1886. Once again the watercolours can be seen to be highly detailed and skilled. One is a case of syphilis and another with a foreign body embedded. Interestingly the *Transactions* also show that Margaret's sister, Alice, possessed a similar capacity, having made drawings of slices of the optic nerve.

Margaret, as well as working as an ophthalmological artist, continued her art studies at the Slade School of Art (possibly with the aid of the Everest inheritance), where she met her husband Edward Ingram Taylor. They married in 1885. Her two children, Geoffrey and Julian, became distinguished in their fields as scientist and surgeon respectively. (Geoffrey is the subject of Chapter 18.)

Carmelita's account of her voyage to England in 1924 with her children mentions the other aunts she met. One was 'Mrs Nettle', otherwise known as Ada Nettleship, and in Chapter 6 as 'Daisy', the daughter of James Hinton. She had made a successful career designing and creating theatrical costumes, some for the famous actress, Ellen Terry. Ada had married John Trivett Nettleship (1804–1902), a well-known painter of animals. Their daughters were Ida and Ursula. The other aunt, 'Maggy', was persuaded to travel to Devonshire on a caravan holiday.

Family photo, Scotland 1924 with the three Hinton children, their mother Carmelita, Margaret, her son, Geoffrey Taylor and someone unknown.

'She put on a warm dress and her second-best bonnet and packed her two botany books in her small black reticule.' Carmelita stalled the battered old Rover in Piccadilly Circus and had to be rescued by a 'kind bobby'. 'Threading every few miles through friendly white villages, doll-like with thatched roofs and rose-twined doorways', they eventually reached the seaside. 'There are no lonely sand shores in England,' Carmelita writes. 'All swarm with people … such an air of contentment so simple and child-like, splashing in the wavelets playing ball, running races, dragging ropes of seaweed.' An excursion takes them all 'on a slow lazy steamer up the Dart River to Totnes there to feast on red raspberries and Devonshire cream in an Elizabethan inn.' The bliss is short-lived: the long train journey to Oban to join Maggy's sons, Geoffrey and Julian, on their boat *Frolic* is harrowing without a sleeper car: 'My English relatives said travelling that way was almost unknown, save perhaps among the nobility.'

It is difficult to create a full picture of Carmelita's other Boole aunt, Ethel (or ELV as she was known), who she met often in the US. Her stern, principled and rather relentless novels portray something of her nature. She certainly did not live the stable family life of Alice and Margaret.

A useful cache of letters to Esther Pissarro, the wife of impressionist painter Lucien (son of Camille) is housed at the Ashmolean Museum, Oxford. The two women maintained an exchange until 1951 when Esther died. Although of a similar age and background, they did not exactly open their hearts to each other. Small details, however, are illuminating.

Both were obsessed by their gardens, forever digging up roots and comparing plants. Letters addressed to Esther show that trips abroad could be afforded by the Voyniches or were necessary owing to Wilfrid's book safaris. Ethel writes from France, August 1905, 'I have got some seeds of the mountain crocus [echoes of *Jack Raymond*] for your and my rockeries.' In May 1907 she's staying in the Alpes Maritime and Chartres but, 'would like to be at home resting under my own dear cherry tree … tomorrow to Paris to meet Wilfrid, then South – possibly book-hunting'. In spring 1906 she's in Paris again looking up animals in the *Jardin des Plantes*; perhaps research for her next novel. There are stays by the English seaside: Dawlish, Cromer, Sidmouth. In 1912 she writes from a hydropathic hotel in Peebles, Scotland that it is 'noisy, rowdy, snobbish and generally detestable [Ethel never pulls any punches], but the treatment seems to be really doing Wilfrid good. Sent hairbells [sic].'

She and Wilfrid seem to have had separate domestic lives to match their separate professional interests but there is obviously a closeness; she refers to him as 'my man', which suggests a devotion but not one of intimacy. She is in contact with her 'available' sisters. The Voyniches had moved to Chiswick circa 1910 and again in 1912 to Richmond, up the hill from the charming bourgeois waterfront by the Thames. There are horticultural vicissitudes, such as in May 1914: 'It is dreadful about the wisteria.'

After the substantial success of *The Gadfly*, Ethel must have been sorely tempted to capitalise on a winning formula. *An Interrupted Friendship* was the result. Like most sequels it is very hard for an author to re-ignite the flame of an earlier inspiration; the 1910 novel asked even more by being a prequel, set in the mid-nineteenth century.

Rene, Marguerite and Henri are the children of a widowed marquis with an estate in Burgundy. With an echo of Ethel's own life the children are sent to be fostered elsewhere. Years later they re-convene: Marguerite is bed-ridden from a hip disease and angry with God; Rene, now nineteen, almost incestuously close to his sister retains a religious inclination; Henri runs the estate. Rene decides to join a scientific expedition to Ecuador to earn money for an operation on his sister's bad hip.

In Quito he meets a starving, half-caste, scarecrow of a man, lame with a deformed arm, two fingers missing and a pronounced stammer. It is, of course, our friend Felice Rivarez, the Gadfly, way down on his luck. Felice joins Rene's party to interpret as they negotiate the Andes jungle. We are now in the midst of an adventure story among the head-hunting Jivaro tribe. Voynich tells it with the air of one who's done her research, even if the episode in which Felice's expert marksmanship saves Rene from a mauling by a jaguar stretches credulity.

Three years later the party returns to France where Rene's bounty enables Marguerite's successful operation, blighted only by her jealousy of her brother's almost homoerotic attachment to Rivarez. The latter has become a journalist in London, meeting Giuseppe (Mazzini) during one of his various stays in exile. He asks Rivarez to help organise an uprising in Italy near Bologna. The insurrection fails but the Gadfly, as he is now recognised, has received a sabre cut across his cheek to add to his Andean deformities. Rene and his sister eventually meet up with our hero at their chateau where she falls in love with him. Marguerite taunts him about his anger towards the world. Two years later, linking up with the Gadfly they read about Felice in a newspaper: 'Atrocities in a Papal fortress. Horrible treatment of a political prisoner'.[8] Brother and sister are left unrequited in their love of Rivarez, 'one of the rare spirits that go through the world like stars, radiating light', but who represent 'a dangerous thing to love too much'.[9]

The plot overall is unconvincing. Without a reading of its parent novel it would make little sense; several reviewers found themselves at a loss to understand the motives of Rivarez. The adjective most frequently used by them was 'morbid'; the novel is yet another exercise in suffering. There is very little indeed of life's beauty and certainly no amorous attachment. At least in Thomas Hardy's world, ruled by the indifferent 'President of the Immortals', human physical love and humour can exist – in Voynich's, ruled by the callous 'Thing', love flourishes only around 'rare spirits' and their nobility in enduring life's ruthlessness.

Ethel writes to Esther on the brink of the First World War from Wales, holidaying with her sisters Margaret and Alice. She has taken on relief work with Polish refugees, writing in December 1915: 'We are all on overtime now. 2,000 kiddies to supply with tea and toys and sweets and games and Christmas trees and cheerfulness …'. Wilfrid is in the US exhibiting the manuscript and cannot get home. In September a year later she writes, 'My man turned up last Sunday unexpectedly. He is badly run down having worked through the great heatwave in Chicago.'

It was during the war that Ethel Voynich made her own contribution to motherhood, perhaps inspired by her work with children. She adopted a girl who she regarded as a daughter but who was never legally so. The child was fostered out. Winifred Eisenhardt, born in 1904, was the daughter of a German interned at Alexandra Palace during the war. Suffering the public opprobrium felt for the 'enemy', she must have touched Ethel's sense of justice and perhaps a latent maternal impulse, somewhat buried. Her own sense of loss in her early life with a difficult mother plus the early separation from her sisters probably accounts for this. It seems typical of her to construct a rational but unworkable relationship – she was hardly in a position to be a proper mother as she moved to Cornwall for three winters until 1919 to recover from whooping cough.

Wilfrid, usually remote geographically, was even more estranged. Much later in 1926 he writes to their friend, Edward Levetus explaining about Winifred, 'I know her very little', but adds with typical generosity, 'I foot the bills but I am not taking part in her education or bringing her up. ELV loves her, and that is the end of it.' Ethel's own emigration to the US made any real closeness difficult, although she maintained an overarching concern. She had, however, during the years of the war made an attachment to a figure who might be seen as the kind of son her sisters Margaret and Alice had produced, that she had foresworn or been denied.

In 1908 Ethel had collaborated with the musician and performer Marion Scott by translating Lermontov's dramatic poem *The Song of the Merchant Kalashnikov*. She combined this in 1911 with a translation of *Six Lyrics from the Ruthenian of Taras Shevchenko*. Shevchenko was a nineteenth-century serf who suffered years of internal exile in Russia far from his native Ukraine. He writes morosely in the first poem, 'Wretched is the fettered captive / Dying, a slave / But more wretched he that, living / sleeps, as in a grave …'. Despite Ethel's concentration on writing, her interest in music had never really gone away. In 1911 she joined the Society of Women Musicians founded by Scott at the Royal College of Music in London. Although Ethel was ten years older, aged forty-five, they must have shared a good deal in common; both were from intellectual families and shared radical unconventional leanings. One thing they certainly shared was a penchant for the twenty-one-year-old, Ivor Gurney, who had won a scholarship there in that year.

Born in 1890, the son of a tailor in Gloucester, he showed early musical promise, winning a place in the cathedral choir there and later as a pupil organist. Alongside another contemporary Gloucestershire composer,

Herbert Howells, he would have soaked up the atmosphere and prestige at one of the cities contributing to the famous Three Choirs Festival. After the first performance of Vaughan William's *Tallis Fantasia* at Gloucester in September 1910, the two young men were so enraptured they spent the night walking the streets unable to sleep.

At the Royal College, although demonstrating a Schubertian talent, Gurney's individuality found him at odds with his teacher, Charles Villier Stanford. Pamela Blevins, biographer of Scott and Gurney, speaks of him as possessing 'a tremendous force within him … manifested in his physical energy and a talent for writing poetry as well'.[10] But alongside this dynamism resided an unstable restlessness plus odd habits in dress and binge-eating. Blevins ascribes it to the onset of bipolar disorder.

Another Gloucestershire friend, J. W. Haines, described him as 'a kind of combination of Don Quixote and D'Artagnan, gallant, intractable, kindly, ferocious and distressingly lovable. With blazing eyes he would pour forth an endless stream of talk …'.[11] The personality was attractive to women. Gurney did not exude sexuality, however, and opinions differ as to his libido. Michael Hurd, another biographer, detects none at all, but his boyish charm would have attracted women wanting to nurture and make sacrifices for him. Marion Scott and Ethel Voynich were two such. Hurd talks of 'a select group of maiden ladies, older by many years, to whom he instinctively turned for comfort and encouragement'.[12] Blevins believes that Scott was actually in love with Gurney; Ethel's attraction may have been motherly but it was, *quelle surprise*, mostly intellectual.

The archives of Ivor Gurney contain a number of letters between them that reveal a good deal about both their natures. The list of subjects they discussed demonstrates a rare ability to be equally conversant around both words and music: Bach, Milton, George Meredith, Robert Bridges, Chekhov, Wordsworth, Beethoven, Hardy. Voynich seems to take on the role of teacher. In a letter to Scott, Gurney reports that Mrs Voynich gave great praise for a poem called 'Steel and Flame' in which the line 'nor steel nor flame has any power on me / Save that it's malice work the Almighty Will',[13] a sentiment that would inevitably have appealed to her dismal view of God and his works.

A year later in January 1917, in a letter to a mutual friend, Gurney mentions rather dismissively that Voynich 'shows distinct signs of making me confidant as to her new cantata'.[14] One wonders what Ethel would have made of an earlier dismissal in August 1913 penned to Scott regarding *An Interrupted Friendship*. He declares that it is 'without form

Framilode on the River Severn, Gloucestershire.

and void but not uninteresting'.[15] He had, however, read *The Gadfly*.
Despite finding it 'awfully fine', he also suggests that it is 'the kind of thing
one would write in cold grey dawn after a substantial breakfast of cold beef
steak pie and porter'; an indigestibility that others had complained of.[16] It
also points to a major difference between them: Gurney writes of the
immediacy of the senses, uneasy or ecstatic.

Gurney volunteered for the war and found himself in May 1916 dug
into a Flanders trench. Here, life made vivid by the possibility of sudden
death, triggered a lingering longing for the sights and sounds of his
beloved Gloucestershire countryside. Gurney is rooted there as Hardy is
to Dorset. Villages become mantras to the homesick soldier, 'Framilode,
Minsterworth, Maisemore …', he incants to Ethel in February 1915 and
again in April 1916, 'Cranham, Framilode, Crickley, May Hill'. Gurney
used to walk from Gloucester city to Framilode, no more than a hamlet
perched on the levee above a great sweeping oxbow in the wide River
Severn. Ethel, a city cosmopolitan in comparison, has no internal

childhood arcadia always calling her back, despite her evident love of nature and the wild. She has no connection with her mother's rural roots at Wickwar a dozen miles south of Framilode. Cornwall is the closest to her soul. As a result, Ethel, apart from her love of flowers, lacks the common touch of small pleasures to lighten the lives of her characters, faced as they are with the meaning of life, the universe and everything.

Love and sex may not be major themes for Gurney either but there is sensuality in his work. His trench letters also show a skittish humour. He thanks Ethel for the 'tray bong', chocolates she's sent him and mentions the joke among his men of life after the war, 'Après le gore'. Discussing Milton with her he contends, 'His spirit was huge, but not generous because he lacked humour, and its attendant quality,'[17] a statement that could be matched to Ethel Voynich. No one could think of her as devil-may-care like Gurney; her devil cares too much.

Gurney is very pleased with a new word his companions have donated to the world: 'twallet', meaning 'one of small intellectual powers'.[18] Ethel is no 'twallet', however, and apart from music and literature there are references to her political interests. She has been sending Gurney information about Russian persecution of the Jews and a whole letter about the news concerning Poland. Gurney writes nostalgically about the home front and English tea times which he has enjoyed with Ethel and her sister Margaret's family in St John's Wood. He is a regular visitor: 'just delightful … Mrs Voynich was as nice as could be, Geoffrey Taylor [their son] ditto'.[19]

The group met up another time to spend Christmas in the Cornish setting of *Jack Raymond*, Gurnard's Head. Ethel had been staying in St Ives since January 1917 composing music and often walking twelve miles over the moors. Cornwall suited her rugged individuality but she is feeling a little bit isolated. She writes sadly in January 1917 (unlike the cosy image presented at Margaret's) to Esther Pissarro, 'Ours is a family apart. If I were dead I doubt if it would occur to anybody to mention it …'. In March 1918 she sends Esther a list of the wild flowers she has encountered on a walk followed by the incongruous, 'Occasionally a submarine comes along and murders a few folk just outside the bay … we hear the bell tolling. It is a filthy business, all of it.'

Ivor Gurney shared Ethel's love of wild flowers and writes on Boxing Day, 1918, 'Violets in bloom (though not many) at Christmas. Primroses glorify all this place in Spring and here the sea-pinks abound. Here also we found a little blue flower … called something like "squininsy wort".

(Mrs V is still out and not to be consulted).'[20] He too must have found inspiration like Ethel in the wild coastline – 'the Armada clouds, huge breakers, wicked-looking rocks and brown and grey moorland ...'[21] – so much so that he literally lost himself on a darkening clifftop, composing the very fine song setting 'Desire in Spring', dedicated to Ethel. Intrepid Geoffrey Taylor and a friend rescued him. A near disaster would have left posterity the poorer was relieved by 'a scrumptious feed at a hotel ... After that we played Musical Chairs and suchlike frivolities till late.'[22] The image of Ethel scrambling to occupy the fast-disappearing seats seems incongruous – maybe she was playing the piano, Chopin probably – not a ditty from the music hall.

The episode suggests a good deal about Gurney: high-spirited but nature-mystical and wrapped up in his own intensity. Three months later in 1918 he told Marion Scott that he had been speaking to 'the spirit of Beethoven'. Two months further on he had sent his friends suicide notes. Wounded in April 1917 at the front and gassed later the same year he was obviously not so incapacitated as to be unable to enjoy rapture in Cornwall. Ethel writes to Esther in June 1919 how she had spent all Whitsuntide sleeping out of doors, 'in an officer's valise by the sweetbriar hedge and Ivor in the buttercup meadow ... on a groundsheet'. The ordeal of the war must, however, have contributed to his latent instability. Apparently Ethel in 1920 was trying to set aside some money in a trust for him. He continued to write a considerable number of valuable songs and poems but after a brief return to the Royal College of Music he was committed as insane and permanently exiled to Dartford, Kent. He died there in late 1937.

The 1911 publication of the Shevchenko/Lermontov translations marked the end of Ethel's literary career for over thirty years. She had presumably exhausted the medium; its link with revolutionary politics had become progressively attenuated. Her connection to the Royal College of Music, Scott and Gurney meanwhile had returned her to music, her first love. It was to occupy her for the rest of her life although the traces of her political consciousness remained. This was hardly surprising given the cataclysm of the First World War. Two events during that period jolted her into putting notes onto staves.

On 3 August 1916 Roger Casement was executed at Pentonville Prison, London for the treasonable offence of attempting to smuggle arms into Ireland to aid the republican uprising of that year. Fifteen other nationalists were executed in May in Dublin, including James Connolly who, although wounded, was shot tied to a chair. Sympathetic to Ireland's troubles, the first draft of a cantata for male choir and orchestra was written soon after. It was entitled *An Epitaph in the Form of a Ballad*, based on the poem of that name by the fifteenth-century poet Francois Villon, written as he too was facing execution. It includes the line, 'Pray to God that he forgive us all', repeated three times. Ethel re-worked the piece many times, the last being in 1948.

The other shattering event took place in her old haunt of St Petersburg. Ethel wrote many years later, '… during the last years of Tsarism, I began to be haunted by music … but I could find no words for it'.[23] She had written out ghostly chords on the piano,

> then one morning … I saw in the newspaper … a headline stating 'The Tsar abdicates' [15 March 1917] and at the moment the missing words flashed through my mind: 'How long, O Lord, holy and true, dost thou not judge and avenge our blood on them that dwell on earth'. These words suited the music almost without change. I thought to myself, it seems I shall have an oratorio and it will be about Babylon.[24]

The settings are from the *Apocalypse of St John (Book of Revelation)*. In a letter to the Levetus family in 1953 she recalled that an initial theme came from hearing the roar of a German aeroplane on a bombing raid in September of that year. *Babylon* was dedicated to Wilfrid Voynich. She would refer to it as 'my major thing', and as her 'royal white elephant'.

The revolutionary news of 1918 from St Petersburg must have taken her back to her days thirty years before, giving rise to another apocalyptic vision, this time taken from a poem by M. A. Dmitriev which seemed 'to reflect a legend according to which Peter the Great and his capital … were cursed': 'Thy city shall be empty, it shall return to the swamp and will be empty, empty, empty.' The cantata for baritone, mixed choir and orchestra, *Submerged City*, was revised again in 1928. In a letter to Edward Levetus, 11 August, 1949, she declared that it took three winters at St Ives to get the sound of the surf onto paper. Ethel had indeed by the end of the First World War marked out her life to be a composer rather than a writer. There was one more novel to come, almost unbidden, to interrupt her compulsion to compose.

# BRINGING HOME THE BACON

When Wilfrid Voynich was resident at his bookshop in Shaftesbury Avenue, life there was, according to Millicent Sowerby, 'an exciting and happy whirlwind'.[1] His tentative English, often minus the definite article, led to sympathetic mirth-making – although it was not always well-received. She recounts how when Mrs Voynich once brought the staff flowers, without thinking, she aped Wilfrid, 'What we have we can put them in?'[2] Mrs V was not amused. In just such an anecdotal way Sowerby remembers 'the treasure that he had found in some ancient castle in Southern Europe … the famous Roger Bacon cipher',[3] later to become the eponymous *Voynich Manuscript*. In early summer 1914 she tells of her sadness at having to leave his employ as he was moving to the US. We cannot know for sure why he upped sticks and sailed on the *Lusitania* for New York on 27 November 1914. The outbreak of war in August might have been a factor but was it purely a Whittington-style move – the cat looking for richer cream in a wealthy, neutral land far removed from an imbroglio possibly long-lasting?

Wilfrid Voynich, antiquarian bookseller.

Ethel at that time was caught up in the musical world of the Royal College of Music, Scott and Gurney prior to her burst of creative composition. Apart from the Polish community, Wilfrid and Ethel had an active social life with the Levetus and Pissarro families and a wider field of eminent personalities. Wilfrid was one of the invitees to the Round Table National Liberal Club New Year's dinner of 1912 that also listed Kropotkin, Belloc, Shaw and Walter Crane.[4]

According to passenger records, he returned briefly to London in February 1916 but re-crossed again in May. Letters indicate that he was in the US in October 1917 and January 1918. In early 1917 Voynich opened an office at Aeolian Hall, 33 W. 42nd Street. Residence in the US was to be a continuing problem for the Voyniches, although their strong alliance with the Polish cause may have helped in some circles. The pair were to spend almost the entire war separated, Ethel mainly writing and convalescing in Cornwall and Wilfrid exhibiting and setting up his book business in New York.

Arnold Hunt maintains that during his residency at Shaftesbury Avenue, Voynich had begun to trade more in expensive illuminated manuscripts in particular, managing to sell one worth £1,200 to the discriminating Bella de Costa Greene, librarian to millionaire J. P. Morgan.[5]

Three years later he tried again with an outrageous price for another that was apparently laughed at by her. This did not prevent him, however, pushing to make a sale with the 'ugly duckling' Bacon manuscript from Italy. By October 1915 he had, among other treasures, as mentioned in Chapter 2, exhibited it at the Chicago Institute of Arts. The *Chicago Daily Tribune* on 9 October reported that 'Hungarian' Voynich had escaped the war to bring works from 'a remarkable tour of Europe in which he spent $8,000,000 in purchasing the most valuable articles in the collection of the royal families and monasteries of half a dozen countries'. Even allowing for misreporting, Wilfrid had been obviously blowing his own trumpet. Alongside one exhibit was the description of 'a work by Roger Bacon in cipher to which the key has never been discovered'.

Hunt also reports that in 1915 Voynich is said to have done very badly in America. Worse was to come. In 1917 he came under suspicion by the Bureau of Investigation, the forerunner of the FBI.[6] The fact that suspicions about his probity had got around book dealers must have been a blow to Wilfrid's own self-esteem as well as his public face. Some mud must have stuck. In 1920 Wilfrid writes to his English friend Edward Levetus, 'Three or four dealers have opened a regular campaign of slander and abuse of my books and my prices.'[7] In 1921 he complained that private collectors were still scared of him. This did not prevent him, however, from continuing his project to prove that his 'Bacon Cipher' was indeed by Roger Bacon. My notion that Voynich himself had at least hoaxed the Marci letter to bolster that provenance does not infer that he did not believe in the actual manuscript's authenticity. What led him to this conclusion?

As we saw in Chapter 2, Wilfrid had made a plausible case for the manuscript being handed on from John Dee to Rudolf II in 1586 and then on to Kircher via Marci in 1665/6. Voynich's contemporary scholars did not scoff that it might have been penned several centuries before by Bacon. A. G. Little, an expert, wrote in 1914 that 'many spurious writings are attributed to Bacon and some genuine ones are hidden under other names'.[8] Many of his works had not even been translated from Latin. The Bacon work could have been an opus of his re-written later by some scribe deliberately in cipher to hide blasphemous knowledge such as his belief in astrology.

One leading Bacon scholar and expert in early books, Robert Steele, had been a firm friend of the Voyniches since 1900. He was a witness at their official wedding in 1902 and also the librarian at the Savage Club, endorsing Wilfrid's membership there. In one of his articles in *The Library* (1917) he investigated the *Secretum Secretorum*, the 'Secret of Secrets' that Bacon himself had translated from its Arabic source. Although without illustration, there is a comprehensive list of plants that could be used as an elixir of life. It is highly likely that Voynich would have consulted him. Steele hoped that the manuscript would prove to be 'something very important indeed'.

Voynich sent photocopies of the manuscript to various experts, one of whom, William Romaine Newbold, Professor of Philosophy at Philadelphia University, was literally to drive himself mad over it. Working from what he assumed was a key partly in Latin, partly in cipher found on its last folio, he stumbled along an almost unbelievable path of decryption. This involved anagrams and the cabala plus the assertion that each letter of the script, when magnified, revealed minute pen strokes in the pigment that could be transcribed from Greek shorthand into Latin letters.

In April 1921 in Philadelphia at the College of Physicians, the staggering details of Newbold's analysis were declared. The naked nymphs were human souls brought to earth and embodied through sexual intercourse. In the illustrations, reproductive details could be seen to be ovaries and spermatozoa that Bacon must have discovered with the help of a microscope – long before its known invention. And from the micro to the macro, Bacon, he declared, had identified, through the equally unknown telescope, the Great Nebula of Andromeda. Robert Steele in 1921 provided a manuscript of Bacon's he had discovered in Paris that Newbold decrypted as medical advice to the Pope concerning the pontiff's kidney stones. Another revealed a process for precipitating copper.

Voynich and Newbold were making extraordinary claims but given the favourable backing of some academics Voynich felt justified in maintaining a price of $160,000 for the manuscript. If it sold, his financial problems would be over and his credentials as a dealer enhanced after his poor start in the US. Over the years his adverse status seems to have attenuated; Bella Costa Greene was again in negotiation for a Justinian manuscript and a psalter for $6,000. Henry Huntington, railway magnate, art and rare book collector had become a client in 1925. Voynich was happy to service a clientele that since the end of the war had been 'growing here like mushrooms …who believe that now they can buy Europe, Universe, God and British Museum'.[9]

By the end of 1921 he writes again to Levetus of his main prize, the Bacon manuscript, 'It is very exciting to be a medium for the recovery of great discoveries … made by one of the greatest men ever lived after their being buried for seven centuries.'[10] Voynich adds that it will mean 'riches for him and most likely a Nobel Prize for Newbold'. He is hoping for at least ten deciphered pages and 'then to conclude deal with Philadelphia millionaire'.[11]

Newbold did indeed struggle on further decoding Bacon's description of an eclipse in 1290 and a recipe for gunpowder. The spiral of the Andromeda galaxy was also revealed leading to its comparison with folio 68r in the manuscript, even though the former wasn't discovered until the twentieth century. Also, somewhat problematically, in reality it turns to the right rather than the left. (This didn't stop one Voynichero from claiming that naturally to the aliens who wrote it it would be seen that way from the 'other side'.) The toll on Newbold's health was enormous. Roland Grubb Kent described him 'half sitting, half lying on his bed, a powerful electric light over him as he examined the text with his reading glasses and his microscope, the latter set in the right side of a pair of spectacles, while the left was closed by an opaque disc'.[12] The strain was too much for him; he died in September 1926.

It is a tale of obsessive self-willed belief with underlying cryptomnesia. Voynich held out to the last believing that his Bacon would save him from a financial smash. He had at least eventually gained some respectability within his profession. Two of his contacts, Le Roy Crummer and Joseph Miller, sang his praises in 1928: 'Mr Voynich is assuredly the Sherlock Holmes of the old book trade … the King … and the most honest man in the game.' His letters to the Levetus family, however, throughout the 1920s are embarrassingly full of his insolvency. Although the Florence office was closed in 1922, suggesting that costs were being minimised, he did keep up his book-finding safaris in Europe. He writes in autumn 1920 that he

Professor Newbold, cryptographer
1865–1926.

has been in Germany visiting eighteen shops in Leipzig in three days, returning to the USA 'penniless but with some treasures and many interesting things'.[13] The following autumn he was in Switzerland and Italy.

Letters also make clear that he relied heavily on the shrewd and skilful services of Anne Nill, his American assistant, who often accompanied him on his voyages. Born in 1894 and much younger than Ethel and Wilfrid, beyond business concerns, she seems to have attached herself to them as a member of a trio. Little is known about how they met; her first listed crossing was with Wilfrid on the *SS Scythia* in November 1921.

Ethel and Wilfrid began their joint stay in the US in February 1921 when their boat docked in New York after a stormy crossing. The last four days, Wilfrid writes to the Levetus family, Ethel was, 'sitting all day on the deck … spending evenings in the most shaky part of the boat gazing at the stars …'.[14] Ethel established herself in Washingtonville, Orange County, northwest of New York City until 1927 at the home of an acquaintance, Mary Marquis. She seemed to be finding her now calmer suburban life a little strange, writing to Esther Pissarro that the area is 'almost comically like Surrey until one sees *sumach* instead of hawthorn'. Wilfrid writes to Esther in 1923 that Ethel 'only comes to New York when very interesting concerts are given'.[15] She had been present, however, in Philadelphia at the April 1921 temporary ascendancy of the Bacon manuscript, writing to Esther of, 'a roaring time … it is probably the beginning of big things. Wilfrid has started a snowball that may end in a really big development of medieval research and the discovery of much lost science.'[16] Wilfrid writes of her ill health but nevertheless that she has had a burst of composition.

Later that year in October Ethel went to the Catskill Mountains to visit her Hinton nephew Ted (Sebastian) and his wife Carmelita. Ted was the youngest son of Ethel's sister, Mary Ellen, and Charles Howard Hinton. Their country home 'Camelot' was on the edge of the Byrdcliffe Arts Community near Woodstock, New York. As a baby Ted had left for

America with his parents in 1887 at the same time Ethel had set off for St Petersburg. It was the first of many visits to the Woodstock area, tying the two dynasties together across the Atlantic. She describes them as 'delightful folk … We drove down to a wood near the village … and there found a real concert hall built of rough pine trunks … but as big as Aeolian Hall …'. They listened to, 'a late abstruse Beethoven quartet and a fine quartet of Ravel to a perfectly silent audience'.[17] (Carmelita and her children will figure very largely in our account later.)

Wilfrid was certainly not excluded from the Boole/Hintons; they seem to have taken a liking to him, but he was often busy elsewhere. He too had struck up an old acquaintanceship that seems to have counter-balanced his arm's-length attachment to Ethel. The connection to Erla Rodakiewicz dated back twenty years. Fortunately, Erla kept a cache of the letters between her and Wilfrid that was handed down to Jean Hinton, her god-daughter.[18] In a letter dated 9 April 1898 Ethel had invited Erla Hittle, as she then was, to have tea with Wilfrid and herself during their stay in Florence, where the latter was convalescing. There was no further contact with the Voyniches until August 1914 when Erla was staying in London.[19]

She was obviously highly intelligent, having studied at Cornell, Leipzig and Heidelberg universities where she gained a doctorate in 1901. Over her life-span (she died in 1964) she nurtured occult interests in psychical research, the *I Ching* and, most notably, medieval manuscripts. Foremost Renaissance scholar Paul Kristeller much later reported that while insisting on being an amateur, Erla herself had developed as a consummate expert.

Before Ethel's arrival in New York she seemed to have also developed an interest in Wilfrid personally. She addresses him in letters as 'Thornbush' for no clear

Erla Rodakiewicz, (d. 1964), friend of the Voyniches.

reason but possibly referring to his life's prickly career, and occasionally too as 'Gypsy Chief'. He, in return, writes to her as 'Early Christian', presumably a convoluted pun on her name. She sent him obvious tokens of affection – poems, an amulet and a tie – earning, it seems, a mild chiding rebuke in a letter dated 31 January 1918: 'I think you are attaching often to my words undue importance or deeper meaning than this contains.' He adds that he is 'glad that my social duties kept me apart from all kinds of personal entanglements, domesticity and parentage'.[20] This suggests a disposition that partly complements Ethel's own feelings.

What Wilfrid's letters to Erla most usefully reveal is a frank evaluation of his own character. In a letter of 1926 he invokes his feline stance again: 'I am a cat walking by himself,' adding, 'I can thrive on this solitary state but I am deeply moved when I meet understanding if not sympathy.' Wilfrid continues, in his halting English, 'It is always a great pleasure to have chat with you in your home.'[21] Erla has fashionable cocktail parties there on the Upper West Side. Wilfrid was grateful for the introduction to her friends this allowed. They also dined regularly together. One undated letter includes Wilfrid's ambiguous comment, 'I have enjoyed the two evenings immensely but at what cost. I am afraid you will be ill as the result of sleepless nights and exhausting days.'[22] Erla had proved a good friend over the years, to Ethel as well; both women had attended the 1921 manuscript presentation in Philadelphia. They also spent vacations together, often staying at Woodstock.

Despite the trials of what he calls 'that beastly thing humanity', the egoism of his early London days shines through. He writes in an undated letter from the Manhattan Hotel in New York, 'I'm always the same, old, young, half-mysterious, very good friend who really deserves all the pampering of friends.' He mentions his 'pep' and how, like Erla, he loves 'music, woods, flowers … dogs, animals, frogs, cats … and everything created by Good God'. But against this yea-saying his hot-bloodedness emerges. 'Purity means stagnation by deterioration. Only sinners are capable of advancing, of improving. What life would be like if we were all sinless; I would devote myself to debauching their souls and bodies.'[23]

Ethel, too, often refers to mankind's beastliness as against the beauty to be found in flowers; her response is to take creative refuge in music and literature. Wilfrid's urge is to engage with the cut and thrust of a glamorous cosmopolitan world, the risks and highlife of the time. 'I am in love with your country to distraction,' he writes to Erla. There is something of a minor Gatsby about him. The loyal but unlikely Voynich pair had to live apart to pursue their separate drives; Erla seems to hover around Wilfrid's life, a vibrant and engaging admirer from the smart set.

The glamorous prospects of the New World may have entranced Wilfrid but the failings of 'beastly humanity' back in the Old World still plagued him personally and politically. He writes in April 1922 to Jill and Edward Levetus, 'My sister died last year from typhoid in Russia. Niece has managed to escape back into Poland but little boy was taken by Bolsheviks and nothing is known about him. Probably dead.' Wilfrid tried to keep in touch with the niece, Zocha. Erla made efforts on his behalf to see her when she was in Europe. A year later he was involved in good deeds again, 'relieve work [sic] in Poland and east'.

Wilfrid fancies himself as a historian and occasionally when business takes a back seat in his letters to Edward Levetus he develops his theories. Two things emerge strongly. One, unsurprisingly, is his lingering distaste for all things Russian. It is the US, he writes in 1918, that has saved the world from 'Russian barbarism and the thin coat like cheap varnish which covered Asiatic souls and Asiatic bodies ...'. Two years later after the revolution the Bolsheviks have improved nothing with their 'famine and universal madness'. A further two years on he is predicting economic and political domination by Russia as imperialistic as under the Tsars. His allegiance lies squarely with Poland, free from Russia at last – but in 1926 he writes sadly how 'imbecility reigns there too ... It is a wonderful germ, thrives on nothing.'[24] He is referring here to the *coup d'état* of his Tunka camp companion, Joseph Pilsudski.

Ethel and Wilfrid returned to the UK on the Cunard liner *Covonia* for a three-month stay in August 1922. She wrote of missing home but her stay in America, despite returns, was to be permanent; she wants to be 'within reach of my man'. The two were in Britain again in the summer of 1925. Ethel is worried about Winifred's education at a Quaker boarding school in Suffolk. She has time for a holiday, writing to Esther of having 'a glorious time in Ireland', while Wilfrid is elsewhere, 'well and as busy as ever book-hunting'.

Her main preoccupation is music, revising her choral works *Babylon* and *Jerusalem* and enrolling in 1923 at the Manhattanville College, Pius X, School of Liturgical Music in northern Manhattan to study plainsong. She moved to Staten Island in 1927 presumably to make travelling easier. This connection with a Catholic institution may seem strange given her atheistic tendencies, but for her, as with party politics, ideology is less important than the pure aesthetics of medieval polyphony. A nun at the college tried to convert Ethel but 'she gave up when she saw that it was useless'. For her, as with her father, George, a church was a good place to hear music. Her librettos were anything but in praise of religion.

By 1926 *Babylon* was finished, of which she un-selfpityingly writes to Esther Pissarro: 'Nobody is likely to perform it I assure you, at any rate not in our life time, but that is a secondary matter.' The disavowing may have been a little over-hasty; one choral part for forty voices had been performed publicly in 1925. She writes to Erla Rodakiewicz in August 1928 that 'orchestration – the most elusive, slyest, wariest of games – has this last year grown tame enough to eat out of my hand'. Well-known musical figures Phillip Hale and Carl Engel had been trying to get a conductor to undertake the work. A year later she writes again that the music publisher C. C. Birchard of Boston, who had Ives, Varese and Copland on his books, expressed an interest in *Babylon*, 'your formidable manuscript'. She had composed yet another large-scale work, 'that has been hanging about since before the war ... a symphonic poem for full orchestra and wordless choir called "The Riders"'. True to form it's anything but flippant, based on the *Four Horses of the Apocalypse*.

Ethel seems to have been removed from the thick of things. In September 1927, for example, she is head down in her music away in the Catskills. Her grand-niece Jean reports that Ethel would spend time wandering in the mountains looking for wild flowers and snakes. 'She used to turn pale and angry every time she found a snake crushed by a car ... especially if it had not been killed but left to die slowly.' While out swimming with Jean's sister, Joan and brother William, Ethel 'interrupted our aimless wanderings by asking us if we noticed all the currents in the stream. Before she was through with us she had brought to light three separate currents moving at different speeds. We were amazed and fascinated.'[25]

Wilfrid shielded her from his affairs where possible. Earlier in that year he had suffered from pneumonia and business was not good; he could never establish a reserve of funds and was always behind servicing bank debt. Although he needed to replace stocks from Europe, he and Ethel had to return every year, at some expense, for US visa reasons. In 1927 a zealous official had reported their overstaying resulting in four weeks' notice of deportation. Voynich wrote to Edward Levetus on 7 January 1928, however, that he had 'used very powerful backing and visited twice Washington', securing a retrospective right of residence.

Wilfrid's own lifestyle, opposite to Ethel's, must have eaten into any profits. He always lodged at the best hotels and while no doubt offsetting any wooing of clients against tax, lived the life of a city slicker. He writes to Levetus on New Year's Eve 1925 that people this season 'will drink as never in its history. Prohibition agents looking that this done decorously and "legally". At his hotel there will be no chance for sleep – 3,000 seats for supper at $15 per couvert.' The week after he is entertaining colleagues, 'so after heavy days work instead of bed will be galavanting'. In an undated letter headed from the impressive Hotel Louvois, Paris, he addresses the Levetus couple as 'Two dear Vagabonds' and suggests, 'after business we can go to any wild Russian or other cabaret'.

Wilfrid's health was deteriorating, however. Headed 'Confidential and Personal' to embargo Ethel, he writes in November 1929 to Levetus, 'Dr's tell me that I am desperately ill … any moment I can get tuberculosis.' He had taken a six-week vacation previously in the sun, to recuperate with a business associate. After making contact with a club of millionaires in Minneapolis he continued a strenuous tour across northeast USA. Fortunately, despite the Wall Street financial crash, 'business is good, all big transactions on long credit and managed by Miss Nill who is brisk, has brains and worked like devil for three persons'. Other news mentions that he had been elected honorary member of the American Medical Library Association. It gave him additional prestige, 'and that means money in the near future …'. The last letter to his stalwart Levetus friends of 25 February 1930 was optimistic as to his health: 'with my constitution and habit to fight I am sure I will win'. He ends, 'I love you two.'

This had been written from a sanatorium in South Carolina. Anne Nill, who had obviously been shielding Wilfrid from their true financial status, on 14 February 1930 had written a begging letter to the Levetus family asking for help to tide business over and 'to relieve Mr Voynich of one worry at least'. Everyone is trying to help out. Erla, who had left for Europe in 1929 for an indefinite period, had returned to see Wilfrid, offering to raise money. Ethel is rushing to finish her translation of the *Chopin Letters* to bring in a little cash. There is talk too of selling the film rights to *The Gadfly*. Erla suggested approaching film maker Robert Flaherty, but Paramount is apparently scared of likely religious controversy.

Wilfrid's cause, however, is hopeless. At first doctors diagnosed a treatable infected gall bladder. Ethel writes of the news, 'I am as cheerful as a cricket … despite ruinous expenses.'[26] Doctors had, however, been completely fogged by a rare and obscure form of abdominal cancer. Wilfrid died peacefully, aged sixty-five, on 19 March 1930. Alongside his bed were

Ethel and three women friends: Anne Nill, Mary Marquis and Erla Rodakiewicz. At one point he murmured, 'All friends, all human beings'.

This is in some ways a suitable epitaph. Voynich was a maverick adventurer of great charm who won people over and bred confidence. Erla had written to him two weeks before he died about their friendship, 'No one truer, more faithful, half so kind and forgiving.' She reminded him of how they argued 'about life and politics and ethics and institutions. But we also knew how to play and laugh. What jokes we made in the days before we grew old and serious.'[27] The exuberance was coupled with what seems to have been a genuine compassion for others and an awareness of the world's ills that had made him a revolutionary hothead in his youth and a continuing loyal supporter of Polish freedom.

He was headstrong, energetic and ambitious but like many of that ilk essentially a loner and not a team player. This emerged way back in the 1890s when he became unpopular while working with Stepniak's Russian circle, who accused him of vanity. Even before that as a twenty-year-old, despite his bravery, he typically claimed unjustified membership of the central group of *Proletariat*, the Polish socialist group. His bragging could make him unpredictable and appear suspect when embroidering the truth. His decent self was shadowed by the cynical *doppelganger* who promulgated the virtue of 'the need for sin to advance'. Millicent Sowerby, who attended the funeral in New York, never forgot her two years working for him in London, believing him to be 'as immortal as any human can be'.[28] I suspect, however, that more about the unreliable side of his character will emerge one day. For the moment his immortality does indeed seem secure for as long as the *Voynich Manuscript* remains unexplained.

What had kept Wilfrid and Ethel together for forty years? A more unlikely couple it's hard to imagine. Ethel is the serious-minded, inner-driven, artistic creator at home in concert halls, wild natural places or rockery gardens. Wilfrid is the flamboyant romantic at home in city bars and his London club, posing and haggling for money and status. Both, however, are outsiders, outside their own families and outside each other in their separate lives emotionally and physically. He was certainly attracted and attractive to women but professed fidelity. Their shared loyalty to each other was born of a sense of injustice in a 'beastly' world that found them both courageous in their younger lives. The years of the Stepniak circle must have welded an inseparable bond, drawn tighter by their mutual expulsion from it.

In his will Voynich left the *Manuscript* jointly to Ethel and Anne Nill.

The 'ugly duckling', however, had still not obtained full plumage and was valued at only $14,900. His estate was divided 60/40 between Ethel and Anne. Nill's allocation was presumably recognition of her loyalty and hard work over many years, but also perhaps a recognition that she had become 'one of the family'.

Ethel tried to arouse interest in the manuscript among various circles, sending a copy to the Catholic University, Washington where several people worked to identify the 'herbal' plants, concluding that some were European and others New World. Conscious of her role as co-guardian of an important item of heritage, she drew up a letter to be opened after her death acknowledging that Wilfrid had found the manuscript at Villa Mondragone, Frascati in Italy and gave basic details of the purchase. The *Voynich Manuscript* was placed in a bank vault far from the gaze of the public.

# THE BIGGLES OF COMMUNISM

Ethel returned to England soon after Wilfrid's death. She was perhaps feeling homesick and missing her adopted daughter, Winifred, and her two remaining sisters, Margaret and Alice. She stayed with Geoffrey Taylor, her nephew, with whom she shared a strong mutual affection. On her return Ethel and Anne Nill took up residence at 333 E. 43rd Street, moving on to 45 Prospect Place in 1933. We can only speculate why she didn't stay on in England. She writes to Esther Pissarro in 1931 describing New York as 'the biggest and most detestable of cities in the world in the worst year on record with banks smashing, businesses at a standstill everywhere, breadlines and unemployed apple sellers in the streets …'.[1] Her deep attraction to the natural world and wild places seems hard to reconcile with living in a skyscraper apartment, even though she kept a rooftop patio garden.

There were business concerns, however. Both the New York and London offices were kept going for a few years despite decreasing book sales generally. Apart from the Bacon manuscript and two valuable Valturius manuscripts, they had sold much of Wilfrid's book legacy to survive. Perhaps Ethel's attachment to Manhattanville College suitably reinforced her musical ambitions. She took up teaching part-time there in 1934 and ran an annual summer school. Her Library of Congress files list, as well as her major works, wide research notes on music from many cultures and periods. In the same year she was teaching immigrant children at the Henry Street Settlement on Manhattan's lower east side, the worst slums in the city. She also offered help to a 'negro' choir in Harlem. Although not avowedly a socialist, Ethel shared the Boole/Hinton dynasties' sympathy with the underprivileged, involving themselves in the Settlement idea.

One other reason for staying might have been her attachment to Anne Nill. As a couple, sharing incomes, the bond certainly lasted their lifetimes after Wilfrid's death. Anne was a rather stern-looking woman, short-haired and heavily bespectacled, some thirty years younger than Ethel. She is always referred to primarily for her hard work. Whether or not there was more intimacy between them than companionship is hard to say. There may have been an element of hero worship on Nill's part. Ethel's sexuality seems to have been a redundant subject – although as we shall see it re-emerged as something of a scandal after her death.

Despite the long period of her concentration on music around 1935 she began work on another novel, dedicated to Anne Nill, one that seemed to come of its own volition. She wrote in the foreword to *Put Off Thy Shoes*, eventually published in 1945, of having being impelled: '... these bodiless offspring ... have come and gone ... I have stumbled after them as best I can.'[2] The novel completes a trio of 'Gadfly' histories looking back even further in time than *An Interrupted Friendship*. I had put off reading this late novel; somehow the old-fashioned title with its biblical reference just didn't entice. I didn't expect it would reveal anything new about the author, especially one writing in her seventies. Despite its continuation of well-worn themes, I was wrong. (Lesson to biographers: mine every scrap of information about the quarry.)

Things deep down in her subconscious must have needed an airing. One can safely assume that the story of the central character, Beatrice Rivers, followed from youth to old age, contains strong autobiographical elements. Anne Nill told Taratuta that 'Beatrice gave her no peace and was constantly talking to her'. Ethel had never known her father, George Boole, a loving and bookish man who died when she was a baby. Beatrice's father, of a similar disposition, dies when she is twelve, his blindness acting perhaps as a transposing sense of Ethel's own loss and unknowingness. Beatrice possesses an older brother and a younger sister. Their mother is fickle and adulterous, marrying in turn the rakish ne'er-do-well, Jack Carstairs. Similar to *The Gadfly*, the novel's three parts would serve as an opera.

Enter the somewhat boorish country bumpkin Henry Telford, owner of a Midlands farm, looking for a wife. During the London season of 1763 he meets and is encouraged to woo the highly reluctant Beatrice. She is a good catch: educated and, like Ethel, 'not a beauty ... but passably good-looking'. In a reprise we learn why there is undue pressure to see her wedded.

While her mother had been in town Jack had attempted to gag and rape his stepdaughter but was repulsed by 'a finger-nail driven deeply into his eye'.[3] The mother, herself prey to evil because of the family's tainted blood, suspecting what has happened, realises she can now keep a hold on wayward Jack via blackmail. Secondhand Beatrice needs disposing of. The episode continues Ethel Voynich's fascination with violence but marks the first explicit recognition of sexuality in her novels and its potential destructiveness. This reinforces her ever-present distrust of a benevolent godhead; Beatrice acts as a mouthpiece. After reading *Gulliver's Travels* she realises that the world is dominated by Yahoos, the verminous, monkey-like creatures that Jonathan Swift adjectivised as cunning, malicious, treacherous, revengeful, insolent, abject and cruel. She concludes, 'clearly whoever had made the world liked it that way. Just a bigger Yahoo.' Despite the relative kindness of her husband, defiled and faithless Beatrice is trapped in a hideous marriage, but, she reflects, 'wasn't all life hideous ... who was she to complain of defilement when everything was defiled'.[4]

There is some respite, however, as Beatrice, like Ethel, finds beauty in her English garden with its 'roses, jessamine, honeysuckle and traveller's joy'. Henry tries to be patient with his frigid wife, feeling as if he is treading on a violet. Beatrice muses despairingly, 'what else are violets for?' Page after page of the novel recounts Beatrice's rage at woman's lot as she dialogues in her imagination with another tainted female ancestor embodied in a family painting. Should she kill herself rather than submit to sex? No, says the visage, 'You'll howl at first but you will be fruitful as your owner chooses.' The image of fertility recurs to Beatrice as a kind of medley from a 'shepherd's calendar': 'a woman flowers from the innocent daisy, the virgin lily, the modest violet, the blushing rose ... and next the fruitful apple-tree'. And after? 'The sour withered worthless old crab. And last, no doubt, the rotting carrion fungus.' The cycle brings her mind back to 'the emblem of every triumphant male: the stink-horn ... the filthy thing'.[5]

There's something highly unsettling about the vehemence of this erotic exposition from the pen of a white-haired woman in her seventies. From what source does it emanate? Her own childlessness, her adoption of a daughter and separate living from the man of her life might indicate an aversion to heterosexual coupling, as too her long association with a woman, Anne Nill. None of her novels feature a tender or romantic attachment between men and women leading to motherhood and family life. Just the reverse: she speaks of the explicit exploitation of women, 'the only two things in female life that were real under the shams: sex and childbirth'. Had she finally given vent in literature to feelings she'd long repressed? Could the sexual repugnance originate in her own intimate personal experience?

Despite Beatrice's deep recoiling from Henry with his 'gross desires and no brains', she eventually bears four children. Her dreams when 'all faces leered and inanimate objects turned into phallic objects seldom tormented her now'.[6] When her finally abandoned, alcoholic mother ends her life by taking rat poison, the stillbirth provoked in Beatrice by the shock provides a medical escape from any future breeding. Convalescing, she is able 'to watch with a new-born reverence the ecstatic marriage-dance of insect, squirrel and bird ... able to manage reproduction without indecency ... luxurious yet not obscene. Although Spring-time would last so short a time... that was a blackbird singing and crocuses were out in the grass.'[7]

This very much echoes Ethel's own feelings about nature and life. The marriage rubs along although Henry now finds a sexual outlet in wenching. His doctor sympathises with this choice because, perhaps in reference to the author, 'clever women are sometimes a little cold'.[8] Beatrice turns out to be a competent mother and organiser of their well-run farm, remarking, two centuries ahead of her time, 'If I were his housekeeper not his wife, he would have to pay me wages.'[9] Rustic tranquility cannot last, however: her favourite son, Bobby, aged nine, while flying his kite is charged by another rampant male symbol in the shape of an escaped bull. Beatrice throws herself over her son to protect him but is tossed aside; Bobby is gored to death.

In the next act the scene changes. The setting again is north Cornwall, a clear autobiographical link to Ethel's childhood and long stay there during the First World War. Beatrice goes to convalesce there and diminish 'the onset of the melancholia that was in their blood'.[10] Ethel's own musical perspective phrases her heroine's recovery as Beatrice absorbs 'the intermittent counterpoint between the ecstatic descant of the skylarks and the endless ground bass of the surf'.[11] Discord is never far away, however – the local fisher-folk peasantry live in poverty, dependent on the fortune of the catch and the rulings of their estate landowners.

Ethel, always a friend to the poor, describes a cottage Beatrice visits with its 'leaking roof and bulging walls; broken windows stuffed with rags ... a fringe of patched and frayed washing which dripped forlornly onto an uneven floor of beaten clay'.[12] She is there to commiserate and help Bill Penwyrne, who, risking his life and losing his vessel, had saved the lives of two of her sons messing about in boats trying to follow the pilchard catch. Voynich deploys a full-frontal use of Cornish dialect. 'Yeur lousy bastards baint the fust lives I've saved. An' wot may yer graatitude be woarth, eh?'[13] rages Bill. Her charitable response is not only to provide the family with a new boat and cow byre, but also to offer their precocious son Arthur a life of advancement with her. The boy is to be her consolation for the loss of Bobby. Once again, the theme of adoption is used by Ethel.

The third act sees a softening of the novel's tragic inclination, consistent perhaps with Ethel's exhausting of her own inner demons. Husband Henry continues his mild degeneracy, but young Arthur strikes up a bond with their daughter, Gladys. In Cornwall, where the adoptee returns for his summers, Bill and a mate are now making out with their new vessel. A complicated subplot involving Beatrice's brother and his disastrous marriage bring us to her last days. She does not reveal to her family that she is dying from cancer but offers her views on life to the children's tutor and, one suspects, Voynich's latter-day philosophy. There is a ranting pessimistic logic that inverts the anarchist utopian sentiments of Ethel's mentor, Charlotte Wilson.

'Man is an ugly animal ... in the bottom of our hearts we do not desire justice or beauty; we cannot breathe in their presence ... our brutal laws and customs are only a reflection of our own minds ... the churches, creeds and moral codes have made humanity criminal ... by making it unhappy ...'.[14] Plans for the regeneration of mankind are nothing more than beautiful fantasies – yet the French Revolution is in the offing. 'The whole world is full of desperate people ... something is beginning everywhere, boiling up from underneath ...'.

Beatrice, aware that in England, like France, big estates have been won by robbery and kept by force, asks, as Ethel did during her Stepniak days, 'Massacre, will that help anyone? Happiness is the only thing that really matters. People's souls flower under it just as grass turns green under sunshine.'[15] Beatrice's own soul has never fully burgeoned, yet as her vitality ebbs there is an acceptance even of the sensual. 'What a fuss we make about the body, as if that were what mattered, perhaps after all it's not a yahoo appetite, a thing of the primeval mud.'[16] Beatrice dies acknowledged by all as a perfect wife – despite her own inner distaste for motherhood. For Beatrice, unlike Ethel and her music, there has been no ongoing passion to be fulfilled. Arthur and Gladys receive a farewell blessing at Beatrice's bedside – the saga will continue into the next century and the dramatic tales of Arthur as the Gadfly.

Put Off Thy Shoes was received by critics with some acclaim, The New Yorker even suggesting it had the qualities of Gone with the Wind. Another praised its intense vigour and skill. The Cornish setting is vividly evoked and could have indeed been conveyed well on screen. But the novel is overlong, despite its narrative drive. Beatrice's visceral hatred of sex and maternity is never quite adequately explained. The novel's professed link to the other two of *The Gadfly* trilogy seems to be, in Voynich's own words, 'scarcely more than names'. The 'taint' of the maternal heredity is redolent of the real-life Boole/Hintons but not worked through. The *raison d'être* of Voynich's last novel lies perhaps in the author's need for catharsis. The essential Ethel, in my imagination, stands at the edge of a Cornish cliff, as the sun sinks, petulantly waving a fist of wilting flowers heavenwards.

At the outbreak of the Second World War Anne and Ethel had moved to an apartment at 450 W. 24th Street, where they were to remain for the rest of their lives. Ethel, at a distance, is horrified by the new hostilities. She writes to Esther in September 1939, 'Well, now the sky has come down on us all. We are reverting ... reverting. Neither war nor any other kind of ruin can make us anything but what we are. The only real compulsion is the inner one. I am glad that my husband died before all this ... and other folk that I have loved.'[17] The compulsion was being realised at least for her in teaching and performing; in December 1941 she was involved in a Christmas concert for 100 singers and an audience of 500 poor children, performing a huge polyphonic mass by Palestrina.

She hears about the war and her nephews: Leonard, Alice's son, is back from Libya and Egypt but there is worryingly no news of Maggy's son, Julian, since the fall of Singapore. Geoffrey, her other nephew, presents himself twice in New York; he is in the US, as we shall see, on secret 'special work'. In January 1946 Ethel wrote to the Levetus family how she and Anne seem to live in a backwater. Ethel is revising and arranging her music scores for publication and posterity; Anne returned to another dull public library job.

They live in genteel poverty but Ethel is not unhappy, writing, 'I never hoped for such good luck in my old age.' Her flat, number 7 on the seventeenth floor (strangely, numbers redolent of her St Petersburg stay), is part of a drab, monolithic block in the Chelsea district. It consists of two small bedrooms, a dining room, a study full of books and a piano. A letter to the Levetus family of February 1949 mentions that a small legacy will allow the piano to be re-keyed and the sofa repaired. She writes to Esther in 1948, 'I have a few pot plants – they stay healthy with a weekly sluicing in soapsuds to get off the incredible NY soot and grime.' The New York winter is harsh but her friends the Davidsons in Florida invite her to stay. The eighty-two-year-old Ethel writes, 'I wade in the warm surf and lie on the smooth sand to watch enormous brown pelicans swooping overhead.' The family run a newspaper, the politics of which she approves: decent government, education and elementary justice to the black population. American friends visit her and some of her English family; Winifred stayed in 1947, not having seen Ethel for ten years.

In 1947 she travelled to visit her American grandniece, Carmelita Hinton, in Vermont, and writes later to Esther with obvious pride about the three Hinton children as

> solemn young crusaders. Joan has chucked a promising career … and is now on her way to China … as she puts it 'to do something besides inventing new ways of killing people'. Bill, already there [is] working … behind the fighting lines … teaching Chinese peasants to use tractors, ploughs etc. Jean has just gone off to a big Pennsylvania town to work as a factory hand and preach her particular form of social gospel to the employed girls.

Ethel spent another week in 1949 with Carmelita at her Putney School in the Vermont mountains. By now her overall health is deteriorating; she needs an escort to leave the apartment owing to her increasing arteriosclerosis. Her hearing is failing. She relies on Anne, of whom she pays tribute to Esther in November 1949: 'She has given me for nearly twenty years a selfless devotion that amazes me.' Her old friend, Esther Pissarro, finally managed to get to New York to stay in 1950 but died the following year. Their letters reveal a strange – affectionate but carping – relationship. Ethel's slight but ironic humour is often characterised in letters: 'You are an exasperating person but I love you. If you don't write … you may expect the rough side of my pen – alas – you cannot have that of my tongue. Aren't you glad?' Eventually housebound, she continued with the main aim in her life: to revise her music and see copies lodged. She was completely unaware in 1955, aged ninety-one, that her sequestered apartment was about to be turned into a veritable shrine.

Her novel *Put Off Thy Shoes*, after its long gestation, had finished a trilogy that must have seemed finally done and dusted. The most successful part of it, *The Gadfly*, had long been forgotten in western bookshops. Its old-fashioned high drama had long been replaced by more modern styles. Triumphalist heroism, however, still reigned beyond bourgeois shores. In tsarist Russia, *Ovod* (*The Gadfly* in Russian) continued to be popular enough to be considered seditious. After the 1917 revolution, however, it was issued by all the main Soviet publishing houses in thousands of copies. The novel's lauding of heroic sacrifice and revolutionary ideals and the condemnation of religion was taken up genuinely in public consciousness and manufactured as propaganda.

The Gadfly became a favourite book, for example, of the fictional hero Paul Korchagin in the hugely popular novel How the Steel Was Tempered by Nikolai Ostrovsky.[18] A film of The Gadfly was begun in 1919 but abandoned. In the 1920s a stage version was produced, two operas were written, one of which was performed in 1929. In the preceding year a film reached the public. Ovod, translated now into many of the languages of the polyglot Soviet Union, possessed a wide popularity inspiring citizens to name their children 'Arthur' and 'Gemma' and their boats 'Ovod'. Even more remarkable than its genuine acclaim was the fact that no one then or in the USSR of the 1950s actually knew who had written it.[19] More stage adaptations were made: another opera in 1957 and perhaps most persuasively, a full-bloodied film by Alexander Faintsimmer in 1955 with music by Shostakovich.

Evgenia Taratuta, Ethel's Russian biographer and devotee, published an article in April 1955 about her, having consulted Who's Who, and controversially suggested that she was still alive in New York. Taratuta approached Boris Polevoi, a journalist there, to find out more. At the same time a UN Soviet member, Peter Borisov, mentioned the article to a friend, Anne Fremantle, who was learning Russian from him. She was intrigued to find that she knew nothing about Voynich or the novel. Fremantle was blown away by The Gadfly when she read it and hastened to make a visit to the author. Borisov visited some time later taking along his young daughter, Irina, who obviously took to the old woman. A photo in a later Look magazine shows a dignified, neat, white-haired Ethel sitting listening attentively to a young girl dressed in her best caped dress with a large bow in her pigtailed hair. On the wall is the picture of the youth in black she had first encountered in the Louvre so many years before.

A more formal visit had taken place on 17 November 1955 when seven Soviet journalists (those numbers again) presented themselves at the apartment with flowers. I leave it to Taratuta to describe the meeting:

What an amazing evening that was, when the sensational, unparalleled fame and love of millions of far-off people gushed into that tiny flat on 24th Street. They gazed with reverence at this small, thin woman who

Ethel Voynich, in her New York apartment, receiving flowers from a Russian girl.

slowly told them about her youth and her life in St Petersburg, about Saltykov/Shchedrin's funeral and of how she used to take parcels to the political prisoners in the detention house … '[20]

Ethel found herself, in her nineties, deified in a land that she hadn't known for sixty years and had changed beyond recognition. Over two million copies of her novel had been sold; millions of people would have borrowed and read a copy. She had assumed that her own 1913 Russian edition of *The Gadfly* had been the last published there. In China too the novel was enormously popular and in the early 1960s Faintsimmer's film was shown widely and loved.

Following Ethel's discovery, the front page of *Pravda* printed the headline, 'Voynich Lives'. Resurrected like the Gadfly himself, she continued to receive a huge amount of tributes, letters and gifts – one rather grisly: a battered copy of the novel found on a dead Russian soldier in the Second World War. Most welcome of all perhaps was a 1956 royalty payment of $15,000, a previously unheard-of transaction. Ethel worried that with a headline in the *Washington Post* on her ninety-fifth birthday, 'Reds Honor US Woman', this might be misconstrued in McCarthyite times. A film projector was set up in her apartment to show the recent film of *Ovod*. When it ended, expecting approbation, Ethel was embarrassingly quiet until, ever truthful, she said, 'It's not like that at all.'[21]

If Ethel rejected the Hollywood treatment in the theatre version of *The Gadfly* back in 1898, she must have felt much the same about its reduction to revolution rousing. The film is shot through with stylised images of the military presence of the oppressors, the opiate glory of the Church, the cloak-and-dagger bravery of the nationalists and, especially, Arthur's execution. These are constituents of the novel but this overlooks the other larger components of love, betrayal and faith. For Soviet pedagogues the political message was its most important attribute.

Taratuta went so far in a letter to Ethel to talk of religion being 'dead'. She must have been shocked at her indignant reply: 'I consider that the most vital requirement of the human soul is the need for some kind of faith. In my youth I was at one time a violent atheist ... for more than half my life I have been ... a kind of pantheist.'[22] This rather woolly term (applicable perhaps to her mother, Mary and father, George) would have meant nothing to Taratuta. Although *The Gadfly* pilloried religion, as with the other novels, the underlying theme consists of how individuals cope with existential suffering, not primarily political struggle. Taratuta has to maintain a party line, however, and insists upon the novel's anti-clericalism. Those of faith evince a 'dusty mouldiness', compared to the rich intellects of the atheists. She compares Voynich to Dostoyevsky and finds him amiss. In the former, 'everything is simpler and more integrated'.[23] Voynich she proclaims has created 'the immortal image of an immortal hero ... our friend and comrade in arms. As long as there are

Ethel being shown a poster announcing the performance of an opera based on *Ovod*.

people who devote … their whole life to the revolution the Gadfly will continue to live on earth.'[24] The rigidity of Soviet triumphalism would have appalled Ethel.

The enthusiasm for *The Gadfly* continued unabated in the Eastern bloc. A third opera premiered on the fortieth anniversary of the October Revolution in Perm in 1957 and *Put Off Thy Shoes* was printed. *The Gadfly* continued to be issued in huge quantities: as of January 1980, Taratuta reports that it had been published 180 times in twenty-four languages selling 11 million copies. Even in the West there was revived interest. The writer of the *Look* feature of July 1958, Laura Berquist, socked it to her readers declaring that it had 'not a trace of Marxist doctrine and as a whacking well-told adventure story … outclasses the turgid propaganda novels that pass for Soviet fiction'. Soviet authorities did, of course, use *The Gadfly* in that role, set widely as a school study text. The highest accolades of all came from the stratosphere: cosmonauts Yuri Gagarin and Valentina Tereshkova cited the novel as their favourite; a minor planet, 2032, discovered in 1970, was named 'Ethel'.

On 27 July 1960, aged ninety-six, Ethel Voynich died peacefully of pneumonia in the apartment on W. 24th Street that she had shared for so long with Anne Nill. Her ashes were spread in Central Park. She left her estate in order of precedence to Anne Nill, Winifred and Geoffrey Taylor. According to a letter from Nill to him she was spared a likely leg amputation as a result of spreading gangrene. Ethel's attitude to the hereafter was summed up, according to Nill, as 'The dead body is only old clothes.'[25] This blunt realism is certainly part of Ethel's make-up: a dour, no-nonsense recognition that we live, however we strive, in a vale of tears.

In a touching part of the letter she mentions that an old friend, Olga Sutcliffe, a botanist, enjoyed a special relationship with Ethel, 'based on their love of tiny spring wild flowers'. If Ethel never quite forgave God for his mess-up in not being able to make an octave pure, her delight in the transitory but beautiful must have been a great consolation.

A very interesting letter dated 5 May 1924 from Geoffrey Taylor to his cousin George Hinton in Mexico says a great deal about her and the Boole/Hinton dynasties as a whole. Apparently George and Ethel had quarrelled, perhaps by letter, about something or other. Taylor states that he has quarrelled with her likewise, sometimes violently, 'but I still like her very much indeed'. He continues, 'she is a very abnormal person … she sees certain things frightfully strongly and is completely blind to others. She has the real fire inside her. Her want of judgement and balance is a thing which she inherits from our grandmother [Mary Everest Boole] … All of us Booles have this deficit in greater or lesser degree.'[26] He describes it as a 'taint' and suggests it is there with the Hintons too. James Hinton and Charles Howard certainly fit the bill. Ethel, of course, had used the same word in *Put Off Thy Shoes* in relation to the portrait of the cynical woman ancestor who literally hangs over her descendants.

Having, I suspect, exhausted herself of crusading to right social evils at the defining time of Stepniak's tragic death, it was only her creative drive, still compassionate, that motivated her afterwards. On a grander scale the wildness of Cornwall and the Catskill forests provided a soothing consolation given that no one person could make her complete. Wilfrid was her mate probably mainly out of loyal friendship. The final ten years of her life, adored at a distance by millions, must have somewhat gratified her need for recognition. The real crowning glory would have been the recognition and performance of her music.

Nill writes, 'It is not I'm sure the end of ELV in this world … one day though not in the near future, her genius will be fully recognised.'[27] Sadly, neither in her five novels nor her many musical compositions has that proved to be the case. In 1967, Raymond Leppard, a principal conductor for the BBC, was asked to look at some. He noted that they were 'very simple, not inexpertly written … the style is restrainedly religious'. This might suggest that they are not major works but motets. He noted how difficult it would have been for a woman composer to make her way in earlier times but thinks she was 'probably more talented' than another Ethel composer, Ethel Smythe. 'They would perform still if a suitable occasion presented itself.'

Sidney Reilly (1873–1925), the 'Ace of Spies'.

Not long after her death in 1960 an appraisal was made of her life and most famous novel completely opposite to that of her lauding in the communist world. It represented one more episode in the Cold War. On page 23 of Robin Bruce Lockhart's *Ace of Spies*, published in 1967, he recounts a story told to him by his father, Robert Bruce Lockhart, who in 1919 had met that most notorious of spies, Sidney Reilly – the putative inspiration for James Bond. Lockhart senior and Reilly were both at that time engaged in Moscow on a British mission to assassinate Lenin. Reilly had told him how, aged twenty-two in 1896, he had journeyed to Italy 'with a woman a few years older than himself who was just beginning to make a name for herself in London as a writer. There under the Mediterranean sun … Sidney bared his soul to his mistress …'.[28] The young lovers visited Elba but in Florence he deserted his mistress and returned to London. Reilly's amour was none other than Ethel Lilian Voynich.

The liaison was attested by George Hill and Sir Paul Dukes who, like Reilly, were working for British Intelligence at the time of the Bolshevik Revolution. Not only did Reilly have an affair, he also claimed that after the First World War he met her again but was disillusioned by the meeting. His third claim was that the biographical details of Arthur Burton, the hero of her novel *The Gadfly*, were based on his life. Like Arthur, he too had discovered the treachery of his father, faked a drowning and stowed away on a ship to South America where he suffered the same indignities that befell the fictional hero.

For the sensationalist media of 1968 looking for ways to embarrass the Soviet Union this story scored very highly. In the *Daily Mail* Gilbert Lewthwaite seized on the essential story as expounded in *The Spectator* (17 May 1968) by Tibor Szamuely:

> The final twist of fate has just caught up with Russia's number one popular fiction hero The Gadfly … the Biggles of communism. Szamuely has established the almost unbelievable irony that the Gadfly was none other than Sidney Reilly the British agent who tried to overthrow the Russian revolution.

The BBC joined in the fun with a World Service Russian language programme on 9 June pointing out that all Gadfly inspirations in the USSR would be now hollow indeed. Many subsequent versions of Ethel's life have paid lip service to the story.

Biographers of Reilly have not been helped by the lack of reliable information about his life. Michael Kettle in *Sidney Reilly: The True Story* (1983) has him born as Sigismund Rosenblum in 1874 of Polish/Jewish stock in Grodno, then a Lithuanian province ruled by Russia. Richard B. Spence's *Trust No-one: The World of Sidney Reilly* (2002) affirms the Grodno lineage and has him fleeing to France from Odessa in 1893 after an incest scandal. Andrew Cook's *Ace of Spies* (2004) repeats this story and traces Reilly's arrival in London from France in December 1895 after being involved in a murderous robbery on a train. What is essentially relevant to the truth of the Ethel Voynich story is whether any opportunities arose for Reilly and Ethel ever to meet, allowing him to pass on the story of his life – true or not.

We do not have anything like complete details of the continental travels of either of the Voyniches. Both of them visited various cities meeting other radicals, for example en route to Switzerland in late 1893. In 1894 they were busy setting up their rival League of Book Carriers smuggling anti-tsarist material abroad. Both Watson and Cook assume that Reilly mixed in such circles at that time, known to the French security services as Leon Rosenblat. Plausibly, Ethel and Reilly's paths first crossed at that time or in early 1895 on her journey to meet Ukrainian activist Michael Pavlyk and back smuggling books. According to her own accounts she left England in mid-April for Florence to work on *The Gadfly* and stayed for four months there and in Bologna. On her return she wrote to Pavlyk, 'I had absolutely no time to see anything in Italy, spending all my time in library archives or writing in my room.'[29] There is no intrinsic reason, however, to assume that she couldn't also have met Sidney Reilly there and to have travelled further afield with him. Most of Reilly's biographers have no information to locate his whereabouts at that time. Based on Lockhart's assumption that Reilly was born in 1874, however,

Ethel Voynich, Florence, 1895.

Ethel Voynich, Florence, 1898, aged 34.

his Florentine twenty-second year would indeed be a year later in 1896. Cook, however, based on British intelligence records, puts it at 1873 – giving the appropriate year 1895.

Is it likely, one wonders, that the couple would have been drawn to each other? Reilly was a notorious womaniser. Ethel may have been somewhat serious but was not unattractive. Millicent Sowerby described her fifteen years later as 'a beautiful English woman with a glorious head of golden hair'. Her attraction to him is more problematic from what we can glean of her sexuality, yet consider her attachment to Wilfrid, another suave, romantic adventurer with radical credentials. In Chapter 10 we noted that there were hints of attraction to underground radical, Pavlyk.

It is perhaps not so far-fetched to imagine her amid the romantic glories of Florence falling for a Mazzini-like figure with the glamour and wiles of Reilly.

Anne Fremantle in 1955 had met and interviewed Ethel at her New York flat several times and published articles on her and *The Gadfly*. Years later in 1978 she was researching a biography of her. In the process she contacted Jean Rosner (Hinton), Ethel's grand-niece, who was also collecting material on her life for a family archive. Jean's inventory listed xeroxing some photos for Fremantle that had come to her from Geoffrey Taylor after he died in 1975. Among them was one taken of Ethel in 1895 by the famous Florence firm of photographers, Fratelli Alinari. It portrays Ethel looking serious but demure in a crisp, high-collared white blouse with ruched sleeves sporting a large spray of roses. She has a far-off look compared with the no-nonsense photo taken three years later in Florence in 1898 when she and Wilfrid met Erla Rodakiewicz. At that time she was wearing a drab apron dress and looking a lot less dewy-eyed. Most importantly, however, on the rear of the Alinari in her handwriting is the legend in Russian, 'The Gadfly's mother to the Gadfly's father from the Gadfly's country.' One can read many things into a photograph perhaps but the message is clear enough: it can surely only refer to Reilly. This does not mean they had a steamy affair – although Reilly claims that they did. It does add up: unbelievably, something passed between them.

The actual photo with the writing seems to have disappeared like so many Boole and Hinton records. Fremantle's copy has not survived either. The fact that Fremantle never mentioned the liaison publicly is presumably because she subsequently lost interest in Voynich as a subject. Perhaps also she felt it etiquette not to do so. In the draft of her biography of Ethel, however, she is clear enough: 'Above all the liaison is confirmed by the photograph taken by Alinari.'[30]

Reilly was, of course, notoriously devious. Was he just bragging about his sexual prowess to his mates Lockhart and Dukes, knowing probably that *The Gadfly* had been a big hit particularly in Russia? Does the 'mother' of the book imply the nurturing relationship of a woman ten years older to an inspiring youthful acquaintance? The substitute maternal 'adopting' role implying an absence of sexuality recurs frequently in her later novels. But 'mothers and fathers' is usually a different game altogether. Perhaps Reilly was her first lover; was her life with Wilfrid up to then just a platonic political engagement? Or, as dramatic as an episode from *The Gadfly*, was Reilly's 'abandonment' of her the lifelong source of her detachment and sourness towards men? Was Anne Nill a safe sexual surrogate? Was Reilly the 'stinkhorn' that she had to purge in her last angry novel? We shall never know for sure.

On the basis of an actual romance or a meeting at least, the other question begged is whether Reilly's life story could have found its way into Ethel's novel? Ethel in later life stated that Arthur's image came from her old interest in Mazzini and the portrait from the man in black in the Louvre. She told *The New York Times* in August 1898 that both plot and characters of the book were purely fictitious; Gemma was the only character taken from real life, based on Charlotte Wilson. Nevertheless, Lockhart, Spence and Cook all agree that in the early 1890s Sigismund Rosenblum as he then was, living in Odessa, became aware that he was the bastard son of an adulterous relationship between his mother and brother-in-law; he left Russia in a hurry. In Reilly/Lockhart's version he feigned suicide, emigrating to South America; in Cook's and Spence's he left for Paris or London.

Following Ethel's return from Russia *The Gadfly* had gestated after its conception at Wyldes about 1889/90. She maintained that in December 1895, although it was nearing completion, she used to make corrections later on. She did have time to append Reilly's 'betrayal' prologue (Act One) to the 'nationalist struggle' essence of the story. As there is no evidence for Reilly spending time in South America, however, as Cook convincingly points out, this part of his tale told to Lockhart must have been culled from Voynich's later novel, *An Interrupted Friendship*. Having read the book Reilly merely annexed its plot to his vainglorious claims.

The Reilly and Florence link makes Ethel a much more puzzling figure. Reilly must have exploded her cherished romantic visions of Mazzini and the black-hatted man in the Louvre.[31] It also adds to an intriguing historical irony. While Ethel's sister's husband, Charles Howard Hinton's, fourth-dimensional ideas had influenced the design of Lenin's cubic mausoleum to promote his immortality, Ethel's lover, Sidney Reilly, had been sent to Russia a few years before to assassinate him.

# OLIVE AND *OVOD*

It was a matter of unfinished business. My sporadic but intense relationship with the Soviet Union in the 1980s and the brave dissidents who had challenged its stultifying bludgeon, had created a link that remained dormant long after the liberating events of 1989. The emergence of my connection with the Boole/Hinton family had started things off again. It felt like wanting to reconnoitre the replacement partner after a divorce and re-assess a previous attachment. I also felt the more immediate incentive of locating Ethel Boole and Wilfrid Voynich in their historical settings and possibly dredge up some more clues about their lives. Did Ethel's novel *The Gadfly*, so hugely famous in Soviet times, known as *Ovod*, still have friends in the new Russia?

It is fascinating and powerful to locate the immediate environs of town, street, room, bed and inkstand to a biography. In Wilfrid's case the possibility arose of discovering more about his unknown early background and the milieu that persuaded him to become a revolutionary. His home country was the first port of call. As a consequence I found myself arriving at Vilnius airport in Lithuania late one October evening in 2009. Arrival was the first tricky moment of three weeks of them. Marvelling at the ease of withdrawing the local currency, *lits*, from an ATM, I was issued a note worth an awful lot of them when I only wanted a little for the bus into town. A charming young woman I approached, however, not only paid the fare but persuaded a local traveller to see me to a hostel. It was a great relief to find, late in the day, a warm welcome at a reception desk and a lower bunk in a musty dorm.

In the morning a short step took me into Vilnius's Old Town through a medieval arched gateway. Under it lay two beggars, one with a medieval-looking, biscuit-sized, red-raw wound to his leg. When Lithuania was ruled by its Communist Party this scene would not have been visible. But at least a mendicant now has the consolation of a fully refurbished Catholicism: next to the gate was a vast, sugary church relaying mass through speakers to one and all. Vilnius is full of fine churches: the history of baroque, St Casimir's nearby encapsulates the slings and arrows, not only of its other religious houses, but the fate of all three buffeted Baltic republics. It has three times been burnt down deliberately and occupied in turn by Jesuits, Napoleon, Lutheran Germans, Nazis and Soviets, who turned it into a museum of atheism. It reverted to the Jesuits after Lithuanian liberation in 1991.

The fall of communism has brought obvious benefits to Lithuania's population, previously unimaginable: common freedoms and consumer goods for the many through a market economy. Vilnius is well-stocked with boutiques and classy merchandise for its inhabitants and tourist visitors. Its cultural life is now nationalist and cosmopolitan. I was amazed to find a performance that night of the first movement of Gustav Mahler's uncompleted *Tenth Symphony*, a work rarely heard even in London. I strolled down later to the wide Neris river and the White Bridge that lies astride. On a plinth adjacent stands a remnant socially-realistic, socialist couple: the husband's legs firmly apart, arms poised ready to make the inevitable future a reality, his mate only just behind at his shoulder. Across the bridge cluster the high-rise, chrome and glass office blocks of the capitalist present.

The modernistic Congress Hall swaggers nearby, a token of communism's benevolent but stuffed-shirt mentality towards culture. The edifice inside is, however, no longer any kind of People's Palace: the wide foyer was scattered with the twee-est, petite round tables decked to the floor in bow-tied, flouncy linen. The only food consisted of pricey, equally tiny bowls of whipped cream and chocolate plus miniature, crustless salmon sandwiches that wouldn't have sated a pygmy shrew. Fortunately I had a Snickers bar on my person that I carry as an emergency ration. I wolfed it and ignored what I perceived as subdued snickering coming from the haute bourgeoisie around.

The orchestra was not quite up to scratch – nor was the audience. It wasn't impressed by Mahler's *weltschmerz* resignation. Towards the end of this threnody of complete despair comes a monstrous repeated chord that I was awaiting with dread. A few bars after the abominable sound came a tinny, Bach mobile-tune from the row behind me, followed by a frantic ferreting in a bag. The offender, I remember thinking, should be immobilised behind bars as well. Afterwards, celebrating the two years since I was last in eastern Europe in Warsaw, resisting menu delights such as hot-smoked pigs' ears and zeppelins with minced meat, I betook the same meal: borsch, carp and a brandy.

The weather, unsurprising for October, was deteriorating; pale sun between piled black clouds and occasional dabs of rain. I paid a visit to the enclosed food market where market values soothe rather than cajole. Here were stalls laden with shoe-shine aubergines, lovely bunches of stubby carrots, pumpkins – some as big as your head – amber pots of honey, cheeses and apricot-coloured chanterelles. Breathing over all was the high note of spiced meats in fatty chunks and myriad sausage shapes sized from delicate fingers to rolling pins. Markets always bring one down to earth, as poor Gustav tried with his rhapsodic *Das Lied von der Erde* – The Song of the Earth – composed shortly before the *Tenth Symphony*, but ending with a resigned fade out. I bought some excellent homemade chocolate biscuits to reinforce my Snickers stash. The friendly vendor said that they were *karasho* – Russian words seem still to be in use. (The only word I'd picked up in Lithuanian was *aciu* ('thank you') pronounced 'a-chyoo'. It stayed in my brain by remembering how awkward it would be to have a bad cold here.) Outside a ragged line of old peasants were selling small heaps of apples and cabbage harvested from their cottage plots before a winter would set in, of a severity unimaginable to our gulf-streamed shores.

I took a train the following day from the imposing Vilnius station, once, in communist days, serving many more routes – at the time not even connected to Riga, next Baltic capital north. Fortunately, Telsiai, Wilfrid Voynich's home town, is still on the line to the coast some 150 miles distant. And a yummy BIG diesel train to take me there! The ticket-collecting *provodnitsa* in a smart uniform doesn't exactly welcome me aboard, but I'm too excited to care. The countryside is almost flat, harbouring woods, trees – birch and pine – trees and more woods. Occasional patches are cleared for hedgerowless, arable fields. Little wooden houses in peasant smallholdings hold the odd cow, a row of cabbages, a covered well. Many dwellings are empty or run-down, symptomatic of the migration to the city.

What am I expecting to uncover about Wilfrid? Probably not much, but the *Voynich Manuscript* is such a puzzle one wants every scrap of information to complete the jigsaw. What we do know is that he was the son of a 'nobleman', Leonard, and his mother, Emilia Vajculevicz, both Poles. Wilfrid was educated for two grades at a *gymnasium* and took a pharmacy exam in nearby Sauliai. Will I find his house? Did he have brothers and sisters? How important is the aristocratic title of his middle surname, Habdank? If nothing emerges there is still the aesthetic pleasure of matching a life to a location.

Telsiai, Lithuania, hometown of Wilfrid Voynich.

I'm wary about getting off at Telsiai because 109-year-old traveller, Pangdora, posted a blog on the net saying that she'd missed its station and had to come back from the next stop. As I was later to commiserate, this was not due to her decrepitude, because, unlike in the UK, the only place-name indication is written on the station building itself. I managed, however, and booked myself into a smart, friendly hotel in the middle of the town. Telsiai strangely reminded me of New England: one wide main street that once held a market, at one end a pretty wooden, spired church on a hill and a few wooden buildings in the centre largely replaced by brick, including a large supermarket. At the other end is the externally rather plain, baroque cathedral, the home of the Bishop of Samogitia. Modern cars throng in between. The whole sits picturesquely on the edge of Lake Mastis.

I dropped into the tourist information centre for any early Voynich leads. 'Never heard of him,' I'm told. 'A-chyoo,' bless you, anyway. Never mind, tomorrow I shall visit the Alka Museum that houses the ethnographic heritage of Samogitia, this feisty region that resisted adjacent invading Teutonic knights and even Christianity until the fifteenth century.

A very helpful employee there taxed my German to the last as he explained that he'd never heard of Michal Wojnicz either. He showed me the Lithuanian Wikipedia site on which there is no reference to him. I in turn showed him the English site on which he is definitely listed as a favourite son of Telsiai. Another Lithuanian website at least recognised him but gives Kaunus as his birthplace. Mystification, but I recalled Wilfrid's own birthplace declaration on entering the US as Telsiai; surely he should know. The more pressing issue is whether Wilfrid's family existed here at all; records it seems will be in Kaunus or Vilnius. The only place where records cannot go missing, I decide, is the cemetery. I hurried off there.

Oh how little we Brits know of the world. Surprisingly not much for a nation that boasts having mothered parliaments, begotten several of humanity's top ten genii, and fixed longitude for the whole globe. It never occurred to me that the beautifully-positioned hilltop graveyard would have few if any pre-Second World War graves. Pressure on space due to Lithuania's misfortunes has meant that earlier sites had been re-used. I was very impressed, however, with the neatness of it all, nowhere overgrown; most sites sported elaborate headstone designs bedecked with flowers. Where next?

Fortunately a local schoolteacher of English, Elena, came to my aid and took me under her wing as we pursued the Voynich goose chase. At the local library I learnt that all pre-1922 records are housed in Vilnius, but we found at least the several coats-of-arms of the Voyniches including the clans, Abdank and Nalecz. Elena then managed to phone the Archives in Vilnius who did indeed confirm that there had been a Wilfrid Voynich baptism recorded in Telsiai church witnessed by a good crowd, suggesting that their notion of nobility carried status. Elena augmented this by telling me that the tsarist authorities had closed the only school in 1832 following the independence uprising of 1831 that had also given rise to the Citadel in Warsaw. Wilfrid must have been educated by tutors, and later in Kaunus, indicating another sign of family wealth.

I learnt too that Telsiai had been ever since the end of the Polish/ Lithuanian Commonwealth in 1795 a centre of resistance to Russian occupation waged by the nobility. Both Lithuanians and Poles fought against the suppression of their languages and the subjugation of the Catholic Church. Was Wilfrid's father involved? Even if not, his son would have certainly grown up in a climate of opposition to the existing order. How his connection to *Narodnya Volya* was made we'll probably never know, but possibly via university. I left Telsiai grateful for some very kind help, but felt in return, at least, that the town now knew it had a favourite son to attract the world beyond.

Retracing the railway line back to Saulei I was obliged to take a bus to Riga in Latvia. My main reason for being there was to join a train onwards to Moscow. I checked in to another hostel right in the Old Town on the embankment of the wide, oily River Daugava and went for a guided tour of the city. Riga is a much more substantial and commercial city than Vilnius, having been a major player in the trade of the Hanseatic League. I joined a party of young people taking a city tour. The rain came down steadily. I quipped that we were all likely to end up with a bad case of 'riga mortis' – without any visible appreciation.

On a happier note, one fortuitous encounter was with one of the hostel workers, Angela, who was the first to volunteer her opinion of *Ovod*. Only in her twenties, she remembered it well, partly because of her background: her grandmother had escaped the siege of Leningrad in 1942 and settled in Riga. Most unusually, her granny had married a German SS officer, but Soviet culture had been handed down in her family. What she admired in the novel was the romantic relationship between Arthur and Gemma, comparing it to *Gone with the Wind*, 'only more serious'. It was thanks to her knowledge of Russian that I was able to make a booking for the hotel Lubliyana in Velikiye Luki, my overnight stop, halfway between Riga and Moscow. This was wholly essential as I would be getting off the train at 3.30am. I was going there to chase up Ethel's past history.

The train from Riga to Moscow left at eight in the evening. The carriage was packed and uncomfortable with hard seats. The bunks too looked un-appetising – but I wouldn't be sleeping. Two fat old *babushkas* played cards opposite; the older had a heavy, almost Stalinesque moustache. It was all a little disappointing after my wonderful Trans-Siberian journey. Sleeper trains have a romantic appeal for Brits – waking up on the Scottish moors or at Penzance, but what can you expect for thirteen lats, about £15, to travel 500 miles? From King's Cross you wouldn't even get to my father's datum location at Peterborough. Nearly everybody is busy attending to a mobile; I attend to the spinach pasties and curd tarts that I provisioned myself with from Riga market. I'm given an immigration slip to fill in for the border control.

What's my father's patronymic? 'Thomasovich'? I suddenly feel nervous when we reach the border and our passports are taken away to be inspected; memories re-form of my awkward comeuppance with contraband documents

in 1983 at the Mongolian/Chinese border and again in 1984 entering
Hungary. The young guard snaps something at me that I can't make out.
I answer 'London' and hope for the best. All eyes are on the foreigner. It
takes the best part of an hour but, thank God, there are no problems. I'd
shown the guard my ticket so he could warn me before we arrived at
Velikiye Luki in the dead of night. But he forgot or couldn't be bothered
and to make matters worse I hadn't realised that we'd crossed a time zone
en route; I'm an hour behind. It was only by pure chance that the station
building was on my side of the carriage and I saw the place name when we
arrived. I scrambled for my rucksack and dived off the train, having
narrowly escaped a 'Pangdora'.

There's miserable desolation outside, but I locate a taxi that takes me
to the hotel where I gratefully expire, back in the USSR after twenty years.
And it is the USSR when I wake up in a room straight from the period.
I'm the last to complain about hotels but the décor is dreary, the toilet
door won't close, there is no bathplug as of old (when we carried a squash-
ball for the purpose) and it's over-heated. Welcome back, I almost feel at
home. Breakfast is two fried eggs (still) and a frankfurter encased in plastic.
The hapless, plump waitress can't be bothered with me, so I serve myself.
A cook pops out of the kitchen and slyly slides two slices of cheese into
her pocket. Ah, the proletarian touch, the sheer *gorkyness* of it all.

Outside, the city of 100,000, mainly a communication and engineering
centre, feels familiar, but in fact has been rebuilt after a battle here in 1942
when 7,000 besieged Germans were wiped out. People look brighter than
I remember from the eighties, the girls in their short skirts and knee-length
boots. A shop is selling fitness equipment, once unthinkable. But making
such sketchy judgements is not why I'm here. It was not far from Velikiye
Luki that Ethel Boole came by train and handcart in the summer of 1887
to stay in the village of Uspenskoye. Stepniak's sister-in-law, Praskovya
Karaulov, asked her to help nurse her weak and ill brother-in-law, Nikolai.
Ethel used the stay as a setting for her third novel, *Olive Latham*, of 1904.
The whole book has echoes of her time in Russia.

Olive is described as a serious-minded young woman given to good
works but also something of 'a sexless water-spirit',[1] characteristics that
Ethel might have had applied to her. She is in love with Vladimir
Damarov, a young nihilist she has met in London. Olive notes that
'nihilist', 'In Russia nowadays ... simply means a person who has the wrong
opinions',[2] and for which he too, like so many, has suffered in prison.
Vladimir is under the influence of a Lithuanian Pole, Karol, who has been

Ilya Repin's 'The Unexpected Visitor' (Tretyakov Gallery, Moscow), depicts a ragged Narodnik returning to his family.

released from Siberia and resembles both Stepniak and Wilfrid with 'a certain massive set of head and shoulders, like a Norse god'.[3]

When Olive receives a letter informing her of her lover's pleurisy she sets off to St Petersburg. From there she too travels by train and then on by cart to Vladimir's dilapidated manor house in the 'wild and lonely district where the Volga rises,[4] ... cursed alike by nature and by men; frozen for half the year, malarious the other half, neglected, chronically famine-stricken, infested by beasts of prey'.[5] Irresistible, I had to see it. I wanted to fit the factual village to the fictional but also to visit somewhere in the countryside; back in Soviet times we had been denied the right to witness its poverty.

It only took half an hour by taxi to Uspenskoye. The village is desolate indeed, a clutch of buildings straddling the road. There are some patches of cleared land but many of the wooden cottages are abandoned, their small, corrugated-roofed barns and scattered outhouses dilapidated. Only the odd apple tree shows any sign of lingering life. The inhabited dwellings seem sodden, with little sign of activity save for small stacks of unprotected hay and a few exhausted Lada cars. There must have been attempts at one time to revive rural life as by the road are several, obviously communist era, two-storey, barrack-like blocks, but they too are dishevelled, the top floor windows often boarded up. Another has no windows at all.

In a single-storeyed building, there seems to be a school, as some scruffy children emerge. The snazzy bus stop is the only suggestion that the village is going somewhere, sometimes. The one obvious sign of a cared-for, collective past is a wheeled, Second World War gun guarding a line of twenty slabs with lists of names inscribed – testament to the suffering the locality bore resisting the German army that fought its way eastwards looking for *lebensraum*. There is plenty of room here now but no one wants it. Conditions in Uspenskoye are probably no better than they were when Ethel stayed here but presumably the wolves have gone. Olive comments harshly that it is a crime to keep people alive who should never have been born. I have no time to explore for either Nikolai's or the fictional manor house as the taxi meter is ticking. Sure enough when we arrive back at the hotel the driver wants twice what we agreed. Depressed by the village, I couldn't be bothered to argue.

Literary business done, I had no reason or desire to stay in Velikiye Luki and booked myself a ticket onwards to Moscow, another 250 miles overnight. Despite sleeping well on the train, my arrival there at the Rigevskaya station at 6.30am was completely bewildering. Everyone scuttled off somewhere but I felt dazed and was unable to orient myself. I needed to contact Dimitry, someone with whom I'd corresponded regarding state archives on Voynich. He'd kindly offered me the use of his mother's empty flat. But it was too early to make contact so, eventually locating a Metro, I took off to Tretyakovskaya station where an internet café was listed, hoping for coffee.

At the top of the steps out of the station I emerged bleary-eyed into the bright new Russia. Velikiye Luki had been a useful foretaste, but here was the real thing: a tiny Rasputin-like monk evangelising, a McDonald's in front of me and to my left, atop an office block, the red banner of Canon where once might have been a red proclamation announcing the road ahead to socialism. I took the road ahead to McDonald's, grateful for a semblance of breakfast. Eventually I made contact with Dimitry who took me back to his large but slightly run-down apartment on Leninski Prospekt.

After black tea and introductions to his family we sat at his computer to look at the material he'd located at *GARF* (State Archives). He showed me 400 pages mostly written by agent Seraev in New York regarding the activities of the Friends of Russian Freedom for which Voynich had worked. I was particularly keen to see if there was anything interesting about the mysterious Mister Stein who'd been one of the breakaway trio of the League of Book Carriers with Ethel. The only evidence seemed to be that it was a pseudonym used by Stepniak – which made no sense at all. I then remembered that I needed to register my presence with the authorities before three days, so we went to the nearest post office. It took an hour to navigate the forms in duplicate and cost 200 roubles. What struck me most was the appalling rudeness of the young, stony-faced woman clerk. Might it have been because Dimitry is obviously Jewish or just a hangover from the noble tradition of Soviet times? Probably both. This was compounded the day after when I accompanied him to the polling station for Moscow's *Duma* elections. This is most certainly a post-Soviet function but it had an old feel: guards outside smoking, *militsia* inside and hardly a festive air. 'It's still a police state,' offered Dimitry, 'nothing much has changed.' Only 30 per cent bothered to vote.

My first adventure in Moscow was not a success. It was cold with rain pouring down, but at least the Metro was womb-like warm. The stations still have their imperial grandeur but the trains are old and clanky. I had some crazy notion that I could pop into the Lenin Library and consult some books, but I never got past the cloakroom and entry kiosk. (I was also reminded of twenty-five years ago when our guide had proudly pointed out the Lenin Library as we passed it. Much to her annoyance, I'd quipped that we have 'lenin' libraries all over England.) Outside in the pouring rain the streets were puddle-flooded so I made a sea-line for another McDonald's: an island of predictable inedible sanity. They are everywhere spreading the cholesterisation of Moscow.

The other task for the day was to visit the Writers' Union and try for an interview re *Ovod*. Eventually I found it near the inner ring road, a fine eighteenth-century, yellow manorial building with gardens. At the entrance someone in camouflage gear (as in the old days, the military are every-where) took me to an office upstairs where I met a writer, Marina Anashkevich, and Ksenia Bogoyavlenskaya, who could translate for me. Marina was born in 1960, the year Ethel died, and is, gratifyingly, a huge fan of the book, having read it ten times. As she was a Christian this seemed surprising but what attracted her was the dramatic conflict between Arthur and his errant father, Cardinal Montanelli. For her the book had 'soul'. On the way out I remembered that this was where, on our END delegation in 1987, Ian McEwan had got blotto on Georgian champagne.

Wandering back into town I stumbled across Arbat Street, immortalised in Bulgakov's novel, *Master and Margherita*. It is now a paved theme park of entertainment with *matryoshka* shops, buskers and beggars, people in bear, dog and cat suits hoping to lure passers-by into shops, once impossible in the Soviet Union. Moscow has always been a city for cars with its multi-laned, broad boulevards but the volume is now so much greater. I've never witnessed so many tinted-windowed 4x4s and BMWs charging about at great speed. Maybe that's why so many buildings are hung with football-pitch sized, gauze adverts, as they'll never be seen otherwise. The result of this car consumerism is inevitably gridlock at any junction. Moscow, fired up by oil and gas and corruption, has been dunked in a thick candy-coating of capitalism, for some anyway.

Musing on all this I found myself at Pushkin Square. Here under his statue, a symbol of freedom, in Soviet days it was considered high treason to doff one's hat on Pushkin's birthday. Today a small crowd of onlookers and journalists are outnumbered by *militsia* and some very nasty-looking special riot units, the OMON, grouped around an armoured lorry. One long-haired demonstrator was just being hauled away as I arrived. I hung around and took pictures unobtrusively. The following day I learnt that about thirty demonstrators had been arrested for showing their displeasure at the weekend's probably rigged elections. With suppressed dissent from inside and terror legitimising media control, Dimitry's speculation about democracy seemed justified. Two reports in an English-speaking paper highlighted abundant corruption and inefficiency. A thousand leather-jacketed, gold-chained mafia hoodlums had attended the funeral in Moscow of a mobster gunned down outside a restaurant. Another told how a hydro-electric power plant's turbine had vibrated to destruction, killing seventy-five in the subsequent flood. At fault said the writer, was 'the complete deterioration of the country's technical infrastructure'.

The new Russia, rich and poor.

Continuing my *Ovod* trail I called in at the *Moskva* bookshop on Tverskaya Street to see what stocks of *The Gadfly* it held. There were no new copies but a couple of used ones in the basement. This was a little strange; in Riga out of curiosity I'd asked two old ladies working in a book kiosk under a railway arch if they had a copy. Delighted, they sold me a brand-new one. I strolled on into Red Square passing at its entrance that well-known figure of world revolution, a costumed Bart Simpson. The

Demonstration in Pushkin Square, Moscow, 2009.

square, I think, is more impressive at night and in the snow. There are now no queues outside Lenin's mausoleum and sadly no arms-bearing soldiers changing the guard with their ridiculous goose-stepping strut that once had made me laugh. GUM, the department store down one side of the square has now had a complete makeover into a boring celebration of the elevated world of the super-rich. On to the only art gallery on my itinerary: the Tretyakov.

Inside today's GUM Store on Red Square.

I purchased a ticket and left my coat at the garderobe, both transactions performed with the same world-weary resignation of the attendants that I remembered from all those years ago. There is still a particular Eeyore quality in Russia. I was more pleased to view a picture by Repin that would have meant little to me in 1983. *The Unexpected Return* is a poignant portrayal of a ragged, sheepish *narodnik* aristocrat returning to his family after a no-doubt unproductive mission to succour the peasantry. Crossing the Moskva river one comes to the major re-instatement of religiosity post-Soviet Union, the huge Cathedral of Christ the Saviour. Finished in 2000, the controversy surrounding it reflects the yo-yo symbolic history of communism and the Russian Orthodox Church.

It is a faithful resurrection of a cathedral on the same site, the original of which was blown up by Stalin in 1931. Consisting of one large, central, copper onion-dome with four attendant shallots, it sits on a square,

Cathedral of Christ the Saviour, Moscow.

off-white monolithic base surrounded on all sides by a high paved plaza. This relative conservatism is, however, swamped by the exuberance of the interior. I am rarely jaw-droppingly overawed but the blaze of golds and reds on the vaulting arches depicting the saints and biblical scenes was overwhelming. The olfactory seduction of the incense was augmented by the amplified drone of a hidden choir. Even for a soul as detached from the Absolute as myself, the emergent voice of the extra-baritone solo, like a kraken from the deep, seemed like God himself surfacing. Fortunately blessed with the saving grace of a trivial mind, I could ward off submission by reminding myself that the Russian word for God is *Bog*. Submitting readily around me was a large crowd including many head-scarved old women crossing themselves, kissing the floor and touching images. At the end of the liturgy two robed priests disappeared mysteriously into an inner sanctuary. To me it was all vain and vulgarly absurd.

Sometimes on one's travels the really pleasantly unexpected turns up. Her name was Alina. Lost in a rush-hour crowd at the end of the Arbat I asked her how to get to Smolenskaya metro. She kindly offered to take me there. Her English was very fluent so I couldn't resist telling her about my mission. Yes, she knew of *Ovod*, it was a book that had meant a lot to her. I met her later at a charming café not far from the appropriately named Kropotkinskaya metro where we were joined by her ex-partner Greg. Alina loved the book, like Angela in Riga, because of the romance of the attachment between Arthur and Gemma; for her it was 'about love not revolution because it appealed to the minds and hearts of young people who dream of something extraordinary'. Among what she called a lot of literary garbage in Soviet times *Ovod* was a celebration. The sacrifice and heroism of the revolutionary cause was in a way just part of how ideals were portrayed at that time, and therefore did not make an especial impact.

Greg was a lot less enthusiastic because the book was 'mandatory'. 'Whatever they said was good I would translate as bad.' The Bolsheviks, as he called them, 'unsentimental, heartless and cold, used the novel to keep the nation under control'. Greg pragmatically saw the book as being about 'systems' and exploitation, compared to Alina's emotional response to two lovers unable to realise their love.

Later that day in Oktyabraskaya Square under the statue of Lenin, I met Dimitry again. It is also the regular Friday night meeting place for a chattering flock of punks. His wife kindly took me onwards to Leningradskaya station for another overnight trip on the *Two Capitals* train to St Petersburg. I was beginning to get used to the making-up of bunks, the round of tea in glasses and the appalling music on the intercom. What was less attractive was yet another turfing off the train at six in the morning onto the empty streets of the city. Deciding not to be ripped off again by a taxi I walked the distance to my reserved room in the entirely un-Egyptian *Ra* hotel in a grubby neighbourhood. En route I stopped off in the only open café for a bowl of warm *kasha*, a gruel-like sweet porridge, which made me feel quite Chekhovian and brave for some reason.

I awoke in the afternoon to a wonderfully warm day and strolled Nevsky Prospekt down to the Winter Palace. Petersburg is indeed a beautiful city with its fine, stuccoed buildings often painted sky blue or ochre. By ancient decree none of them could be built higher than the spire of the cathedral within the Peter and Paul Fortress. This rule has now been breached but fortunately not in the city centre. The presence of the river, more like a sea, and the two crescent-shape canals it feeds create soothing Venetian vistas and long, low perspectives.

But that's the tourist bit; I wasn't here to go to the Hermitage or marvel at the Bronze Horseman. The next day, with an arctic rain sloshing down, I went in search of the Museum of Political History to set up an interview. I took a tour. One wing is rather haphazardly devoted to odds and ends of tsarist life and nineteenth-century history. I asked an attendant who had some English why there was no mention of protest and the *narodniks* who figured so large at the time. She explained their absence as being because they were terrorists and fought against the Tsar's best intentions at reform – a similar rationalisation as in Warsaw regarding *Proletariat*. She informed me that Princess Diana had been one of our own great leaders. The other wing was a badly presented but exhaustive account of Soviet history with letters, posters and artefacts. Nothing hidden here: full chilling tales of the gulags, even Trotsky made his goateed appearance. Sakharov, Brodsky and other more recent dissidents were acknowledged, although no mention of the Trust Group I'd once tangled with.

While truthful, the museum still exhibited the lack of panache that typified Soviet times. At the sales kiosk I decided to buy the grandkids some Soviet lapel badges and pointed to them in their glass case. *Zavtra*, snapped the over-lipsticked, chubby matron, 'Tomorrow'. I expostulated pointing to the Russian family she'd just sold some stuff to. *Zavtra*, she snapped again. Maybe she's an intentional *tableau vivant* reminder of days past. The snappiness of ordinary people bugged me the whole time I was in Russia, the inability, for example, to hold open maliciously swinging doors. I looked up the word for 'rudeness' in my dictionary – *groobost* – and when riled muttered it as loud as I dared. I had, however, at least set up an interview for another time with young and friendly museum worker, Maria.

I stayed in a nearly empty hostel on Tretskaya Ulitsa, Third Street, just a stone's throw from the Moskovskaya station that Ethel and I had arrived at in the very heart of the city. I had a room to myself and cooked up packet borsch soup and noodles. Despite the supermarkets I was unable to find fare I knew what to do with. I did, however, find a basement café nearby where an omelette breakfast could be had. It was here that I met Sergei, a very voluble Petersburg guide, for an interview about *Ovod*.

He too knew the novel well and confirmed everyone's view that books, in general, once had a central role in people's lives, almost gone now in the young, apart perhaps from *Harry Potter*. He told me, 'People once were queuing to get a copy of *Ovod*, and the last film of it in 1955 made it even

The author outside Ethel Voynich's apartment block in St Petersburg, 2009.

more popular.' He also suggested that part of its attraction lay in the fact that it had different levels –the revolutionary, romantic and universal – especially the struggle between the priest Montanelli and his son Arthur.

As a professional guide he was happy to accompany me to find Ethel's apartment where she had spent her two years in St Petersburg. I knew the address according to the 1889 Police Department as flat 17, number 7,

Seventh Rozhdestvenskaya Street – now Seventh Sovietskaya in the area known as Peski. We walked up there, another four streets on the grid system from the hostel. The five-storey blocks are still handsome but run down, although there are signs of gentrification: a sushi bar on 7th and a basement café opposite number 7. The dilapidated, bare inner courtyard was reached through an archway and tall metal gates that were fortunately unlocked. We decided not to ring the bell of number 17 but the atmosphere was enough: dark and graceless. It wasn't difficult to imagine Ethel, nervy, thin and upright, dressed in black, sombrely emerging, perhaps with a leather case of music and books to go and teach or meet other political rebels. It's her haunt.

She must have found some relief from the hardship surrounding her by a short walk to the Tavrichesky Gardens; we followed in her footsteps. It is indeed a charming park with rounded paths and a wandering lake that contrasted with the regular geometry of the streets nearby. Maybe she passed through it, however, on her way to the gloomy assignment she kept, to the detention centre on Shpalernaya Street, a few yards from the Neva river. It was here where she took food to Vasily Karaulov who was awaiting his sentence to Siberia. The Big House, as it is called, was built in 1871 and held many political prisoners, including the 193 *narodniks* before their despatch to the Peter and Paul Fortress. It held Lenin himself in 1895 in the apposite cell number 193. It is arranged around a large courtyard, which cannot be seen beyond the monolithic constructivist façade which was added in the 1920s by architect Noah Trotsky. It later housed the offices of the security services as their acronyms morphed from *OGPU* to *NKVD* to *KGB* and today's *FSB*.

In her novel, *Olive Latham*, Ethel described the scene as Olive enquires about the fate of her lover, Valodya, after they have returned to Petersburg from his country home and he had been arrested:

A continual stream of figures in uniform passed in and out, hurrying, lounging, gossiping, rustling papers and slamming doors. Beside Olive sat a poverty-stricken woman with a child leaning against her knees. From time to time a few tears trickled down her cheeks and were brushed away mechanically by the rusty black sleeve of her jacket.[6]

Valodya's friend, Karol, tells her in a precise echo of Mark Reitman's story about a society used to big stones raining on them every day: 'Once you're accustomed to these things, you can manage to stand them; the whole

В условиях одиночного заключения единственным способом общения узников друг с другом было перестукивание. Для этого использовали так называемую «тюремную азбуку» из 28 букв (исключены наименее употребительные: ё, й, i, ъ, ь, ѣ, ѳ). Буквы алфавита располагались по порядку в таблице из шести строк и пяти столбцов. Каждую букву выстукивали в два приема: сначала номер строки, затем — номер столбца. Стучали каблуком об пол; ложкой, кружкой, пуговицей по стене или ножке кровати.
Акустика тюрьмы Трубецкого бастиона позволяла слышать не только звук, доносившийся из смежных помещений, но и из камер другого этажа. Постоянно царившая в тюрьме тишина болезненно обостряла слух.
Перестукивание считалось нарушением тюремного режима и строго пресекалось охраной. В наказание узника могли лишить свиданий, книг, перевести в другую камеру где не было соседей, заключить в карцер. Никакие суровые меры не могли заставить узников отказаться от перестукивания; по словам В. Н. Фигнер, «борьба за стук... это — прямо борьба за существование».

| №   | 1 | 2 | 3 | 4 | 5 |
|-----|---|---|---|---|---|
| I   | А | Б | В | Г | Д |
| II  | Е | Ж | З | И | К |
| III | Л | М | Н | О | П |
| IV  | Р | С | Т | У | Ф |
| V   | Х | Ц | Ч | Ш | Щ |
| VI  | Ы | Ю | Я |   |   |

Tapping was the only way of communication between prisoners in single cells. They used so called "prison alphabet". Letters in alphabetical order were accumulated in a table of six lines and five columns. Each letter was tapped in two steps: first, number of a line, then number of a column. Prisoners "spoke" tapping the floor, walls or legs of a bedstead

A tapping table for making contact between cells.

trick is to get accustomed.'[7] This seems to be the rule for survival in every extreme of deprivation. Olive has to withstand the deliberate, Kafkaesque rigmarole of false information from the police. 'To the end of her life she would see hell as a labyrinth of white corridors, of false directions, of smiling faces.'[8] Eventually she discovers that Valodya has been sent across the river to the fortress. In leaden despair Olive stations herself in a niche of the parapet of the granite embankment directly opposite the prison staring across the ice-bound river. The spiteful cat and mouse game of officials continues until she is able to learn of his death for sure by a compassionate gendarme she meets at the cemetery where Valodya has been buried. Her life is shattered; she returns to England.

Church of the Spilled Blood, St Petersburg.

I left Sergei at the Big House and strolled on through the Summer Gardens, now autumnal, past the eternal flame in the Field of Mars by which a couple of hippies were trying to warm themselves. The Church of the Spilled Blood nearby was built to commemorate the exact spot where Alexander II was assassinated in 1881. It's an extraordinary confection of gold and sky-blue domes spangled with stars and all sorts of cones and barley-sugar twists. If it wasn't so ideologically oppressive it would be even more wondrous. I couldn't take my eyes off it but couldn't

Tombs of the Tsars, Peter and Paul fortress.

quite take it in. I settled for trying again to choose some presents for the grandkids at the nearby souvenir market. Never put off until *Zavtra* what you can do today. Despite the freezing cold, the stall-holders warmed to their free-market patter. I bought a *matryoshka* doll, inevitably, a glass snow storm and some Soviet badges, 'forbidden' elsewhere.

The other place I felt I ought to visit in homage to Ethel was, of course, the Peter and Paul Fortress; we had both visited it separated by a century. Built on an island in 1703, it was the first building Peter installed as a defence against foreign invasion but was used exclusively as a defence against his dynasty's subjects. The six bastions enclose a large courtyard accommodating barracks, a mint and in the middle the baroque cathedral distinguished by its pencil-thin, gilded spire, a spiked reminder of the Tsar's temporal might.

Like the new Saviour cathedral in Moscow, the building is plain on the outside but obscenely lavish inside. All of the Romanovs are now buried within after years of forensic wrangling over the authenticity of the remains of the family of Nicholas II, executed by the Bolsheviks at Ekaterinburg in 1918. The uncertainty led the Orthodox Church to hold back on canonisation, but in 2000 a fudged sainthood was conferred. I joined a crowd of schoolchildren to gawp like them at the entrance to the dedicated chapel and wondered what the nineteenth-century revolutionaries would have made of the rehabilitation of all they had fought against. My act of dedication was to visit the Trubetskoy bastion prison where many of them had been locked up and tortured.

A two-storeyed pentagon, it was rebuilt in the 1870s to contain mainly political prisoners on remand, under investigation and serving sentences, all usually kept in solitary confinement. The cells, as at the Warsaw Citadel, were not cramped, but poorly lit and furnished only with a bed and side table. Despite the felt, sound-absorbing layers between walls the inmates managed to communicate, as in Koestler's *Darkness at Noon*, by a process of tapping or stamping on the floor from an agreed coded alphabet. Touring the corridors revealed details outside each cell of celebrated inmates, some of whom I'd come to feel as historical pen pals: Mikhail Bakunin, Alexander Ulyanov (Lenin's older brother) and Kropotkin.

Ethel must have come to know some of them during her stay in the city. Many suffered torture, went mad, were shot or sent off to Siberia like Yakubovich. The Peter and Paul Fortress must be one of the few places in the world where the reified lives of one group of human beings can have been so exalted and another so debased and made wretched in such close proximity.

I returned to the Political History Museum to meet a Russian historian to ask, via Maria, about *Ovod*. The conversation rambled on; he certainly knew the book well and held it in esteem – too much so. According to him the revolutionary message within it was still potent and might be a rallying cause one day for a future generation. I thanked him and Maria warmly; they reciprocated and honoured me in that typical awkward style I remember from Soviet days by offering me a large box of souvenir matches.

One final visit, almost touristic, was to the last St Petersburg apartment inhabited by Dostoyevsky on Zagorolnye Lane. Unlike the Political

History Museum it had a dignified post-Soviet flair, not matched by my own composure. I finally lost my grip on *groobost* when I was challenged briskly while taking a photograph for which I had duly bought a permissive ticket. Russian has very few words that begin with 'f'; English carries one probably understood everywhere. Rather ruffled, I didn't overstay my welcome but sufficient to observe the well-appointed gentility that had come to Dostoyevsky in his last years before he died in 1881.

As Stepniak noted of Ethel's condition on her return from St Petersburg, Olive Latham in the novel returns after Valodya's death to her parents' Sussex home 'a hollow-cheeked and hard-eyed ghost',[9] unable to communicate with them and given to hallucinations. In the garden she experiences, as a kind of embattled spectator, 'her familiar surroundings, all its joys, its loves, its gods [as] but painted films'.[10] Her friend Dick tries to persuade her to come back to life and work for socialism, but Olive scornfully asks, in what may express for Ethel her distance from her days at Wyldes farm and English politics, 'The Hampstead variety, all economic statistics and afternoon tea, or the Bermondsey sort … with the beer and banners? I think if I were going in for anything, I'd try anarchism … there's something satisfying about that, anyway.'[11] Her Polish friend Karol arrives in London and comes to see her, to whom she confesses that, although she doesn't believe in answering violence with violence, she has a nihilistic *Raskolnikovian* desire to kill someone. Karol recognises it as the need 'to destroy something that has a tangible bodily presence, and satisfy yourself that it is solid'.[12]

In answer to her alienation he suggests that she should come to London and help with a project once supported by Valodya: to look after peasants fleeing the Russian suppression of the Uniate Catholic Church in Poland and Lithuania. She agrees and her life gradually revives until she is called to rescue Karol, wounded while trying to help refugees in Galicia. He recovers but still suffers from a progressive, possibly terminal, spinal paralysis as a result of his Siberian years. This had prevented him from professing his love for Olive. An understanding is reached between them, however, and the novel ends, cautiously optimistic about their future together.

*Olive Latham* is told in a strange mixture of animation and passion yet curiously flat and removed, but its historical truthfulness and examination of principles conveys an essential honesty that is admirable if somewhat relentless. Like *Ovod* it is lacking in sentimentality. *Olive Latham* is, of course heavily autobiographical, based on Ethel's remembered St

Petersburg stay, but it also draws, naturally enough, on her more immediate life. Karol is Wilfrid writ precisely: the revolutionary Lithuanian Pole damaged forever physically by his political activism, suffering but undaunted. Both Ethel and Olive Garnett looked up to his type with deep respect.

It is not in Ethel's basic character, I suspect, to commit to another from sexual or familial desires. This is shown by her rendering of Valodya, the other male character in *Olive Latham*. He too has suffered for his beliefs like Karol but is a lesser man physically and morally. Olive looks upon him as her offspring, despite Ethel's undefined use of the term 'lover'. She tells him, 'I've carried you back in my arms from the very edge; you're just as much mine as if you were my own child'. She will comfort and stay with him but although he had won her utmost devotion … 'he had not awakened the woman in her.'[13]

After three weeks I was dizzy with trying to grasp the changes I'd witnessed in Russia. Despite the lack of any stunning revelations about Ethel and Wilfrid Voynich, I did feel I knew them better and had cemented a kinship. I was fairly exhausted and all too ready to get back to an England overflowing with marmalade and sprouts, where zebra crossings are sacred and people say sorry when *their* toes are trodden on. These are a few of my favourite things; not much to write home about but felt fondly when away. None of my postcards from Russia made it back.

# A CQR LIFE

The name of Geoffrey Taylor had cropped up regularly during my investigations into the Boole family, not only in connection with his mother, Margaret, the second Boole daughter, but also his aunt, Ethel. He seemed to have maintained an especial bond with her. As a leading Cambridge scientist he was often in the US lecturing or at conferences; while there he visited her regularly in New York. The transatlantic visits also allowed him to keep up with Ted (Sebastian) and Carmelita's three Hinton children: Jean, Bill and Joan. In 1924 he entertained them, as we have seen, on his boat *Frolic* off the coast of Scotland.

For the purpose of the biographies of the Booles and the Hintons he is an important character not just because he acted as a link across the Pond, but also, conscious and proudly aware of his antecedents, he acted informally as an archivist. Hardly any of the letters, diaries, etc. of the other Boole/Hinton *dramatis personae* seem to have survived. Sometimes I've found myself hatching conspiracy plots to explain this. His high recognition in 'respectable' society meant, however, that what he had kept found itself lodged later in a public institution.

There is something Lewis Carroll-like about the environs of Trinity College, Cambridge, once *alma mater* to Sir Geoffrey. The building that houses the Great Gate is big enough to be a college itself. A statue of King Henry VIII stands oddly in one recess with an orb in one hand, a table leg in another. Some wag many Michaelmas-taking terms ago stole the sceptre. Going through one enters the largest square in Oxbridge (they compete not only in rowing boats), arranged in rectangles divided by paths that should have sported wonderland flamingos. On the far side up some stone steps one enters a corridor off which leads the grandiose hammer-beamed refectory where Trinity students tuck in. My memory jogged to Leeds University days when, on a tray of refectory sandwiches a biroed notice offered, 'Reduced to clear'. At the end of the passage is a tiny door.

Carmelita Hinton and Geoffrey Taylor, 1938, holidaying in Canada.

There were no helpful instructions saying 'eat me' next to an inviting cake; stooping through it led not to a ground-keeper's shed but, curiouser and curiouser, another huge grassed square. Clinging to the cloistered edge I arrived, up marble steps, at the entrance to the magnificent library. Completed in 1695, a Wren copy of a Venetian library, its one room is enormously high and long. I offered up my certificate of worthiness and was directed to the distant far end of the building. I'd have appreciated a fanfare as I set out along the marble black and white tiled floor that would have been perfect for the Olympic hopscotch finals. Past rows of rope-cordoned bookstacks sheltering venerable tomes, past a statue of Byron rejected by Westminster Abbey owing to his naughtiness, I arrived eventually at a raised dais where I could fill in the forms to request Taylor's archive. The be-ribboned folders were magically summoned up.

The authoritative source concerning his life is G. K. Batchelor's book, *The Life and Legacy of G. I. Taylor* published in 1994. An Australian, he joined G. I. as a research student in 1945. Much of it, inevitably, as a fellow scientist, concerns the description of his career development and experimental details. Much of it I'm afraid, therefore, I had to gloss over. Nevertheless his account reveals interesting facets concerning the role of the scientist in society. Any conflicts that there might have been are adumbrated by Batchelor's unstinting praise of Taylor as 'a truly simple, good man'.[1]

Taylor was very proud of his connection to the Boole family, delivering an *Address to the Royal Irish Academy* in 1954 about his grandfather, George Boole, and writing several articles about his other aunt, Alice. Geoffrey Taylor's father, Edward (1855–1923), showed early artistic promise and studied at the Slade School of Art, later undertaking commissions in stained glass and decoration of large public rooms including some on ocean liners. He and Margaret, the second Boole daughter, one of his pupils, were married in 1885. Geoffrey was born on 7 March 1886, his brother Julian three years later.

The Taylors' family house at Blenheim Villas, St John's Wood, London was not far from where his Aunt Ethel used to visit Sergei Stepniak in early 1887 to learn Russian. Geoffrey and his brother Julian kept in contact with their widowed grandmother, Mary Everest Boole, in her house in Notting Hill. He wrote, 'I visited her very frequently and enjoyed her conversation, particularly when she described her childhood in a Gloucestershire vicarage and in Paris.'[2] Otherwise, however, he found her eccentric and had little interest in the things she wrote about. Through

Ethel he remembered once sitting on her anarchist friend Kropotkin's knee. The kindly prince, 'who shined in my early childhood memory',[3] had made him a model boat and explained the rigging.

His parents had a considerable influence on Geoffrey, fostering his love of the countryside and wild flowers (Edward's drawings of them are to be found in the British Museum). The outdoor life, sailing in particular, later became a passion. But it was science that grabbed him, aged eleven, when attending a children's Christmas lecture on wireless telegraphy at the Royal Institution. Here he met the scientist Lord Kelvin, who had mentioned his friendship with George Boole. Soon after Taylor built an electric machine to generate x-rays, with which they diagnosed his mother's arthritis. Less constructive experiments involved the proverbial schoolboy fascination with explosives. He and his brother blew a large hole in their garden door. At Kensington Gardens Round Pond a model fireship built by them was detonated, destroying a close-by more genteel flotilla. It was reported that 'their owners were led away sobbing by indignant nursemaids, gleefully watched by the two small buccaneers from an adjacent clump of shrubs'.[4]

At University College School in 1899 he sailed through his studies to land an 'exhibition' at Trinity College, Cambridge in 1905. The nautical analogy is prompted by the first arousing of his interest in boats. When at school he had built a dinghy just smaller than his bedroom that could only be lowered into the back garden after the removal of the sash windows. He launched it on the Thames and took it out to sea. In 1908 Taylor graduated and was awarded a scholarship at Trinity to undertake research. Here at the Cavendish Laboratory he remained anchored for the rest of his life – a remarkably linear career.

His first project idea on quanta, tendered to his professor J. J. Thomson (and I quote from Wikipedia), showed that 'Young's slit diffraction experiment produces fringes even when feeble light sources such that less than one proton was present at the time'. His lab was set up in his childhood playroom at St John's Wood where for three months he used varying intensities of light from a candle screened by smoked glass to throw a shadow image from a needle. (Strangely, the grandmother he found eccentric used shadow-casting to help explain geometry to his own mother, Margaret, and her sisters.)

To a non-scientist like myself what is fascinating is the incongruity of the impenetrability of the task and the everyday settings and materials used in problem-solving. It's the Professor Brainstorm stereotype found boyishly charming by many. 'From the outset,' according to Batchelor, 'Taylor exhibited the engaging curiosity of a bright child and retained that fresh enquiring attitude throughout his life.'[5] As if building a soapbox cart, his experimental attitude was 'to test some ideas or calculations after he felt he understood what was going on rather than to acquire data from which understanding would come later. He usually knew in advance what to expect.'[6]

Just before midnight on 15 April 1912 the SS *Titanic* on its maiden voyage to New York collided with an iceberg off the coast of Newfoundland. There was heavy loss of life. Its vaunted unsinkability in theory, owing to its sixteen watertight compartments, was disastrously holed when five of them were pierced. The calamity led to the first International Convention for Safety and Life at Sea and the recommendation that an International Ice Patrol be established to give warnings. Taylor was recruited as meteorologist to the 230-ton wooden whaling ship *The Scotia*, on which he put his practical skills to work flying kites and balloons to gather data to understand and model atmospheric turbulence. Not long after, during the First World War, the same combination of theory, practice and adventure directed him into another emerging field, that of aeronautics at the Royal Aircraft factory at Farnborough.

Although, according to him, it was 'an asylum for theoreticians',[7] his adventurousness ensured he signed up as one of the magnificent men in their flying machines. Like many, he diced with danger while parachuting. His last jump in 1921 was less risky, descending into the then women-only Girton College at Cambridge. More dangerously, he narrowly escaped death when the plane in which he was testing stability crashed and killed his designer colleague. The flimsy, fabric-covered creations held together by piano wire fell from the sky only too regularly and needed great skill to keep them aloft. Taylor reported that on cross-country navigational trips kindly station masters would spell out their station names with whitened stones on their flower beds. Navigating in cloud or at night presented similar problems that provoked early designs for suitable cockpit instruments.

Another military application involved the almost Heath Robinson idea of dropping darts or flechettes onto soldiers from above. The problem was to make sure they dropped without whirling. Taylor experimented dropping prototypes down a disused factory chimney and then needed to check whether they had a terminal velocity enough to inflict sufficient damage. A rifle was bored out and the darts were shot at a leg of mutton to simulate the effect. The successful design, however, was not used as they were regarded as 'inhuman weapons that could not be used by gentlemen'.[8] The niceties of inflicting death can be strange; presumably the medieval practice of propelling darts with a device made from wood, gut and an archer's skill could be considered 'civilised'. A moral concern for life on a galactic scale, compared with these examples, was to face him in later life.

After a period at Cambridge experimenting alongside Sir Ernest Rutherford on fluid mechanics, the years of risk-taking brought new gambles with womankind. Although his gentle manner probably pleased women, he did not find it easy to form close relationships. Early in the 1920s in his late thirties he became enamoured with Ada Nettleship's daughter, Ursula, a singer and choral trainer. In 1924 she was another of the party voyaging on *Frolic* off the west coast of Scotland.

The whole family nexus of Hinton/Boole and Nettleship came to a halt where Ursula was concerned. Appalled by her sister Ida's Parisian ménage à trois with the rumbustious painter, Augustus John, and his bohemian partner, she vowed never to marry. Poor Geoffrey reaped the harvest of others' loose living. Perhaps on the rebound Taylor soon struck up a friendship with Stephanie Ravenshill, a teacher in Birmingham. After many exchanged letters he popped the question somewhat ungallantly in one dated 21 May 1925: 'Will you marry me? It's a risk of course but … if you really care for me as I care for you, it's just the risk … more worth taking than any other in the world.'[9] She reciprocated; immediately after their wedding he took her sailing on *Frolic*. Fortunately she enjoyed seafaring too, including a passage across the North Sea to the Lofoten Islands in 1927 and circumnavigating Ireland in 1931. She accompanied him in 1929, en route to a conference in Japan, on a tough six-month journey through Borneo, braving jungle trials in dug-out canoes, deadly snakes and crocodiles. On firmer ground they planned and had a house built, 'Farmfield', in Cambridge in which they lived until he died.

In 1923 Taylor was appointed to a Royal Society Research Fellowship that he elected to undertake at Cambridge, continuing at the Cavendish Laboratory. This allowed him to relinquish teaching – fortunately for his students as he had apparently little aptitude for it. The post was occupied by him for another forty years and made possible by his faithful technician, Walter Thompson, who made all the required apparatus for G. I.'s main fields, researching the mechanics of solid materials and fluids.

His inventiveness paid off with his production of a revolutionary, lightweight but strong anchor suitable for yachts and seaplanes. The CQR ('secure') marketed by him in 1934 is currently still obtainable. Batchelor

Geoffrey Taylor and Stephanie aboard *Frolic*.

describes it as 'a triumph of geometrical and mechanical imagination'.[10] Indeed its remarkable holding power for its weight suggested its use on the floating 'Mulberry' harbours from which the Normandy landings were launched in 1945. This is also the year that takes us into deeper and more clouded waters.

According to his dutiful biographer, G. I. Taylor could be counted among 'the three giants in mechanics during the first half of the twentieth century'.[11] At the outbreak of the Second World War, 'there was no one in Britain with an equal skill in sizing up a novel scientific situation, uncovering the essential processes at work and formulating the real problem'.[12] Inevitably G. I. would be looked to as an adviser. Graham Farmelo, in *Churchill's Bomb*, describes him as 'a brilliant applied mathematician and unrivalled expert in fluid mechanics …'.[13] Practical defence applications included underwater explosions and their effects on structures, the fragmentation of bomb casings and the dispersal of fog from aeroplane runways using lines of burners.

His most important involvement was with the committee set up early in 1941 to consider the production of nuclear weapons. As part of a small but crucial group of British scientists, Taylor spent three visits from the summer of 1944 at Los Alamos, New Mexico, engaged on the Manhattan Project to develop two types of atomic bomb. His role was to act as a general consultant advising on air shockwaves and aspects regarding detonation of the plutonium version. Doubts whether it would go off at all were allayed when on 16 July 1945 it was successfully detonated in the desert not far away. Following the dropping of the uranium-based bomb on Hiroshima on 6 August, it was exploded over Nagasaki three days later. (We shall return to explore the background to these events in greater detail in Chapter 21.)

His description of the bomb's detonation was repeated soon afterwards (too soon, felt some) as part of a BBC series of radio talks on *The Genesis of the Atomic Bomb*. It was published in *The Listener* under the incongruous title *Trying Out the Bomb*. He concluded:

It has been my good fortune … to see the important contribution which the mostly young … British scientists have been making to the success of this project. I should like to say that the American, British and continental scientists … were fully aware to the effect which this work might have on human destiny… [14]

Taylor concluded a lecture in 1971 with the statement, 'It is now the fashion to blame scientists for social evils which may arise from the use which others make of their work. I must confess to the old-fashioned belief that it is the business of our government and not that of scientists to control that use.'[15] It is ironic indeed that his grandfather, George Boole, who he so admired, thought that 'The authoritative Moral perception when wedded to the discipline of true science … is favourable to a sound morality. If it exalts the consciousness of human power, it proportionately

Trinity College, Cambridge, from a seventeenth-century print.

deepens the sense of human responsibility.' Albert Einstein would have concurred. While he would have also agreed with Rutherford about the 'urge and fascination of a search into the deepest secrets of Nature', he also asserted, 'When men are engaged in war and conquest the tools of science become as dangerous as a razor in the hands of children.'

Batchelor believed that G. I.'s response to the wonderful physics of the bomb was 'innocent and natural, like his enjoyment of aeroplanes. He was not reflective and moral and philosophical issues did not often engage his mind.'[16] His ability to connect with children was commented on by members of the Hinton family with whom he kept in close contact. It must have saddened him to have had none of his own. To be blunt, however, there can have been no more hell on earth than the inferno that engulfed thousands of innocent Japanese children in 1945. Like other scientists, Taylor probably believed that there were mitigating circumstances. His detachment from the moral dimensions of war sits uneasily when compared to a letter he wrote to an American colleague in October 1940 about events in Cambridgeshire: 'The thing that makes me really see red is when … a nazi aeroplane … find[s] a village or town where they are momentarily free from our air-raid defences and come[s] down and machine-guns the people in the street, women doing their marketing, children coming out of school.'[17]

On the one hand we encounter the person who took such a delight in his garden at Farmfield, exemplified by a 1954 letter from Ethel in New York to him: 'Do you know the Korean rhodosa … don't plant it next to jasmine or forsythia, their colours fight', and on the other, the man closing the gate, briefcase in hand, set to visit a group in London considering weapons of mass destruction. We all share contradictory natures and complex values; as his Aunt Ethel remarked, 'The human mind is a queer jumble.' The moral responsibility of scientists in the post-atomic age was an issue much aired throughout the Cold War. In 1955 Einstein, with Bertrand Russell, wrote a manifesto that led to the Pugwash Conferences that brought together scientists and others to consider issues of arms and global security. For Taylor there was a curious postscript to the Japanese episode.

Thomas McMahon, a Fellow in bio-engineering at Harvard University, published his first fiction in 1970 under the ironic title, *Principles of American Nuclear Chemistry: A Novel*. Narrated by the teenage son of a scientist named McLaurin, it aped the personal lives of some of the scientists at Los Alamos. Enrico Fermi became Ferrini, Niels Bohr – Orr, and Lisa Meitner – Selina Meisner (although she actually refused to work on the bomb). Taylor is nowhere named but can be inferred to be McLaurin senior through accurate and explicit references to his papers. Most particularly the novel pinpoints a remarkable piece of later research that had allowed Taylor to estimate the energy of an atomic explosion by comparing a sequence of photographs of a fireball. McMahon fictionalised Taylor exactly in McLaurin: 'The way he had used his intuition to lead his mathematics was nothing less than elegant [he] knew the answer to the problem before he started.'[18]

So far so relatively uncontentious, but McLaurin is having an affair at Los Alamos with a young dancer who appears throughout the novel dallying with an unflatteringly louche collection of scientists. The steamy tone alludes to Selina Meisner as a tease and pictures Orr in a pond with McLaurin's *ingénue*. In reality, the cloistered, dedicated Alamos community was bound to let its hair down. Taylor asked for the withdrawal of the book but had to put up with just an apology. What seems incongruous is his private self-censoring of comment on the vast moral issues surrounding the bomb and his public censoring of a novel of very limited significance.

G. I. was both knighted in 1944 and given the Copley Medal of the Royal Society and awarded the US Medal for Merit in 1946. One commentator on his life suggested that the only reason he was never given a Nobel Prize was because no one could be sure whether he was chiefly a mathematician, physicist or engineer. His fame in the mechanics of fluids and solids drew to him at Cambridge a 'Taylor school', one of whom was George Batchelor himself. Although formally retired in 1951, Taylor carried on at Cavendish Laboratory, producing between the ages of sixty-five and eighty-six no less than forty-eight academic papers. Some of them demonstrated his 'kitchen sink' study of phenomena and experimental ingenuity. After he'd been shown the thrusting of bull sperm under a microscope he was inspired to open up research into a new field of hydrodynamics, ironically akin to Thomas McMahon's own interest in adapting the secrets of animal locomotion to human needs.

Taylor was clearly an original thinker, preferring concrete situations and specific problems to generalisation and the abstract. His enthusiasm for understanding the mechanical came over in the simplicity of his research. One 1964 item was concerned with the instability of adhesive tape during peeling. Often research into the apparently trivial would lead to new insights.

One field of study lay in the propulsion of unusual organisms through water, including snakes. His last paper of 1969, when he was eighty-three, concerned the behaviour of water droplets in strong electric fields such as a thunderstorm. It led to the theoretical properties of a cone-jet, named after him, which found application, for example, in the thrust of spacecraft. Howard Everest Hinton, the son of his Mexican cousin George, as we shall see, demonstrated a similar perspective in his field of entomology. The sense of awe at the natural world is reminiscent too of Taylor's grandfather, George Boole.

His wife, Stephanie, died in 1967 after forty years of a contented marriage. Batchelor comments, 'Stephanie was brisk, efficient and warm-hearted … kind and hospitable', and recalled 'the very English atmosphere of Christmas at Farmfield with punch and hot mince pies'.[19] She must have contributed a great deal to his wellbeing and thereby enabled him to use his scientific ability fully – perhaps much as his mother, Margaret, had done. Somewhat lonely in the last seven years of life he struck up a friendship with a woman twenty-eight years younger. As she represents one of the few face-to-face recent contacts between my Boole family and his, I may be allowed a small digression in acknowledgement.

At Lincoln for the 100th anniversary of George Boole's death in 1964, my aunt Muriel (she with the Everest middle name and a strong interest in lineage) had attended with her cousin, Gabrielle Boole. (A photo in

Chapter 1 shows them both standing in a group with G. I.) I was intrigued to find in the Taylor Trinity archive that the meeting with Gabrielle had led to some correspondence over the years. I too remember her calm intelligence and charm, which must have impressed Geoffrey Taylor as well.

There were continuing contacts; correspondence shows that he'd asked her to come to Cambridge over several years. They certainly met again in February 1972. She wrote after attending the Candlemas Feast at Trinity: 'I enjoyed every moment and feel most privileged to be one of the few women to participate in such a historic occasion.'[20] There was talk of organising an exhibition of Taylor's father's wild-flower paintings. Not long after, however, G. I. suffered a severe stroke, paralysing most of one side of his body and affecting his speech. Gabrielle continued her weekend visits. On 1 May 1973 she writes: 'I look forward to seeing you again before long.'[21] Three years later, just short of his ninetieth birthday he experienced another stroke from which he died two months later. Farmfield was left to his house-keeper, Gladys Davies, who left it in turn to fund a permanent Fellowship at the University in fluid dynamics, a highly suitable continuation of Taylor's work there.

Geoffrey Taylor's remarkable life seems to have belonged to a past age. His securely tenured position at Cambridge enabled him to pursue his interests and instincts without undue reference to commercial or corporate pressure: a pursuit of knowledge for its own sake. Batchelor maintained that this enabled him to recognise the essential aspects of phenomena or a problem that would lead to fundamental and wide applicability. His sequestered academic life perhaps could only flourish where there was no real regard for the political and social aspects of his endeavour. Away from the isolation of his laboratory the real world was encountered by his love of the outdoor life among the elements. Above all, beyond his 'razor-sharp mind', Batchelor concluded in his biography: 'I have never known a man to fit so naturally into the pattern of his life.'[22]

# THE HINTON GENUS

Charles Howard Hinton not only bequeathed a challenging set of ideas relating to the fourth dimension but eventually a generation of kin who strove both to set the world to rights and catalogue some of its natural wonders. His and Mary Ellen's exile from England in 1887 to Japan and the USA seems to have established a Hinton precedent of restlessness. Their Boole relatives, in comparison, mostly based themselves closer to home in England. The Hinton family branch that became associated with Mexico combined an inclination to wander, even courting danger, with an intellectual curiosity.

To continue our narrative we need to follow the dynastic trail back to the Hinton children: George, Eric, Billy and Sebastian ('Ted'), who had followed in their parents' footsteps to Washington DC in 1901. By 1908, within one year, the boys had lost both their father and mother, while still only in their fifties. This loss of family stability possibly added to any existing tendency to wanderlust.

George Boole Hinton, the eldest of the four, had developed an interest in mineralogy when young. This had led to studies at the Minnesota School of Mining and the universities of Columbia and California. Summer vacation jobs in Mexico assaying for mining companies had not only helped pay for his and his brother Ted's education, but generated a sense of adventure regarding such a completely different and exotic culture.

In a somewhat more sylvan clime at the Byrdcliffe Arts Community in Woodstock, New York, where the Hintons were regular visitors, in 1910 he had met and married Emily Wattley, the daughter of a rich owner of a grocery chain. The following year they moved permanently to Matehuala, north of Mexico City, where he worked as a mining engineer. Copying his parents' proclivity in breeding boys, three followed in quick succession: Howard (1912), George (1913) and James (1915). The move from the US was either a brave or foolish thing to have done; the country was in the middle of violent turmoil. As with brother Ted's children, revolution seems to have been a strong attraction.

George Boole Hinton (1882–1943), mineralogist and plant collector, lived in Mexico.

After a period of expansion through foreign investment, Mexico had emerged in the twentieth century as a typical modernising state, replete with the prizes of a national health service, increased life expectancy and lower illiteracy. The downsides of class inequality and a tradition-bound, overbearing Catholic Church were catalysed in 1910 by a rigged election that propelled the country into a decade-long virtual civil war. Power struggles and assassinations racked the nation. Nearly a million died during the tumult but eventually a more egalitarian, anti-clerical democracy emerged.

Before his marriage, George had already set up a mining consultancy in Mexico City. (We shall designate him as 'George Snr' to distinguish him from his son, George.) The family moved around the country as was the nature of prospecting in the industry. In 1913 they found themselves at the Rincon gold and silver mine in Temascaltepec, southwest of Mexico City, 5,000 feet up in the mountains. 'Pretty bad country', as his son George described it in his unpublished and sketchy 1993 memoir *The Life-story of a Hard-rock Miner*. Its worst problem was the violence from the ongoing revolution. He recounts a story reminiscent of his cousin Joan's later exploits in Mongolia.

According to George, his father had gone there to help build the plant at a time when bandit factions were swarming everywhere. In order to obtain money they often ransomed foreigners, obliging any such employees to hide by descending into mine shafts for safety. When the family arrived at the mine George's mother was heavily pregnant with him. Fortunately, finding no one suitable to extort, the rebels quit. He was born four days later, a suitable introduction to a nation seemingly permanently in crisis.

From December 1915 until February 1917 the site was eventually shut down owing to its occupation by the anti-government Zapatista general, Inocencio Quintanilla. He had stripped it of any assets, often torturing workers in the process, and turned its machine shop to making bombs and cannon. With evident bravery George Snr returned there once again, thankfully without the family, to attempt to get the place productive again. A bargain with the general was struck that the company would pay taxes to him on extracted ore. This was thwarted by the arrival of government forces, who, 'spent a day looting and burning, killed many children, carried off a lot of women and left'. The disaster was somehow attributed to George Snr, who was jailed with another company man for ten days by Quintanilla, with a ransom to pay. George writes, in a mammoth understatement, 'For ten days he had a most unpleasant time.'

Not surprisingly, such events prompted George Snr to move operations to the relatively safer locale of Mexico City. In August 1919 he wrote to his brother-in-law, Ralph, 'Business is beginning to move quickly. I now have the best laboratory ever put up in this country.' His faith in his adopted *patrie* is less sure: 'It is a strange country this; a nation with an acutely developed sense of personal pride and dignity but without the capacity for ... truth and honour.'[1] Doubtless few revolutionaries would have agreed with this estimation, but George Snr is living with political instability alongside the macho world of metal extraction and all its accompanying riskiness. World price changes and speculation over viable ore deposits combined with the huge dangers of the mechanical processes: exploding, crushing and grinding rock plus the use of hazardous chemicals such as cyanide. It is all a long way in every sense from the Princeton, ivy-bedecked halls once frequented by his father, Charles Howard.

In 1919, owing to George's and his mother's tuberculosis, they moved to Cuernavaca, a more salubrious small town south of Mexico City at a lower altitude, better for their health. George notes, 'All the main streets were cobbled, the rest were dirt ... local transportation was by horse-drawn carriages and mule-drawn street cars.' If mining in the wild landscape of Mexico was a man's world it was also an environment for boys to be ebullient. Compared to Orwell's recounted boyhood of bird-nesting and blowing up toads there was more plentiful and exotic fare to marvel at and tease. The surrounding countryside was a treasure-trove where the boys could 'run in the fields and turn over rocks to find all sorts of insects ... that we had not seen before. Lizards and turtles ... and beetles of all descriptions.'

Some could be captured and brought home to play pranks with – a bat was kept under Howard's hat and released in the dining room to frighten their delicate mother. George Snr writes in 1924: 'There is a live rattle-snake on the mantle and live animals in every room, also trays, bottles and drawers of them all over the house, and the garden is a zoo.' Alongside the familial fauna to engage with, as with Geoffrey and Julian Taylor, there are boyish antics experimenting with gunpowder: 'Howard's toys are fire-arms, microscopes, boxing gloves, carpenters', masons' and pipe-fitters' tools.'[2] The microscope foretells the direction of his later life.

Beyond the natural world that inhabited their bedroom lay a real social world of incidental and casual cruelty. In *The Power and the Glory*, Graham Greene wrote how in Mexico, 'a little additional pain was hardly noticeable in the huge abandonment'.[3] George remembers being in a warehouse where he witnessed caged rats doused with alcohol and set alight: 'To this day ... I can still hear the agonised squeaking as they burned to death.' Such cruelty to life was complemented by life's cruelty. Their mother Emily was weakened by malaria, which, as a Christian Scientist, she was not prepared to treat except by fasting. She died in March 1923. George, aged ten, remembers his father's grieved outburst: 'Jesus Christ is a rotten potato.' He is stoical, however, writing to her brother in May 1923, 'Three boys to nurse may seem quite a job but the load is shaped to fit my back.'

When Geoffrey Taylor wrote to George Snr, his cousin, in May 1924 about a Boole/Hinton 'taint' (as we first mentioned in Chapter 16) he also expressed 'a strong desire to meet him again after many years'. He writes about 'the faint but pleasant recollection of our early days together', before the family left for Japan in 1887. According to the family group picture of 1897 with the two boys included, they had met again during a return home visit by Mary Ellen to her Boole sisters in England. Now as adults, Geoffrey wanted to invite George Snr and his three boys to come and stay in England. He writes too about current romantic matters, 'there is something exciting in the idea of you me and Ursula [Nettleship] meeting again'. We don't know when the three of them had met, but like his cousin after Emily's death, Geoffrey was thinking of finding a mate. He continues, 'it is only in the last few weeks that I have known her well enough to appreciate what a fine creature she really is'. As we documented in the last chapter, Ursula was set to reject *any* man's advances.

In April 1927 George Snr corresponded with her in wooing mode. He too didn't know that she was not going to be a runner in any man's chase. The hapless contender's style is as woeful as cousin Geoffrey's, revealing his blunt character. He writes, 'My Dear Ursula, A simple little proposition; I want you; what is your reaction?' He canters on honestly but is lacking finesse: 'All your life you have lived in the centre of an effete civilisation ... all the pretty components of your London environment would be like luxuries here.' He is honourably concerned about his financial provision for her but the odds against him mount with his final leap: 'Unconditionally and regardless of the sacrifice you might have to make – will you marry me?' The fated answer was, of course, 'no'. In 1927 Carmelita Hinton (Ted's wife) and her children first visited George Snr and the boys in Mexico. She wrote to Ursula later agreeing that she had made the right decision about marriage: 'It's a fascinating country, but I'm afraid from what you say George would have been a bit trying. He is certainly a queer, eccentric person – but with a large generous heart.'[4]

George Snr had given up his business in 1923 as metallurgist and obtained a post as chief chemist with a cement company. Owing to financial problems half of any wage had to be taken in cement, provided it wasn't resold. Considering Mexico's recurrent earthquakes he put it to good use building a reinforced-concrete house to the south of Mexico City. Its success and the great demand for building materials in the burgeoning economy encouraged him to start his own enterprise. From the expertise gained, in true Boole/Hinton fashion, he invented and patented 'floating cement', a lightweight block-building material made by frothing a mixture of water, cement and chemicals by blowing air into it. A million dollar company was launched in the US in 1929 to exploit the patent but sank owing to the stock market crash. George Snr returned to mining. The worldwide economic turmoil and its implications for Mexico compounded its continuing internal social problems.

After the revolution the Constitution of Mexico of 1917 placed severe restrictions on the Catholic Church, declaring its separation from the state and secularising education. With the election of President Calles in 1924, atheistic principles turned into anti-religious campaigns. Property was seized, religious institutions closed and day-to-day limitations placed on the Church's presence, such as the public wearing of priestly garb. Passive Catholic resistance escalated to actual rebellion in turn leading to draconian government methods that included depopulating areas and concentration camps. After episodes of renewed horrific violence a truce was mediated by the US in 1929 and a limited freedom returned to the Church with strict secular boundaries. Some 900,000 people had died in the Cristeros War as it was called. Martyrs were produced. In a drama reminiscent of *The Gadfly*, the priest Miguel Pro was executed without trial in November 1927 on assassination charges. Posing with his arms spread as a cross before the firing squad, only a point-blank bullet eventually killed him. Photos of the execution were produced to encourage an anti-clerical response but were taken up as objects of devotion.

Another echo of *The Gadfly* is realised in *The Power and the Glory*. Despite the ending of explicit religious persecution, the vendetta persisted. Greene set the novel in the 1930s in the northern state of Tabasco where he had spent some time. The whisky priest on the run, who has, like Mont-anelli, sired a child, is hunted by a socialist lieutenant prepared to kill and torture to eradicate the Church's power. In order to hear confession from a dying man the priest is trapped into meeting the lieutenant and martyred for his beliefs. The moral victory of Ethel Voynich's atheist Gadfly is reversed.

The aftermath of the insurgence was later to involve George junior but during this period the Hinton family were relatively safe. George père had built his concrete house in Colonia del Valle in the countryside on the edge of Mexico City. Once again, his son recalled, 'It was an ideal place for us to collect toads, frogs, snakes, turtles ...'. Back in their city apartment they kept them in the bathtub. One snake escaped down an overflow to emerge in a woman's bath below. The three boys' boisterous life continued unruffled.

Even making allowance for the loss of a mother at the tender age of eleven, it is Howard who emerges as the dominant sibling of the trio – and the most flawed. Howard seems to have developed a begrudging inner resilience. The resentment, for example, over physical punishment from their father for misdeeds was borne by him overlong. George writes of his brother as 'husky and strong in contrast to me ... he was a real bully and made us his slaves ...'. He recounts how during boxing matches Howard could be very assertive. When once the reverse happened and their instructor delivered a too-powerful punch, Howard rushed upstairs to get his .22 rifle to threaten him. Less able to defend themselves, stray cats were given cyanide-laced milk and snakes were beaten.

Howard showed the most academic promise, however, and was the first of the brothers to leave Mexico for the US. (We hear very little about James the youngest.) His education from

Howard Everest Hinton (1912–77), entomologist, eldest son of George Boole Hinton.

1926 involved high schools on the west coast before taking a science degree at the University of California, Berkeley, graduating in 1934. George Snr was able to finance him, writing in May 1929 that he is in a sounder position than most foreigners, 'despite the fact that the revolution messed things up for everybody'. Howard had begun on his career as an entomologist, asking his father for money to make beetle cases for his collection. Next to his photograph in a California newspaper of January 1931 is the heading, 'Junior Beetle Collector' and underneath, '20,000 beetles! That's the ambition of Howard Hinton ... who has already ... identified 2,000 species.'

320

The amassing is a result of a number of summer vocational shoestring trips involving riding freight trains. Robbery was an ever-present risk. On one occasion he writes that having lost his wallet he had to live on rice, carrots and squirrel. The last ingredient 'tasted horrible'. By 1933 we learn that he has mounted 2,500 specimens of moths and is selling some that his father has sent. His career is galloping, by the same year he had had eleven species named after him and written ten papers. He writes to his father in April 1934 in his characteristic bombastic way, 'Not that it means anything, but I have written more than any undergrad. has.'

The collecting bug had also overtaken George Snr, his speciality being flora, especially orchids. He sent seeds to the Royal Botanic Gardens at Kew in England. Mexican species were little known because of the difficulty of collecting in forbidding mountainous areas. His experience of the terrain and people was a major asset. He resumed his senior post at the Rincon mine at Temascaltepec, writing in September 1933: 'We have got out over a million pesos of silver and gold in the last few months.' On the domestic front he had married again, to Erica, a doctor. George and James had continued their education, the former entering the University of California in 1933, the latter finishing high school a year later. Familial contact between the Booles and Hintons was re-established when Geoffrey Taylor met up with the boys on a conference trip to Vancouver.

Both George and James, true to form, joined forces in Canada over a footloose vacation full of scrapes: canoeing, hitch-hiking and riding freight trains, ending up in New York city. James began a degree at the University of British Columbia but was more interested in writing poetry than studying, so returned home to help George Snr gather plant specimens. By May 1935 young George had flunked out of college and returned to Mexico to take up mining at the Rincon mine. The routine was almost as dangerous as when he had been born there in 1913. His father was understandably very worried about the bandit situation as George had written how 500 government soldiers had been killed nearby: 'They stopped the convoy ... shot the drivers and burned the trucks.'

George nevertheless took entirely to the underground life of a mining scientist. Above-ground life, when it wasn't countering banditry, was almost medieval in its ways. Suffering from nosebleeds, miners at the camp referred him to Don Abundio, a witch doctor. Diagnosing from a drop of blood, George was told to drink a glass of his own gore as a cure. 'I took a deep breath but I just couldn't do it. So I called the cook who added butter, salt, pepper and onions. I ate it with great gusto, but my nosebleed did not stop.' His stepmother recounted how in the wilds of Guerrero state a mother had brought her daughter for treatment, 'at the point of death stricken with diptheria; nothing could be done'. A local witch doctor disagreed. Pointing to several dogs outside he commanded that one should be castrated. In not more than a minute the testicles were brought inside on a plate. One was forced into the child's throat cutting off breathing and causing immediate death. The witch doctor merely retorted that the wrong dog had been chosen.

Life in Mexico was dangerous and cheap with death unsurprisingly celebrated in its popular folk art. While working in 1936 at the El Orito mine in Aguascalientes, 300 miles northwest of Mexico City, the area was troubled by factions of the still-active Christeros movement that had turned to banditry. Clashes with government troops were common and extortion a regular practice. George, yet again, was in the thick of it, presented with a demand note for 100 pesos. Eventually at night the bandits turned up for payment brandishing large-calibre revolvers and rifles. Taking exception to his attitude they decided to shoot him and dragged him to his office. Protesting that if he was shot he would not be able to hand over the ransom, he cheekily asked for a receipt. The leader complied. 'I was so nervous on the typewriter that my fingers were hitting between the keys and every other word was mistyped.' All the bandits made their mark on the document and left satisfied.

Howard's life, now settled in Europe, was a good deal less threatened but uncomfortable in its own more staid way. 'I am much disappointed with England,' he writes to his father in April 1934, 'it is a has-been nation. Most English people are born snobs.' He no doubt suffered something of a handicap for someone studying for a PhD at Kings College, Cambridge, hampered with an oddly-tinged Hispanic accent. More importantly, he had visited the British Museum and although unimpressed with its insect collection, having more and better-mounted specimens of his own, he decided that he would spend his life working there. Howard does, however, find a homely anchor having supper every Saturday with his first cousin once removed, Geoffrey Taylor, a senior member of Trinity College. He plans also to meet Taylor's brother, Julian, now a successful surgeon. The connections continued to cement the transatlantic Boole/ Hinton dynastic bonds maintained primarily by Ethel and Wilfrid Voynich. At one gathering he met Geoffrey's mother, Margaret (Maggy) Taylor, the second-born of the five Boole sisters. She remembered George Snr visiting England when he was fifteen in 1897.

Another visitor to Cambridge was Carmelita Hinton passing through on her way to Germany with her children Jean, Bill and Joan, cementing their own familial ties. Howard comments in a letter that 'Joan has grown up to be a remarkably good-looking girl and that Carmelita is very popular with all the family here'. Her offspring are imbibing their mother's radical perspectives, which will impel them to engage later with revolution in China and left-wing politics in the USA. It's not so clear, however, where Howard's own left leaning views originate, not obviously from his father, George Snr.

The atmosphere at Cambridge during the 1930s and the rise of fascism that sprouted the likes of Burgess and MacLean generated a naïve faith in the Communist Party and the Soviet Union. Stephen Korner, in his obituary of Howard, notes that 'his convictions were deeply and firmly rooted in a revulsion against any kind of oppression or exploitation … strengthened by his own observation of grave social injustices in Mexico and elsewhere'. Korner adds that Howard had been greatly influenced as a young man by reading Robert Tressel's *The Ragged Trousered Philanthropists* (as was my own father). The novel describes the exploited lives of a group of painters and decorators in Edwardian England. Howard's daughter, Charlotte, whom I met at her London home, maintained that his Marxist convictions were academic and entirely separated from his personal life. Even blunter, she was happy to assert that her father was a 'stroppy, intelligent, ferocious outsider'. This smacks perhaps of the dynastic 'taint' yet again.

It was at a Cambridge dinner party around 1937 given by Henry Morris, promoter of the communitarian village-school movement, that Howard met Charlotte's mother, Margaret Clark. Her pedigree from 'decayed gentry', according to Charlotte, cannot have helped Howard's self-confidence when he became labelled as 'Margaret's Mexican'. This may have been unflattering but reflected truthfully his origins and his love of an outdoor life, foraging among raw nature, now wedded to a career as an entomologist. In 1937 he joined an expedition to Lake Titicaca in Bolivia in that capacity, plumped up by his knowledge of Spanish.

In July he set off on his own to the east coast, visiting French Guyana and Trinidad. He writes on a cargo plane crossing the Andes, 'my first view of the Amazon region was extraordinary. It seems one vast plain of green with numerous broad rivers meandering through and many lagoons.' His exploits often read like a Graham Greene novel: encountering a gang trying to spring convicts from Devil's Island and narrowly avoiding arrest at a border check carrying gold. Waiting for a steamer to Trinidad he met some English miners and got drunk for two days. 'They more or less carried me onto it,' he reports. Animals and insects could be a constant irritant as well as an object of study; he writes of experiencing a rat crawl his body-length and being badly bitten by bedbugs.

His South American notebook announces very precisely on page one, '3/6d will be given to or sent to the person returning this (if lost)'. It records both his 'rootin' tootin' school-boyishness and his academic interests collecting specimens: '28/7/37 Shot an alligator about nine feet long with a 32 dum dum. Had two shots at a capybara and missed. Termites were taken in rotten palm log on ground. 30/7/37 Got large adult mayfly in vial at light also large species of cricket.' 'From tyrant fish took 3 lice, eggs and many mites. 25/8/37 Most of morning was spent collecting in cow dung. Shortly after I found what may be a new genus *aphadidae*.'

One incident gave him 'the most prolonged fright ... of my life'. Paddling his own canoe up an inlet, 'there was suddenly an enormous roar only a few feet away from me in the jungle, but the bush was so dense I couldn't see the jaguar'.

The animal was obviously following me. I started to paddle back for the village as fast as I could. I put the .38 on my knee and I thought

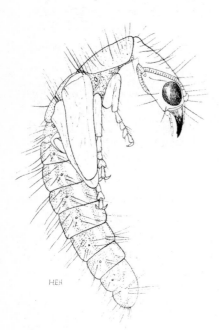

HEH

that if I had a chance I could take a shot at it, otherwise I would over-turn the canoe and try to come up in the air-pocket inside, although ... the inlet was full of alligators. Nothing happened and when I got within 50 metres of the village the sound suddenly stopped. I beached the canoe and shouted to the nearest man 'tiger, tiger in the bushes.' He laughed at me and said 'mono' which means monkey. That was my first experience of a howling monkey and completely fooled me.[5]

Back in the quieter waters of Cambridge, now a family man, he completed his PhD in 1939 on the classification of a genus of beetles *Dryopoidea*. Firmly ensconced in the

Insect drawing by Howard Everest Hinton.

institutions of marriage and academia, his life became increasingly remote from his father and brothers, George and James, back in Mexico. George rebukes him in a letter of 1939: 'Why the hell don't you write more often … I hope to get an answer in about a year if I'm lucky.' Both of his brothers had been living and working in the rough-and-tumble world of the mines, moving around the country.

In 1937 the two were working as supervisors at the Real del Monte mine near Pachuca in Hidalgo state northeast of Mexico City. It was a peak time, aided by US investment, producing 4,000 tons of metal from silver-bearing veins. Mining history had always relied on migrant workers and innovatory technology. In the early nineteenth century the mines were revitalised by an influx of out-of-work Cornish miners bringing with them the Trevithick steam engine – and leaving for posterity the *paste*, a crescent-shaped pastry now filled with hot peppers as well as lamb.

George's account of their lives, when not describing the different ways of extracting ore, recalls anecdotes of dangerous competitiveness and horseplay, often made worse by the drinking of *pulque*, a milk-coloured, plant alcohol that lightened the gruelling working conditions. Tricks were played letting hoists freefall; one persistently sleepy hoist man had a fuse lit and placed in his shoe. George found himself the butt of canteen bravado after the cook, Lilly White, had put chillies in his marmalade sandwiches. He retaliated: finding a flattened rat, he trimmed it to size, rubbed it in mayonnaise and placed it in the cook's sandwich. George himself was sacked for falling asleep on his job signing time cards. He moved on to be a shift boss at the Cananea mine owned by the US Consolidated Copper Company close to the western US border state of Sonora. The mine exemplified the perennial, ubiquitously problematic nature of the industry.

In 1906 a famous strike had taken place there. The Mexican miners were opposed to the differential in pay between them and foreigners and the deplorable working conditions. Reciprocal violence led to the deaths of over twenty men and the calling in of 300 'volunteer' strike-breakers from across the border in Arizona. The continued bitter struggle at the plant was considered to have been a contributory factor in the downfall of Diaz in 1910 and the beginning of the Mexican revolution. In 2010, a century later, the Cananean saga renewed itself: 600 police were called in to oppose Mexican miners' longstanding protest against safety hazards.

Appalling accidents have always been endemic. George recalls witnessing a particularly grisly scene in 1939 at a silver mine at Parral in Chihuaha state. A slab of fallen concrete a metre thick had imprisoned one worker from the waist down, unable to escape.

> He was alive and digging his fingers into the waste trying to pull himself out, until all the flesh was worn from the end and the bones could be seen. When the man was finally dug out it was found that he was completely cut in two; the only thing that joined his torso to his pelvis was his backbone.

George Snr's health had deteriorated owing to malaria, the blood-vomiting symptoms of which he was treating by eating ice. In 1936 he had decided to devote himself fulltime to collecting plants, finances permitting. George writes: 'My father could ride a mule from morning to night, day after day, but because of a heart condition could not climb the highest mountains on foot … [it] was left to James.' They travelled as frugally as possible. The riding mules were treated well and personally shod by George Snr. James was so attached to his own mule, Lenina, that he asked plant expert Carl Epling to name a salvia plant he had discovered in its honour. Continuing the fine tradition of the Hintons, knowingly or otherwise, upsetting the US state, during the stubborn era of McCarthyism, Epling had to prove that there was no intentional approbation of V. I. Lenin.

Their trips took them into the wildest territory, inhabited by seemingly universal bandits. Father and son conspicuously carried pistols and relied on the goodwill of the peasantry to survive. George Snr writes to Howard in 1939 that he'd discovered 'a new orchid as pretty as a snowdrop, *Hintonella Mexicana*'. Whether pursuing insects, exotic plants, exquisite ideologies or hidden dimensions the Hinton family was always single-minded and intrepid. After one stint, working away from home at a silver mine, he returned to find some of his collection had been destroyed by insects.

After expeditions George Snr would return to his ranch at Aguililla in the US border state of Nuevo Leon to lodge and classify his specimens, many of which would be sold or sent to institutions abroad. The lifestyle probably took its toll on his now fragile health. On a trip to Mexico City in 1943, his bruised, unconscious body was found by a chambermaid in the Hotel Palacio. He died shortly after, aged sixty-one, of extreme anaemia and coronary thrombosis. The British Museum, according to his son George, considered him, 'the most important botanical collector of the twentieth century'. His collection included over 300 new species and four new genera. Over 50,000 specimens were later sent to the New York Botanical Gardens and redistributed to forty other herbaria.

George had also married in 1936 and had two children, Steve and Joan, but found himself, unlike his brother Howard, stuck in the macho rut of mining with little other interests. In 1943 he moved with his family, his daughter only a year old, to the San Carlos mine in the wild, arid and treeless part of the northern state of Chihuahua. His account of his stay is peppered with details of mining practices and rare mineral samples that he collected and sold on to specialists and museums. At the mine a *manto* was being excavated, 'a body of ore which is usually lying flat ... in contrast to a vein that is generally upright'. The room and pillar method of ore extraction for silver could often be dangerous owing to cave-ins.

Throughout the whole period George had to contend with the undermining authority of the superintendent in charge, leading to disputes and bitterness. In his memoir he writes enthusiastically, however, of the beauty of fragile clustered crystal formations such as the rare 'brilliant ruby red to orange to yellow brown of vanadinite'. The physical isolation of the camp and the internal clashing wore him down; he talks of quitting and wants to make toys for a living.

By this time James the youngest had found some success writing and selling some pieces to *Mexican Life* at $25 a time. He took charge not only of the plant collection following his father's death, but also the ranch. He married Helen Hart in 1942 and began a family, yet another George, Patricia and Lorna. In 1943 he took a job at Saltillo in the neighbouring frontier state of Coahuila working for the US Goodyear Rubber Company investigating the use of the native latex-bearing shrub *guayule* as a substitute for rubber. James's passion for plant collecting continued, as did his stories. In 1945 he sent an unappreciative Howard a collection of them, writing back to him crossly in December of Howard's 'massive contempt and gargantuan disdain' for his efforts. Undaunted, in 1946 he writes again that he has embarked on a panoramic novel of Mexico, 'unwieldy, like some giant ill of glandular disease'.

Howard's dismissal of James is commensurate not only with his sense of superiority but a reflection of his own grander established status. Since 1949 Howard had joined the scientific establishment like Geoffrey Taylor, in his case as lecturer in entomology at the Zoology Department of Bristol University where he was later professor. The two Boole/Hintons exchanged queries based on their own specialities. Taylor had sent Howard a specimen of an insect that was causing a minor plague: 'they seem to stick themselves

James Hinton (1915–2006), writer and plant collector, son of George Boole Hinton.

onto white paint'. It is a larvae of the moth *Luffia lapidella*, comes the reply. Howard wants to know about the impact of raindrops on the insect eggs he has sent him plus the spinning motion of a top. G. I. replied on 9 January 1951: 'its action did not impress me because I have always believed that a hard-boiled egg will do the same thing if you could spin it

328

fast enough. The trouble with the egg is that it has no little projection which fits between finger and thumb.' If this seems to be an example of boffin egg-headedness there was also a more serious side to their concerns.

An undated letter, probably from mid-1952, that Taylor sent to Howard Hinton concerned weightier matters. Relating family news first he reports that he had met George's son Steve, who has, keeping it in the family, enrolled at Carmelita's Putney School in Vermont. Howard must have also courted Taylor's opinion on whether he should, as a renowned entomologist, presumably on the communist side, join a fact-finding, scientific investigation regarding accusations of US involvement in bacteriological warfare during the recent Korean War. The North Koreans and the Chinese had maintained that outbreaks of cholera, plague and smallpox were the result of US planes dropping containers of infected insects. A number of captured US pilots had apparently admitted this.

The episode was a very early statement of the ideological Cold War that came to dominate the world until 1989. A World Health Organisation and Red Cross-instigated group of scientists sent by the UN that failed to confirm the allegations was countered by a rival group set up by the Soviet-inspired, World Peace Council. As with Hiroshima not so many years before, the objectivity of science was being called into question. Taylor suggested that a scientific group could only decide if the feasibility of particular insects carrying particular pathogens was consistent. He repeated his formula of 1945, 'I think it would be exceedingly unwise to give expert advice on material supplied by politicians in the hope that it will further their political interest.' Taylor, taking his own loaded standpoint, believed that the accusations were, 'phoney and utterly infantile. The reasonably well-trained US scientists would have used methods 1,000 times as effective if they really had made up their minds ...'.[6] The whole international debate is still unresolved seventy years later.

Howard's credibility as a witness attested to his emerging professional stature. During the Second World War he had convincingly earned his wings as an entomological authority by his important work addressing the problem of storing food products to counter the nuisance of moths and beetles. Legendarily, he camped out next to his workbench during the Blitz to provide a 350-page monograph on beetle depredation. Later at his home in Bristol his four children, Charlotte, James, Geoffrey and Teresa also had to put up with living on the job. Howard, as when a child in Mexico, kept a miniature zoo of mammals and rodents in the household including a snake pit for specimens collected in the nearby Mendip hills. There was also a gas chamber with which pesticides could be tested. The proceeds of the latter device helped finance the children's education.

Charlotte continued the dissident tradition of the Boole/Hintons by being chucked out of her Wimbledon Junior School for asking the wrong questions. She later enrolled at Queen's College in Harley Street, London where her ancestor, Mary Everest Boole, had become librarian in 1864. She did not feel at home with its upper echelon finishing-school character, taking exception, like Mary before her, to what she saw as anti-semitism. Her heightened awareness of Jewishness must have been in part fostered by her father's virtual pro-semitism. Miriam Rothschild, a member of the famous Jewish financier family and a fellow entomologist, wrote that she 'never discovered along what strange side-road he became so captivated by the Jews'. Was it, she asked, sympathy for the persecuted? He had taken an interest in their Nazi suffering and the case of the Rosenberg atom spies, but she couldn't fathom his 'fruitless search for Jewish antecedents'.[7]

Howard became increasingly proud of his Boole/Hinton forebears. He, after all, by entering into academia in England had re-convened back across the Atlantic the loftier intellectual life of the Booles. It had been sundered by his grandfather, Charles Howard Hinton's fleeing the Old World with Mary Ellen Boole to settle in the New. Family ties had been maintained, however, enabled by modern transport. Ethel and Wilfrid Voynich had crossed and re-crossed the Atlantic from the US on their annual book-buying visits; Carmelita Hinton came and went for her European educational projects. In reverse direction Ethel's sisters, Margaret and Alicia, took occasional social visits to the USA. Geoffrey Taylor had done so too for both social and academic reasons, maintaining contact in New York with Ethel and Carmelita in Vermont. It was as if the Booles had been away on an extended gap year for two generations exploring new frontiers but had returned to take up more serious, adult endeavours.

In North America intellectual thought was complemented with a greater stress on the 'doing'. George Snr and his sons George and James continued the gutsy life of the Mexican ore industry, augmented academically by prospecting and classifying plant life. On their land at Rancho Aguililla in Saltillo, George and James planted wheat, a crop then gaining high prices on the market. They irrigated it by drilling wells to supply water via an ingenious system of tanks and drainage. It produced enough to pay off a mortgage but yields were subsequently poor. George discovered that the fertiliser they were forced to buy from the government was in fact mostly sand. Other natural hazards included hailstorms that could completely flatten crops. They persevered, however, growing potatoes, beans, apples and later keeping bees.

Howard, now far removed from a world of sowing and reaping, would have been on the side of collective agriculture. He writes to James in 1953: 'Don't get bitten by sentimental twaddle about the land. It can be done by exploiting others or by full mechanisation in a socialist country where the people own the land …'. This didn't go down well with George and James. The latter replied in an undated letter:

Mexico is short 400,000 tons of wheat a year … the second biggest drain on her foreign exchange. We have gone into the wasteland, taken great gambles, poured the sweated saving of a lifetime into the desert … to exploit the masses? Hell no! If the *agraristas* seize the farm … let them take it I will find another desert to make into another garden.[8]

In November 1959 he and George had indeed given up the ghost and reverted to childhood, collecting beetles. Howard rubs salt into their impecunious wounds writing in 1956: 'I have a large Humber that does twelve mpg around town … Eden's attempt at burglary [Suez] has now made the cost of running it prohibitive – 6d a gallon.' He adds that Geoffrey, his third-born, is 'top of his form in maths – so perhaps it has only skipped one generation'.

Howard's views on the 'twaddle' about land was put to the test when from August to October 1960 he toured China, visiting the model Red Star commune, south of Beijing. His notes exhaustively list fulfilled production quotas and its social achievements. In Beijing his first impression is of the absence of flies, dogs and cats, birds and litter; also, approvingly, an absence of nightclubs and Teddy Boys. One wonders, as an entomologist, whether he made the connection between items one and four on the list. During the Great Leap Forward from 1958 the 'four pests' extermination campaign was directed against rats, flies, mosquitoes and the sparrows that consumed grain. Peasants joined together to make enough noise to stop the birds settling, causing a fluttered death from exhaustion. By allowing the expansion of insect numbers previously predated by the birds, things were made much worse.

At Beijing Agricultural University he noted the importance of feeding the Chinese millions and the institution's growth from 400 students at liberation to 3,800 with 500 teachers. The majority of the former pay no fees, live rent-free with food and clothing provided. The vexed question of the Great Famine of 1959/60 he addresses by noting the drought in some parts of China and the floods in others that had resulted in rationing. Howard had met Edgar Snow who told him that if it had not been for communist control such bad years might have meant fifteen to thirty million deaths from starvation.

Back in England he wrote warmly to his host Professor Chu Coching, in December 1960, 'My visit to China was certainly the most interesting event that has happened to me.' What impressed him most was 'the way in which in such a short time so many of the contradictions between individuals have been resolved in a way that is unknown in the West ... we have been told so many lies by our newspapers and radio'. On a personal note he thanks him for bringing his cousin Han Chun (Joan Hinton) from Si'an to meet him in Beijing with a child (Karen) speaking only Chinese. Howard and Joan shared very similar views, as we shall see. Years later Joan and her family were drafted to the Red Star Commune themselves.

Howard's tour was ghosted in a very strange way by two other westerners on a fact-finding tour: Pierre Trudeau (later to be Canadian Premier) and Jacques Hebert. Tongue-in-cheek, in *Two Innocents in Red China* (1968) they told of their meetings as their itineraries crisscrossed the country. They named Hinton mysteriously as 'McIntosh'. In Beijing at a dinner for foreign visitors they speak of this character, 'who always has a dozen stories for us and quotes Groucho Marx and Shakespeare'.[9] In Shanghai they meet again at breakfast. Howard has been collecting tortoises on his travels, which, as in his youth, he keeps in his hotel bath and transports in dampened cotton wool. At a Hangchow hotel, 'Who should be awaiting us but McIntosh, alone as always and delighted to re-discover the *ersatz* Britishers that we are.'[10] At Canton: 'It is with the greatest pleasure that we rediscover this man of sound judgement and delightful humour. He tells of his discoveries including a snake-market and after a glorious feast, in which McIntosh proves himself a remarkable gastronome, we shake hands warmly, he to disappear into Kwantung.'[11]

Having left China en route to Paris, Trudeau was forced to stay at Copenhagen airport where, to mutual surprise, they met once more. Howard had been stranded at Omsk in Siberia where his contraband tortoises had nearly died of cold. After a handshake as Trudeau left at a run for his plane, 'a very British voice called after him, "You know my name isn't McIntosh it's ..." a sound that might have been Hinton was drowned out in the roar of the aircraft'.[12] Many appraisers of 'McIntosh' have noted his sense of humour; what was his reaction to this portrayal, one wonders.

Howard's faith in scientific socialism, as for many another adherent, like E. P. Thompson, was put seriously to the test. Writing in 1956, after Kruschev's secret speech condemning Stalin had been leaked around the world, Hinton suggests the need to retain perspective: 'There were mistakes but I am thoroughly satisfied that they have not made an undue number. The only alternative that has been ever offered to me is a lot of hypocrites, gangsters and deliberate liars.'[13] Late in October of that year following the invasion of Hungary, his daughter Charlotte reports that he stayed in his room.

Charlotte and her husband, Terry, followed the Chinese trail themselves, arriving in Beijing to teach at the onset of the Cultural Revolution in 1965. They also met up with Joan Hinton several times and on one occasion a slim blonde child who could only speak Chinese: Karen again. Charlotte and Terry were holed up in the 'golden ghetto' at the Friendship Hotel and told not to get involved. The Red Guards, however, could not be entirely avoided; the couple had to smuggle candles to their students, past rival groups vying with each other to cut off the college electricity. In the second year of their three-year contract all education was given over to political meetings and the school effectively shut down. As events spiralled they managed to leave in September 1967 – the last British family to quit.

Howard Hinton continued to establish his position as a leading entomologist, founding both the *Journal of Insect Physiology* and *Insect Biochemistry*. In 1961 he was elected a Fellow of the Royal Society. Like Geoffrey Taylor, he continued to pour forth papers on all manner of related subjects, over 300 in all. His definitive work, *Mongooses*, was published in 1967. He continued to attend conferences worldwide although relations with his peers were often acerbic and intemperate. A long spat continued between him and a colleague, delightfully named, V. B. Wigglesworth, over terminology relating to insect metamorphosis. Miriam Rothschild much later wrote that his attitude to him was 'delightfully childish and unreasonable'.

In keeping with the Boole/Hinton tendency to theorise grandly and shake foundations, Howard also made his own contribution to the theory of evolution. Over many years he had experimented with a species of fly native to sub-Saharan Africa that had adapted to its environmental rigours by being able to completely dry out as a larva in mudholes and be re-animated by rain. In the laboratory he pushed this cryptobiosis to the limits by freezing samples to near absolute zero and then heating them above the boiling point of water. Desiccated, they revived and continued to develop into the adult fly stage. In an article in *New Scientist* in October 1965 he suggested that contrary to the idea that life developed from the sea, complex molecules formed in the earth's atmosphere might have found mineral-rich crevices on land conducive to evolution even if dry for long periods. Such forms might well have been washed down to the sea where they evolved further. Theorising from the far past into the future he also speculated that dehydration might be the answer to space voyages that would last long periods of time.

Such speculation was no doubt taken with a pinch of salt; his desire to shock also had a humorous schoolboy aspect he had retained from youth. On occasion at formal dinners he was not unknown to introduce a horned toad from a sock kept in his pocket, or at lectures to delight in drawing attention to scatological aspects of insect mimicry. He recounted how some caterpillars, appearing as a juicy attraction to birds, have evolved the unappetising disguise of looking like their winged foes' turds.

In a letter of 10 November 1975 to Miriam Rothschild, Howard inveighs against a colleague who has, 'of course ... misinterpreted the results of my experiments'. More placatingly he reports that after an operation for bowel cancer, 'Everything has gone well and the surgeon says I am good as new – magic at the age of 63.' Sadly for him the magic had not worked; he died in August two years later. Suspecting the worst, characteristically, he worked even harder to complete a long-planned three-volume treatise on the biology of insect eggs.

In his obituaries due attention was paid respectfully to his achievements, even if 'he scorched those nearby'. Rothschild had commented on his competitiveness, that he 'flew off the handle at the slightest suspicion of criticism'. Another obituary stated that 'he never minded taking on the Establishment'. The catalytic quality of his thought was duly noted alongside the sense of humour. Rothschild noted that, 'endowed with a pair of vibrating antennae ... he became more delightfully insect-like ... every day he made an incredible discovery and was immediately burning to impart the news'.

From what one can gather from the biographical hearsay regarding Howard Hinton it seems that he, like many another, did not metamorphose much from the psychological imprint of his childhood. His boyish fascination and involvement with the appealing and colourful natural world of Mexico remained with him all his life. Perhaps the absence of a nurturing mother, however, left a bitter taste of rejection that he could not camouflage in later life. The desire to compete and win, to shock with pranks, hints at an emotionally cold inner space that was never adequately warmed. His adherence to a fundamentalist political theory like Marxism, itself competitive, objective and cold, seems congruent. Or was it just the ancestral lopsided extremist 'taint' urging through? There is a positive carapace, however, that perhaps relates to the Boole/ Hinton genus.

The prompting to discover and explore boundaries is one he shared with his family antecedents. Howard's interest and kinship with the dynasties suggests that he must have recognised it in himself. Both his great-grandfather, George Boole, great-grandmother, Mary Everest, and his other great-grandfather, James Hinton, possessed it fully. His grandfather, Charles Howard Hinton, demonstrated his own particular boundary concerning the fourth dimension. Geoffrey Taylor, a fully-fledged Boole, spent his life in experimentation with the same enduring boyish spirit that probably cemented his kinship with Howard.

The family impetus seems to have lingered in Howard's own children. The no-nonsense approach that Carmelita personified found resonance in Charlotte's professional life. As head of a school in Newham, London she reacted to threats of its closure by marching her girls to the Town Hall singing protest songs and giving out the telephone numbers of councillors so that they could be contacted. The campaign was successful. His son James, as a historian, has written about trade unionism and the Peace Movement of the 1980s. Lecturing at Warwick University he was allied to grassroots campaigning with the peace group European Nuclear Disarmament. During the period 1982–83 when I was visiting the Soviet Union and living in Shropshire I must have met him at meetings in the West Midlands without any knowledge of a Boole/Hinton connection. Geoffrey, the younger son, has it seems inherited the Boole mathematical gene, having become a foremost authority on artificial intelligence. Teresa, the younger daughter, lives in Tasmania and is a social policy researcher.

Some acknowledgement of the Boole/Hinton threads seems to have been woven in the choice of male names: George, James and Howard re-occur with regularity. Howard's brothers, George and James, despite the cultural distance of Mexico and enclosed lives within the mining industry, continued the fascination with the native plant world passed on by George Snr. James added numerous specimens to the Hinton genera list, most notably that of a giant tree lupin named in his honour, *lupinus jaimehin-tonanus*. According to his daughter, Patricia, James's first love remained writing. Under the pseudonym of Andres Mendoza he eventually had published 100 short stories and six novels, the last in 2004, two years before he died. James's son, George Sebastian Hinton, born in 1949, after studies at Edinburgh and Wisconsin has taken up the mantle by exploring the plants of the gypsum outcrops of Nuevo Leon. Beginning in 1984 he has added sixty-five new specimens, five of which were collected with his own son, George Boole Hinton, satisfyingly completing a cycle reverting to the genus appellation of his grandfather. Three generations of plant collectors in Mexico have appended the nomenclature Hinton and Boole to their discoveries.

# UNCOLLAPSIBLE HINTONS

The dearth of correspondence relating to the lives of the Booles and Hintons applies equally to the life of Charles Howard Hinton. There are references to a collection but it has never been found. The only small coherent cache that sheds light on his four sons, George, Eric, Billy and Sebastian, comes from the youngest, Sebastian, known as 'Ted'. His letters to Marie Little, a member of the Byrdcliffe Arts Community at Woodstock in upstate New York, are therefore very useful.[1]

A young Sebastian (Ted) Hinton (1887–1923).

Woodstock was a gathering point and focus for the Booles and Hintons. Ethel Voynich was a regular visitor to 'Camelot', the white clapboard house acquired by Charles Howard's family on the edge of the community. (We encountered her there in 1921 when she visited Ted and Carmelita Hinton, not long after she had arrived in the US.) George, the eldest boy, (as we have seen), had met his wife, Emily, at Byrdcliffe. The colony and its history illustrates the utopian community aspirations much in evidence in the US in the nineteenth century, both rural and city-based.

Ralph Radcliffe Whitehead, the founder of Byrdcliffe, born in 1854, studied at Balliol College, Oxford in 1871 where he met Charles Howard, beginning a long friendship. From a northern factory-owning background, Whitehead's ideas gravitated towards creating a community of artists –

'White Pines', Ralph Whitehead's Byrdcliffe home.

a contained settlement living a simpler rural lifestyle – the same themes discussed by the Fellowship of the New Life at Wyldes farm on Hampstead Heath. It was one of the first of its kind in the US. In 1902 Whitehead bought a 1,200-acre site below Overlook Mountain in the Catskill mountain range near Woodstock and named it Byrdcliffe, combining the middle names of Jane Byrd McCall, his wife, and Radcliffe.

The existing farms were amalgamated and new buildings erected for individual members. Whitehead built his large timber house, 'White Pines', and a boarding house, 'Villetta', for visitors, plus studios, forges and workshops. While the fine arts of painting, etching and pottery were encouraged, the emphasis was on manual labour resulting, à la William Morris, in the commercial making of heavy furniture with floral panels in a simple style. Attracted by the jovial and communal *esprit*, kindred spirits were welcome to stay, including Charles Howard Hinton's family. It was Ted, the youngest of the boys, who struck up a close relationship with one of its resident artists.

Not much is known about Marie Little. Nancy Green who has written about Byrdcliffe gives her dates as 1866–1949, but scant information other than that she studied art in Brooklyn, 1895–98. The community would have offered her a degree of freedom in a supportive social setting,

Marie Little's home at Byrdcliffe, 'The Looms'.

although Anita M. Smith talks of her solitary life in which she 'vibrated to refinements that common folk ignored … her tastes too delicate to be exposed'.[2] Green provides a picture of Marie as a youth studying music in Italy, 'her poplar-like figure and quavering voice had been much admired'.[3] One imagines a tallish, wispy rather fey character plucking at her loom in the shack-like studio, 'The Looms', where she created her hand-dyed rugs from local materials. It has been suggested that Whitehead, apparently a womaniser, was infatuated with her so much that his wife Jane had initially refused to move to Byrdcliffe if Marie was there.

Some mutual rapport or fascination did, however, exist between Marie Little and Ted, judging by the number of letters from him, bundled together, that she bequeathed. He addresses her for some unknown reason as 'H. Duck' and often refers to her as 'little sister' – slightly surprising given that when the letters begin in 1904 she is thirty-eight and he only seventeen. A running joke is his pidgin southern speech: he signs off, 'Yo "bigger" brudder, Taid'. The tenor is skittish and teasing but strangely ardent. He writes in September of that year, 'I look every mail for a letter from you …', and Christmas 1905, 'I have been lying around and think about you all morning.' They have 'scraps'. He is anxious to have her 'coy' photo and wants Easter 1906 to hurry up to spend five days with her in her house. In February he writes, 'It was just a year ago today that I decorously shook your hand on coming back from England. Wish I could do it now.'

Ted is studying at Princeton. His letters are full of college chit-chat about his classes in Greek, Latin, French and Maths. His father's ability in the last subject seems to have come through; Ted talks of coaching it to other student 'duffers'. He is selling beer mugs, presumably to help with his meagre funds. Food is important: H. Duck is sending cake, apples, chestnuts. 'Dad took me out yesterday and gave me a swell breakfast,' reports Ted. Eating clubs as socially exclusive institutions had been emerging at the university; Ted had found his way into one despite a complaint that most cater for fellows, unlike him, who have been 'sheltered and protected all their lives'. He proposes to start one of his own to make money.

Despite his meetings with the Dewey family and his father's friendship with William James, he felt socially at a disadvantage, although his connection to George Boole, he writes, could easily get him into 'Hahvahd'. He tells Marie that he treasures George Boole's copy of Newton's *Principia* that was presented to one of the Hintons years before. Like his father, his intellect does not prevent him horsing around. 'It's in full blast, Yo' big brudder had to run along with his hat in his mouf today. We made one fellow roll up his trousers … and put his garters around his neck then run up and down the street yelling "I'm a devil".'

The Hinton boys were all by now in their twenties and had not turned out very much like their mother, Mary Ellen's, assessments of them in her 1891 letter from Japan to her sister, Lucy Boole. George was studying at the University of California; his 'lack of strength' at that time is coping well with his vacation trips to wild Mexico. Eric's 'thoroughly good, sound brain' seems to have faltered. Ted sends him a little money while he's 'bumming around' in West Virginia living on a dollar a day. Billy's 'sturdy, cheerful and healthily contented nature' seems to be in the process of an alcoholic deterioration. In April 1906 he is planning to take a correspondence course to be a fireman on a railroad or a lake boat. Like Eric, he travels coast to coast aimlessly looking for work. Only Ted, Mary Ellen's favourite, is living up to expectations. He makes regular trips to Byrdcliffe to see Marie. That winter Ted writes to her, 'I wish I was up there now by the little Looms fire loafing some more.'

Ted, surprisingly, does not mention his mother's death in correspondence to Marie but on 8 February 1909 he writes that he has finished his last exam, thanks her for the malted milk tablets and stuffed dates and commiserates that she must be lonely with the Whiteheads away. Billy died in San Francisco that same month, possibly from epilepsy or drinking, the third member of the family to die in two years. George is now making regular trips to Mexico to sound out the mining industry there. That summer as usual Ted spent time at Byrdcliffe but is clearly unsettled, like his two remaining brothers. The three met up in September 1909 in San Francisco where Ted had taken an unrewarding job teaching at a military academy. He is now thinking of going into law. George and Eric have fallen out. Ted writes, 'The best thing he can do is go back to Mexico and leave Eric to me', adding with an accurate forecast regarding the latter, 'I don't think there is much chance of seeing him again.'

Eric's life is an uncharted mystery except for two official documents. His passport, dated 11 January 1906, records that he had been a student for four years in Washington DC. His disappearances may have meant that the passport had enabled him to continue 'bumming around', possibly in Canada or Mexico where George had been living. The other document, over a decade later, is his Registration Card for the Draft dated 12 September 1918. His details listed suggest that as an 'unemployed labourer' he must have continued his unsettled Kerouac-style existence.

The Draft had been introduced in May 1917 after mounting pressure stemming from the sinking of US neutral shipping by German submarines. Eric, with presumably no dependants and obviously not in a reserved occupation, may well have been one of the two million sent to France, over half of whom saw active service. The war ended in November 1918; if he was sent to fight he may have been unlucky enough to have died, unremarked, in the remaining few months. Or maybe he just evaporated intentionally leaving behind his highly unbalanced family – an obsessive father, a mother who had committed suicide, a brother dying young at twenty-four and two other none too stable siblings, George and Ted, one of whom was also to take his own life.

Byrdcliffe as a community was not a continuing success. Although it drew in many artists and craft workers to reside in the houses scattered on the hillside, there was no over-riding ideology strong enough to integrate it. Whitehead's notion 'to live under conditions that would be healthy for minds and bodies, without political overtones' is hardly a rallying cry. Artists are notoriously often insular and egoistic. His ownership of the Byrdcliffe estate flew in any case in the face of equality. Charles Ashbee who visited from England in 1915 noted Whitehead as 'an English country gentleman transplanted to the Catskills'. His philandering and questionable social attitudes – he was accused of racism – augmented the dominance of his personality.

The furniture-making enterprise did not last, out-competed by factory products. Gradually the colony became almost a holiday resort as residents migrated elsewhere because of the harsh winter or travelled abroad like the Whiteheads. Only Marie Little at her looms, as in some fairy tale, stuck it out through thick and thin. Whitehead died in 1929, his wife Jane in 1955; the experiment has ended up as a registered 'Historic District'. The ghost of the ethos remains in that Woodstock village has become an artistic extension of Byrdcliffe as a haven for artists and in its turn, a symbol of free living (even though the legendary 1969 music festival was staged miles away).

Shortly before the First World War the Boole/Hinton narrative took a shift away from the European highbrow themes of politics, literature and mathematics towards a more grounded focus on the transmission of ideas through education. In an example of the powerful transference of concepts, as with the fourth dimension, in 1889 Jane Addams, social reformer, pacifist and feminist, founded a Settlement community in a working-class area of Chicago. Its inspiration came from the Christian Socialist, Toynbee Hall in the East End of London at Whitechapel that she had visited the year before. (James Hinton as we have seen was an earlier inspiration.) Here male Oxbridge undergraduate students could live, learn and offer support to the proletariat by giving lectures and running courses in arts and crafts. Hull House, its Chicago twin, converted from a dilapidated mansion, was run by university-educated women. It concentrated on 'women's concerns': healthcare, domestic activities, playgrounds, nurseries and child labour, combined with classes on history, literature and art. Over time the project was visited and supported by a number of influential social reformers: John Dewey, Clarence Darrow, Paul Kellogg and, from England, H. G. Wells and the Webbs.

A mature Sebastian (Ted) Hinton.

By 1910 Ted Hinton had moved to Chicago to practise as a patent lawyer, like his father. One of his friends was the educationalist, Ed Yeomans, who was drawn into the orbit of Hull House. A fervent believer in the transforming power of good teaching, he recognised that schools must 'heave prodigiously at the structure overhead, to move it forward a microscopic little … yet be able to pour out men and women of such incandescence of soul and intelligence of mind and heart that two generations would purify the earth'.[4] Over dinner discussions at Hull House such enthusiasm cemented a friendship between Ted and another visitor, the diminutive but fiery, red-headed figure of Carmelita Chase.

Her family background was squarely centred on an American ethos of practicality and realising the possible. Omaha in 1890, where Carmelita was born, was enjoying a boom time based on the intense development of prairie land for cattle across the state of Nebraska to the city's west. Waves of migrants, supplanting the already dispossessed native Indians, were swelling its population to service the meat-packing plants and stockyards. The Chase pedigree was genteel: her mother, Lula Belle, came from an old Kentucky family; her father, Clement, ran a 'social' paper, *The Excelsior*, and two financial papers. His father, Chapman, had been Paymaster General of the Union Army during the Civil War and later Mayor of Omaha. Carmelita's mother's aspirations to steer her towards a fitting social position were frustrated by Clement's belief in women's equality and the love of the outdoors that she came to enjoy too: camping, riding and playing tennis. This alliance with her father had prompted her to take up a place in 1908 at the women's liberal arts college of Bryn Mawr, near Philadelphia.

Founded in 1885, a decade and a half after Girton and Newnham colleges in England, it aspired to give a rigorous academic background to a new generation seeking equal opportunities with men. Carmelita's life at college seemed to be more active than scholarly: plays, choirs and sport. Her stockiness and grit made her into a champion shot putter. Putting your back into what you do was, in fact, her essential motif, echoing a precept of William James – 'If you find yourself deeply stirred by what you read and see … do something' – a sentiment that also resonated with the Boole/Hinton families she was to be allied to. It was her energy and commitment to a 'lived' education that motivated her and her eventual family. It would lead later to the establishment of a highly influential progressive institution in Vermont.

The notion of combining schooling with doing had found favour in the philosophy of John Dewey while at the University of Chicago from 1894 to 1904. Learning should be an interactive relationship between tutor and pupil, a pragmatic process of discovery and self-realisation. Echoing this as an institution, the school should see itself engaged with the wider society around it, promoting social reform. Hull House was an ideal setting to test and put these ideas into practice. Carmelita joined the settlement in late 1913 to work as secretary to Jane Addams. Although the community around was served with practical hands-on activity, Addams told her, 'You can't just bathe babies, you've got to go deep down and really change things whether people like it or not.'[5]

Social transformation was underway in any case with the rapid industrialisation of America. Byrdcliffe was all very well for a few, but how was the mass to acquire skills and live democratic, useful and creative lives? Hull House served, like Toynbee Hall, in the anarchist mould, as a nucleus to empower people to be able to control their own lives. In a poor, oppressed, working-class setting this could lead to confrontation. Some members were avowedly Marxist and supported trade union rights. In January 1915 unemployed workers started a protest march from the house and were beaten by the police.

Carmelita, however, was not then drawn to political activity; her interest focused on taking a course in playground activity. For her, harnessing children's natural curiosity through play and involvement with music or carpentry would furnish society with creative, useful citizens, centred enough to be able to make their own decisions. This practical approach must have appealed to Ted Hinton with his own love of the outdoors, tools and engineering. A career in law was not going to suit his temperament in the end.

In April 1916 Ted and Carmelita were married. He had written rapturously about their prospects in an undated letter sent from his law office: 'Oh sweetheart we are going to have a wonderful life together – you and I. Just you wait and see. Better and better, just as our love gets purer and sweeter all the time.' Carmelita's father, still living in Omaha, had received endorsements for his son-in-law. One from a Princeton college-mate spoke of him being a man of 'unusually fine mind with a high sense of honor and good steady habits', although he had suffered a nervous breakdown from overwork. The married couple promptly opened up their own nursery school in their apartment using a park opposite as a playground.

They decided to leave innercity life in 1917 for the more rural setting of Winnetka, a small town fifteen miles north by Lake Michigan. Carmelita's love of the outdoors was stifled in Chicago. Both had resolved the repeated conflict: whether to seek social change where the masses toil, or live by example in freer air. Carmelita wasted no time: 'I started a school in my own back yard in our carpentry shop with my own and my neighbour's children for pupils.'[6] Jean was born in 1917, William (Bill) in 1919 and Joan in 1921. One attendant at the nursery school later recalled how Carmelita would 'come into a room and it was an explosion … Noise, I associate her with joyful noise.'[7] By the third year there were two more teachers attending thirty-six children. Her husband, Ted, took to the enterprise as well, enjoying the space to keep animals that he'd experienced during Princeton farming vacations. Echoing his Byrdcliffe background, he put his carpentry skills to use making furniture. One particular product of his imagination was to revolutionise children's play.

Convinced of the value of the early instilling of four-dimensional space, Ted's father, Charles Howard, back in Japan, had built a large multi-cube framework of bamboo in their garden. Howard would shout out the co-ordinate letters and numbers where the poles intersected and expect the children to occupy the right place. They, of course, enjoyed the rough and tumble of climbing and hanging about. Sometime in 1920 Ted explained the idea to Ed Yeomans, who enthused about using it simply as a plaything. Ted's patent of 1920 refers to the 'monkey instinct' in children by which they 'swing head downwards by the knees'. The Jungle Gym 'monkey bars' are now found in one form or other the world over. Ted's legacy to posterity has been rather more lasting than the gunpowder baseball pitcher of his father. The prototype still stands in Winnetka.

Despite this success, which would provide an income for some years ahead, a vibrant partner and three children, Ted appears to have suffered badly from bouts of depression, enough to have him admitted in 1923 to a psychiatric clinic in Stockbridge, Massachusetts. There on 29 April he hanged himself. The suicide, however, was not publicly admitted.

It is not really possible to attempt conjecture about this. Outwardly he appears to have shared his father's intoxication with life, more whole-somely an outer one than that of a cubic introspection. His mother, Mary Ellen's, own suicide might have been one factor. Charles Howard and Mary themselves were very complex characters; by modern standards they were begetters of a highly dysfunctional family. How deeply was Ted affected by the early death of his brother Billy and Eric's disappearance? Did the four boys know of the bigamy trial back in England? Had that dark history

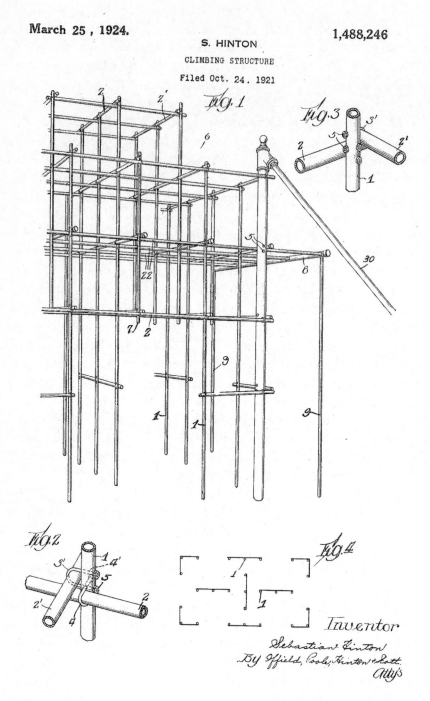

**March 25 , 1924.**

**S. HINTON**

CLIMBING STRUCTURE

Filed Oct. 24. 1921

**1,488,246**

*Fig. 1*

*Fig. 3*

*Fig. 2*

*Fig. 4*

Inventor

Sebastian Hinton

BY Offield, Poole, Hinton & Latt. Atty's

Sebastian (Ted) Hinton's patent for the Jungle Gym.

followed them around? Were they all agitated inwardly by the complete changes in their surroundings: England to Japan to various places in the US. His three brothers had all lived a footloose life. One who remembered Ted as 'a genius who never found his location' was obviously not referring to a position on a jungle gym, but the image may be apt – Ted's own co-ordinates were never secure enough.

No clues to the tragedy come down to us from Carmelita. It was not until 1983, the year she died, that Susan McIntosh Lloyd, her biographer, revealed to Jean, Bill and Joan that their father's death had not been due to pneumonia. Whatever the cause, his demise in 1923 left a wide network of relations and friends stunned. Ralph Whitehead wrote to Carmelita of his 'fine brain which made him distinguished among men'. Warm-hearted Wilfrid Voynich sent a message to Carmelita: 'I would do many things to cheer you up and buck you up in your sorrow.' The Booles sent their condolences from across the Atlantic re-uniting kinship. Ada Nettleship sent hers: 'I can hardly believe it – he always seemed to me the practical head of the family … from whom great things were expected.'[8] The two must have met on Ted's journeys to England.

The parallel with the life of Carmelita and Mary Everest Boole here is very striking. Both women were left with a very young family owing to the untimely, strange deaths of their husbands. With no respite in their personal lives to absorb the tragedies they single-mindedly pursued a lifelong commitment to education, centred on the notion of learning through doing. In their later years both became more politically committed. Carmelita was known as 'the Mrs', Mary Everest as 'the Missus': two indomitable matriarchs. One wonders what would have transpired if they had ever met.

In 1925 Carmelita took the three young children east to Cambridge, Massachusetts to become a teacher at Shady Hill School. Its motto today states that 'the process of figuring something out is as important as the right answer'. It expresses a principle that Carmelita would have also entirely endorsed. She would have added her own style, introducing a sheep into class, for example, shearing it and using the wool to make clothes. Or by staging a construction project to build a habitable miniature town, incidentally encouraging multi-digit addition and subtraction. The Hinton children, of course, took part, supplemented with vacations up at Camelot and enterprising trips abroad. In 1924, as we saw in Chapter 14, they voyaged to England to meet their Boole aunts and sailed with Geoffrey Taylor in Scotland. In 1927 they went to visit Ted's brother,

George in Mexico. Ed Yeomans, in awe of the family's energy, suggested later that a book title to suit them would be *The Indestructible Hintons* or *The All-Weather, Heavy-Duty, Uncollapsible Hintons*. Bearing in mind the recent history of Mary Ellen, Eric, Billy and Ted, any biography would not want to go too far back.

In April 1931 the Hintons moved to a run-down farm near Weston, fifteen miles west of Cambridge, to set up a school providing space for six boarders, some chickens, a pig, cows and horses. It is easy to see how Jean, Bill and Joan developed their involvement with rural life and love of adventure. Carmelita's own desire to travel continued to take them all into often very foreign territory. Internationalist in outlook, she had met Donald Watt in the early 1930s, the founder of the Experiment in International Living, a precursor of the Peace Corps. Its aim was to foster 'peace through understanding, communication and co-operation', via home-swap stays.

A European trip to Austria and Germany in 1933 was followed a year later by another biking and hostelling tour. At one point they visited and naïvely helped out at a Hitler *Jugend* camp, the Nazi paramilitary youth movement that had replaced the banned Boy Scouts. Carmelita could be forgiven for being unaware of the threat behind the swastika flag that was hoisted daily. She rather admired the sense of purpose and discipline on show. The Hinton *jugend*, however, rebelled: ordered to cycle two-by-two at the behest of the Nazi leaders they disobediently piled into each other. The group later found themselves, dismayed, among several thousand others roaring their approval of the Führer himself at a rural rally.

Partly worrying about her own children's high school education, Carmelita in 1934 decided to make plans to found a school of her own. 'If I was going to struggle with my own three children, I might as well struggle with fifty,' she was reported as saying. The essential progressive outlines would have to avoid mere preparation for college and the competitive ethos of sport prevalent elsewhere. Instead she wanted to engender a co-operative spirit based on hands-on labour, music and crafts in a co-educational rural setting: a kind of utopian village for young people. At a time of severe economic depression and with few models to emulate, this was a heady enterprise.

A plan was accepted by Mary and Sarah Andrews to adapt their Elm Lea farm near Putney, Vermont, in the foothills of the Green Mountains. A one-year trial at a peppercorn rent was begun in September 1935, if successful leading to a generous offer to buy for $20,000. The estate consisted of a large, central, stylish house, barns and an assembly hall already in use as a theatre, plus 300 hilly acres of pasture and forest. Carmelita mortgaged her Weston house and launched her own settlement, to plant and sow, repair and build, ready for the fifty pioneer boys and girls later that year.

In 1935 Putney School's first teenage pupils, mostly drawn from middle-class homes, must have found the regime a shock.[9] Rising at six from the spartan segregated dormitories, manual work began by attending to the pigs, poultry and the dairy herd or gardening. A healthy life aiming towards self-sufficiency was the aim. Radios, alcohol, coffee, cigarettes and sex were banned or frowned upon. The inclusiveness and order was tenderised by the complementary illicit – the out-of-sight cigarette, grope or pleasing prank. The emphasis was on self-discipline and collaborative work. Lights out at 10pm was a reward for being tired out, not an invitation to libidinous exploration. It is not difficult to see how life on a Chinese commune would later suit Joan Hinton. In return, inmates received academic lessons from stimulating teachers and generous access to art studios and theatre. There was also the great outdoors, skiing, camping and horse riding giving individual fulfilment but also comradeship. True to Jane Addam's pronouncement at Hull House, Putney School was not just to help people but change the world.

Five years later, despite a barn fire and a hurricane, enrolment reached 120. To offset the relatively high tuition fees, scholarships were offered. Running costs were mitigated by simple living and the dedication of staff. Traditions had been established such as the opening and closing of the school year by the ascent of nearby Putney Mountain. There was a distinctly radical slant to the place. The student-run magazine worried locals a little with its articles on the Spanish Civil War and Bill's exploits in China or Joan's views on Hiroshima. Noted leftish parents of pupils included Owen Lattimore and Alger Hiss, both held to account for communist sympathies.

In the late 1940s her dominance at the head of the school became a *cause célèbre*. Like Ralph Whitehead's control of the Byrdcliffe community as a benevolent dictator, Mrs H maintained a suzerainty over Putney School. In an artistic community, really an association, the everyday details of life were only an individual matter; a school, however, requires some routinised institutional planning. Carmelita's own inimitable powerhouse drive and charisma relied on the willing acceptance of her authority for the sake of the vision. Goodwill from the teaching staff was waning, however, given low salaries, no pension plan, long hours and little overall democracy.

One teacher, Edwin Smith, hired in 1947, was to prove a catalyst. His rather wooden style of teaching did not appeal to Carmelita's notion of class discussion, so she asked him to leave. He in response was instrumental in the drawing-up of a Faculty Association arguing for greater participation but actually becoming a trade union. This led to a strike. Issues were resolved but Carmelita was wounded by the wrangle. Nevertheless, six months later 200 alumni gathered to share a huge harvest festival.

A great deal had been achieved, however, before the season of discontent. Farm income continued to increase, new machinery was bought and food had become 50 per cent home-grown. A large Kitchen Dining-room Unit (the KDU) had been built, a library housing nearly 10,000 volumes established, plus new dorms and a school farm bought at Cape Breton in wild Nova Scotia for summer camps. A second generation of students was now attending, including more Jewish and black students. The co-educational nature of Putney had, it seems, been successful; one later entrant to Harvard commented on how appalled he was at the way men talked about women there in comparison. The perennial problem of individualism versus collective needs had been writ large in terms of leadership at the top. For the students their individual self-realisation had best been served by a disciplined mutualist community. One student commented, 'Putney taught me that I could do more than I ever thought possible.'

Mrs H finally retired in 1955 aged sixty-five and bought a farm near Fleetwood, Pennsylvania worked by her son, Bill. She told a *Time* reporter when she left Putney, 'I hate to leave but I have so many things before me that I'm boiling over.' Her internationalist leanings soon set her in motion, leading to inevitable friction with the US authorities. Continuing her pre-war forays she took a group to the World Festival of Youth, an obvious Communist Party celebration. During McCarthyism she had been under suspicion anyway because of Joan, Bill and Jean's activities. Marni, Jean's daughter, told me that she had learnt early to recognise FBI agents with their 'creased suits, shiny shoes compared to the family in jeans'.

In 1956–7 Carmelita travelled through Europe with five others to Turkey, Iran, India and eventually Kenya then north to Cairo. At home she worked for the Women's International League for Peace and Freedom and took on the role in 1960 of Executive Director for the centennial celebrations of Jane Addams. Despite an official embargo she managed to

Carmelita Hinton (1890–1983), with giant cabbage at Putney School's harvest festival.

Carmelita Hinton at Dazhai commune in China.

get to China via Russia in 1962 and spend a year there. Now in her seventies, she was as active as ever, still skiing despite breaking a leg in 1962 and her hip in 1968 from a fall rock climbing. She later broke the other one walking backwards as part of an anti-Vietnam demonstration.

In 1971 high level moves began for Nixon's visit to China. Bill, with his strong connections there, arranged for her, now eighty-one, with a renewed passport, to lead a fifteen-strong group of young people to experience the country first hand. The visit included a month at the model village of Dazhai working in the fields with peasants. The *Boston Globe* reported that Carmelita worked alongside a woman with bound feet.

351

A rush need to husk the mountains of harvested corn was enthusiastically fulfilled. She recollected: 'We were so joyful that we started dancing.' The collectivist Putney spirit seemed to have been written large. Another month was spent in a factory near Shanghai, but as Marni wrote to Geoffrey Taylor in May 1972, 'the endless hours of note-taking, climbing mountains and late night discussions had taken a toll'. Carmelita suffered a series of mini-strokes and was hospitalised, attended by a team of seventeen doctors. She pulled through.

Between 1973 and 1978 she continued to live at the Fleetwood farm but remained politically active and connected with Putney School, attending the annual summer gatherings at Cape Breton. One final trip abroad with Jean was to stay with Geoffrey Taylor in Cambridge. Her 'uncollapsible' spirit met its match with his housekeeper, Gladys, who evidently didn't want them there mixing up the place mats, 'finding their frontier ways quite gauche'. Geoffrey was in tears, however, when they left.

On my first ever visit to the US in February 1983 as part of my 'peace tour' I coincidentally ended up just a stone's throw away from Putney School having crossed the country eastwards by bus from California.

23 February 1983

Arriving at LA airport from Tokyo the US customs official wanted to know why I'd been visiting the USSR. I explained that it was just a tourist trip and certainly not part of gathering an overview of grassroots activity mobilised against the Cold War. After my naïveté about taking Mark Reitman's document into China on the train I was feeling anxious to hide it more thoroughly. Who knows what some official might make of sheets of paper with mathematical formulae and Cyrillic script? I levered off the top and bottom plastic 'stops' on my rucksack's aluminium tubular frame and secreted as much as I could inside.

Images of the Soviet Union stayed with me on my travels. I found myself often following in my head the paradigm offered by E. P. Thompson that the two superpowers, branding themselves as champions of 'Peace' and 'Freedom', despite their sworn hostility, were in many ways identical. One essential difference, relatively true to the US banner proclaiming 'Freedom', was the ability to organise and speak out unscathed – unlike my beleaguered dissident friends in Moscow.

Whenever possible I met with a great variety of peace groups and

San Francisco demonstration against the Queen's visit, 1983.

recounted my Soviet friends' bravery and plight. In LA I spoke to Alliance for Survival, Women Strike for Peace, Nuclear Free Pacific, the Unitarian Church and a Freeze Movement brunch of 100 people. Several days later in San Francisco I joined a demonstration at the Panhandle, Golden Gate Park that was protesting a visit by Ronald Reagan and the Queen. Despite 'Freedom' it was shadowed by the heavy presence of SWAT special police. There were two big fibre-glass models of the Queen and Reagan, and several other 'queens' dressed in ball-gowns carrying a poster 'Queens against the Queen'. I talk to an oldish guy – a printer, while we march as night settles in to a jazz band playing 'When the Saints'. We are ushered to a cul de sac where from a floodlit stage speakers proclaim oppression against Ireland, Palestine, El Salvador, and solidarity with the handicapped, social welfare, etc. Holly Near sings. Helicopters occasionally swoop down and a light plane tows an illuminated message as it circles: 'Hail the Chief, God Bless the Queen, Long Live Rock and Roll.'

7 March 1983
With my Trailways $99 coast-to-coast bus ticket I set off sleeping on board en route. My first stop was Salt Lake City, HQ of the Mormon Church, high in a mountainous lunar landscape; not a place I

warmed to. Visiting the disneyesque Temple and the Tabernacle, like an upturned boat, I learn of Joseph Smith, the Great Trek and how Jesus Christ, depicted like Rock Hudson, visited the US after the Resurrection. The whole panoply of pseudo-religion is ridiculous and dangerous – racist, sexist and narrowly conservative – anti divorce, alcohol and abortion but pro family and frugality, i.e. Protestantism – lauding self-righteousness, quiet living, rewards for effort, tight-lipped, tight-arsed and morally superior. The more I think about it the more it reminds me of somewhere else – the Soviet Union. The same family values: sacrifice to the ideal, restricted behaviour, censored art, the whole tinged with a military sense of order and urgency waiting for the Millennium like the other's awaited dawning of true communism.

On through Denver and Kansas. I keep thinking of the Wizard of Oz and looking for a twister. Next comes St Louis and Pittsburgh, a big bustling city on the Ohio river, and then New York after Washington – all the while marvelling at the openness of the landscape and the people but finding its mainstream culture as difficult and bizarre as it is portrayed in the UK. The US is a giant squawking fat baby that bangs its podgy hands on the table and shouts for more and more expecting its distended appetite to be satiated.

15 March 1983

My taste for the Big Apple is fulfilled by the usual spectacular perspectives, but there were contacts to be made at Nuclear Times, the Riverside Church, War Resisters League, Pentagon Women's Action, WAND, SANE – there are so many. I contact Cathy Fitzpatrick of Helsinki Watch, the first person to contact the Trust Group in Moscow last year. At one Euromissile conference I find myself proposing that New York's skyscrapers are analogous pretensions to the Stalin-Gothic buildings of Moscow. The Chrysler Corporation = Gostrans, the Soviet Ministry of Transport. This annoys a corps of pro-Soviet members who propose in turn that the Trust Group is a CIA front. Later in the day Reagan announces his 'Starwars' programme, violating the ABM defence Treaty of 1972. He ludicrously advocates that if both superpowers could destroy incoming missiles nuclear stability would prevail – knowing full well that the USSR could never match the technology. My more modest proposal is to take the bus to Vermont to further the three-way twinning of Soviet, USSR and UK peace groups.

Thursday, 31 March 1983

Arrange to meet Ethel Weinberger at Hartland, Vermont. She has organised local people to contribute articles for a box to be sent to the USSR containing mementoes, tapes, quilts that convey a picture of the place. She's hoping for a Soviet town to return the idea. We travel up miles of mud track to a beautiful house her husband has built. We chat, eat and rush off to an Allen Ginsberg evening nearby. He is accompanied by a young guitarist and an outrageously gangly lover called Peter Orlovski. Ginsberg sings in a low drone and plays a filing-cabinet-like squeeze-box.

Friday, 1 April 1983

Ethel drives me around the back lanes; the houses are very evocatively ship-lapped with beautiful barns. Farming has collapsed since the late nineteenth century. One local industry still flourishing is making maple syrup. A plantation is called a sugar-bush from where the sap is taken to the sugar-house where it's thickened into syrup. In one a backwoodsman is stoking a boiler and drawing off the sweet liquid into drums. He offers up small cups which we politely drain despite feeling sick. I buy some maple syrup candies for the kids and we drive back. I bid Ethel a fond farewell, feeling we've made a good starting contact.

Carmelita Hinton would have definitely approved of our twinning project. Had I known of her and Putney School on my 1983 trip, just down Route 91 from Hartland, I would have wanted to meet up with her. She had, however, sadly died just two months before. In March 2011 I arranged to go back to Vermont and acquaint myself with Putney School. Chapter 24 contains an account of my visit.

My narrative meanwhile must make a radical shift to embrace the history of a culture that came to play a dominant part in the lives of two Hinton generations. George Boole Hinton's son, Howard, as we have seen in Chapter 19, had been very struck by his tour of China in 1960. Carmelita Hinton, his aunt, had also been impressed by her visits in 1962 and 1972. Two of her children, Joan and Bill, continuing the Boole/Hinton tradition of globetrotting, had much earlier in the 1940s become immersed in China's turbulent history.

# THE GADGET

The internet is a bit like panning for gold; among all the detritus one very occasionally sieves out a bright and valuable speck. The stumbling on the life story of Ethel Boole and her connection to Wilfrid Voynich had glittered brightly. I eventually pursued both of their lives as far as I could. A steady stream of material about the *Voynich Manuscript* continues to flow, however, mostly on the web via the Voynicheros: the global collection of *aficionados* endlessly sifting archives for clues from any source. What will happen to them if ever a solution to the enigma is found, one wonders?

It was during my research for my book on the *Manuscript* circa 2004 that another fleck from the same Boolean lode had gleamed invitingly. An internet article in New York's *Village Voice* from 4 June 1996 had made the connection between the Boole family and that of a Hinton lineage. While exploring Wilfrid and Ethel I had not paid any attention to the Hintons posted on one side of the battered family tree; they appeared as just a somewhat common name. The article was a comment on a film, *The Gate of Heavenly Peace*, released in 1995, made by Carmelita Hinton, Bill Hinton's daughter (known as Carma). It concerned her controversial take on the events of 1989 at Tian'anmen Square in Beijing when hundreds of protesting students were mown down. What intrigued me greatly was the collateral mention of her radical background. It was Bill Hinton's name linked to Ethel's that grabbed my attention especially.

It was a book of his that had lurked in my mind for thirty years; the notion that I was distantly but directly related to him was pretty extraordinary. In 1966 he had published *Fanshen: A Documentary of Revolution in a Chinese Village*, set in 1948 before the final establishment of Mao's China. The Penguin edition was given to me as a thirtieth birthday present. Hinton was clearly even more radical than his great-aunt Ethel. *Fanshen* had become something of a bible to my generation, interested in both revolution and rural life.

More consequent dipping and diving with Google had come up with further revelations: a reference to the life of his equally fascinating sisters, Joan and Jean Hinton, Ted and Carmelita's offspring, the grandchildren of Charles Howard Hinton and Mary Ellen Boole. Joan too had been living an amazing life in China and was even more intimately involved in Mao's revolution. The Hinton discovery was too late for me to connect personally with William (Bill), as he died in 2004, but I was eager to meet Joan. The internet revealed the bare bones of her story. Somehow she had in her twenties found herself working on the project to build the first atomic bomb at Los Alamos. Badly disillusioned by its actual use on Hiroshima, she had left for China and committed herself and her family to Mao's ongoing revolution. Every one of the articles that came up pointed out her unrepentant praise of him and his works.

In 2005 Joan was in her mid-eighties and still managing a dairy farm just outside Beijing, where she had moved in 1983. That was the year when I had briefly visited Beijing myself on my global peace circumnavigation. If I had known then anything about the Booles and the Hintons other than the mere existence of ancestor George Boole, I could have visited her. We would have had a good deal in common, especially concerning the issue of nuclear weapons. I had tried to extend my visa at that time but was turned down. Ironically, given her honoured position in regard to the Chinese authorities, it would have probably been easy to get it extended at her request. Regardless, this exciting diversion from the *Voynich Manuscript* prompted me to fly to Beijing in 2005 to explore her fascinating life face to face. It was the first time I had been back to what was then 'Peking' in 1983. I still have diary extracts from that earlier sojourn.

23 January 1983
My fellow travellers and I stagger off the Trans-Siberian after six days on board in what seems to have turned into an old people's home. Three of us take a minibus to a hostel that charges about £7 each for the room. A much-needed bath is enjoyed and a 'masses' restaurant visited. The Chinese are amazingly noisy and disgusting eaters. They hold the rice bowl to their mouths and shovel the food in (in so far as you can with chopsticks). They throw the bones all over the table and seem to eat bowl after bowl. But unlike Soviet restaurants, the place is alive, noisy, friendly, proletarian, and the food is cheap.

25 January

Breakfast at the vast and imperial Beijing Hotel – delicate egg sandwiches and coffee. In Europe such scruffs as us would be chucked out pronto – here by virtue of our Caucasian features we are inviolable. We wander into the Forbidden City, a vast, sprawling citadel of gates, throne rooms and courtyards. It's very alienating. Later we taxi to a vegetarian restaurant. We order a selection of dishes and receive a fish, (made of soya) followed by three other dishes with more soya and bamboo shoots – and more soya disguised in meat form. It's delicious and filling. Two Chinese opposite us are eating a bowl as close to heated sink-trap emptyings as you could get (even down to long, thin brackish hairs). They eat it with gusto (their dog). Home by bus – two sweet kids refuse my chewing gum, which I forgot to give away in the USSR.

26 January

Beijing Hotel again for breakfast. Around the corner is Wanfujing Street, the main shopping centre, which is very modest for a giant city. Low-rise shops and a Friendship store where I buy jade earrings for W. The shops seem to be no less endowed than the USSR. Noticeably more transistor radios and Japanese cars, goods everywhere. Advertising is now allowed in the streets. Clothing is still very martial but not unattractive, apparently non-uniforms only allowed since last year. Shop assistants are polite, everyone likes to practise their English. On the bus I'm regaled with 'Sunk yew'. I reply correctingly, 'Th-thank you' several times like a moron. 'Sunk yew' comes the response. Vegetable shops stacked high with Chinese cabbage looking not very crisp. Celery, turnips and radishes also evident. In the restaurants a liking for all sorts of seaweed – it tastes and digests like rubber bands. Most disgusting habit everywhere is the clearing of throats and spitting. Spitoons, shaped like bedpots without the handle, are common in buildings. Who in the name of Mao considers his debt to the motherland to consist in a working life emptying them? Is it the dust from the Gobi (Gobbi more like). Certainly the air is rank, many people wear masks. The little houses burn round briquettes and pieces of consolidated coal-dust cut up like fudge on the pavement outside. The 'hutong' houses are very ramshackle and only reachable by bike, they are being replaced by 6–7 storey apartment blocks. An improvement?

Bikes everywhere. Along the main street to Tian'anmen Square are bike lanes occupying about half of the sixty yard wide road. There are a few soldiers on patrol, but unlike the USSR it doesn't seem to be a siege state. To acrobat show in the evening in a very dingy theatre. Suzie Wong-like character introduces acts in a deadpan voice. They are amazing – girl on back turning a tea-tray and four carpets on feet and hands. Man produces five or six alarm clocks from previously empty soft hat.

27 January
Up early for trip to Great Wall and Ming tombs. Two-hour ride into the mountains alongside the railway we came in on from Mongolia. The Wall plunges up and down as far as one can see. The labour to build it must have been staggering; this stretch is well-repaired. It's bitterly cold. A young American woman I've been talking to says that the Wall is 'neat'. Neat? The only human feature that supposedly can be seen from the moon is 'neat'? We obviously have a language problem here. Back to hotel, pick up gear and taxi to Beijing Hotel. What a life – the Chinese stagger around with several tons on their bikes and we can't even be bothered to carry a rucksack a mile. Have apple pie and ice cream again.

The author at the Great Wall, 1983.

28 January

Breakfast with two girls and donate two of my three poached eggs – do eggs always come in threes in the East? Bus to airport and get to wrong terminal. (This seems to be a habit). Mutter, shout and curse until some nice English teacher comes to my aid and directs me to the new terminal. Heart sinks on espying JAL airlines accident-prone DC10, not my favourite aeroplane, looking monstrous on the apron. We leave and soon China is far below en route to Osaka. Air hostesses are amazingly solicitous – hot towel, fruit juice and peanuts, magazines, earphones and large lunch tray. Welcome back to the capitalist world.

Twenty-something years is a long time in the life of any city. By 2005 Beijing had witnessed a complete transformation. Since the Cultural Revolution the economic powerhouse that is capitalism had emerged, guided by the Communist Party: a somewhat anomalous turn to Marx's historical predictions. For me there were immediate miscellaneous pointers of change. Wanfujing Street is now the shopping showcase of Beijing, the low-rise shops have gone, save one: the Friendship Store where I'd bought the jade earrings. Dwarfing it are the towering post-modern blocks glorifying all the usual suspects: KFC, Olympus, Sony and Adidas, boldly announced by its heroic 'Buddhist' aphorism, 'Impossible is nothing'. In the broad pedestrianised thoroughfare roller-blading smart kids with ice creams pass electronics shops blaring house music under neon hoardings for products graced by western-eyed women.

Outside the giant Oriental plaza/mall complex someone dressed in a sponge outfit to look like a mobile phone leans bored against McDonald's. Inside a desultory troupe of half-clad girls gyrate as part of a Phillips electrics promotion. One could be in any western city. What makes it all more surreal is that literally behind this fashionable façade, that looks as if it's been letrasetted into place, you can still step into the crowded alleyways of the *hutongs*, the traditional patchwork of low, coal-burning houses and courtyards. It's rather like finding back-to-back terraces immediately behind Oxford Street.

I find the *hutongs* congenial, however, and it's where I dine out most nights on bean curd and broccoli in the clutch of houses given over to tourists. The narrow lanes are only really suitable for bikes and act as sanctuary from the horrendously polluted highways and flyovers that are now gridlocked by private cars. As if Wanfujing Street never existed, here one finds the colours of ordinary living: quilts hanging out to dry, tiny

Above: Wanfujing Street in 1983.
Right: A mall on Wanfujing Street, Beijing, 2005.
Below, left: The old and the new in the smog of Beijing.
Below, right: In the hutongs, Beijing, 2005.

one-room shops selling gaudy plastic goods, vegetables and sweets. Dogs loll by the chairs of diners waiting for woks to dish up lunch. It's easy to over-romanticise; until recently, when communal toilet blocks were provided, sanitation must have been grim. The authorities are meanwhile pulling down acres of the *hutongs* every year.

On the corner of Wanfujing and Dongchang'an Avenue stands the now enlarged and modernised Beijing Hotel. The wide steps now ascend past black-tinted, chauffeured limos into a marble-floored palace that is silent save for the high heels that clack echoingly. It demands cathedral whispering; a flunky with serving tray glides by heading for the diners seated at the atrium's cascade. My Caucasian features nowadays count for nothing, the inhabitants are *nouveau riche* Chinese bourgeois. I waft around looking supercilious but am secretly intimidated by the wealth.

A half a mile away Tian'anmen square has not changed, it's still the biggest public space in the world. Mao's body is here in the giant mausoleum in the middle of it. A huge poster of him by the reviewing stand tries to remind all that he hovers yet in the middle of Chinese life. For the crowds gathered at sundown to witness the ritual lowering of the national flag he still no doubt serves as a rallying icon.[1] To get away from the crowds I decided to wander up to Beihei Lake where, pleasingly, people are more intimately receiving head massages, flying kites and playing chequers. After several days sightseeing it's time I made contact with Joan. She was expecting my call.

I was slightly surprised by her 'down home' granny voice. I somehow imagined the cultivated tones of an ex-nuclear scientist, whatever that meant. She seemed sort of jolly, however. I hoped that the family connection would open doors but felt a little anxious about the Maoism. We arranged a visit to the farm accompanied by an old friend of hers. She Jenping is the tiny daughter of a top Party official in Shanxi between the 1950s and '80s. We met in her well-appointed flat in an apartment block for ex-officials in the eastern suburbs, high above an elevated spaghetti junction of roads. We set off on a bus towards the farm about twenty miles away on an expressway, past very contemporary high-rise housing.

At the bus stop where we alighted we waited for Joan's car and driver – she has privileged perks recognising her status. Her accommodation though was in line with what I expected: a meagre, barrack-like-single-storey building with no frills, the spartan rooms leading off a one-sided corridor like an old-fashioned railway carriage. As I entered her room I was greeted by the homely sound of BBC News blaring from her huge

Joan Hinton (1921–2010) at the dairy farm, Beijing.

TV, tuned in by satellite to uncensored global stations; another small privilege. (When she retired to bed, the cook and her dapper driver Lao Zhao – who looked remarkably like Cato in *The Pink Panther* – would instantly change channels looking for lighter entertainment.)

She welcomes me, dressed like an Appalachian backwoods-person in jeans, sneakers, a padded jacket and a flat cap over her unkempt grey hair. Her cook perk rustles us up some lunch and we get down to perusing the crumpled Boole family tree I've brought with me. We trace along the tracks of her illustrious forebears back to her grandmother, Mary Ellen, eldest daughter of George Boole, and pick up mine stationed in more suburban zones. She's amiable with a charming Good Fairy cackle of laughter. Over the next four days we chronicled that part of her life that interested me most: the Manhattan Project and her coming to China.

All of the three young Hintons, Jean, Bill and Joan, had been hugely influenced by their mother, Carmelita, after their father, Ted had died aged just thirty-six in 1923. Joan the youngest, then only eighteen months old, started life much as her great-aunt Ethel had done, not really knowing her father. Indomitable Carmelita brought them up emphasising learning by doing and treating life as one long adventure where possible. They learnt to ride, hike, ski and live the outdoor life, encouraging independence, self-sufficiency and endurance. Being relatively well-off, they had travelled abroad, partly to meet Boole/Hinton relatives.

Both Bill and Joan attended their mother's Putney School, with its daily regime. After milking cows at 6am before lessons, the evening cultural pursuits such as making music or photography could be pursued. An all-round education with an emphasis on a work ethic and the value of labour was a preparation for living a life of curiosity and achievement. Looking back at Joan's life it is not difficult to see its anti-conformist and explorative influence on her and her own children.

At Bennington College, Vermont, which Joan entered aged seventeen, at that time women only, there was yet another dairy farm and an emphasis on the liberal arts. Her interest in physics led to the building of a cloud chamber, used to detect particles of ionising radiation. She forged the metal container parts and turned others on a lathe. The last two years of her degree were spent at Cornell University in upstate New York where Bill was studying dairy farming. Here she got to know some of the other physicists who would later assemble at Los Alamos, New Mexico. When Cornell refused to accept a woman for postgraduate study she enlisted at the University of Wisconsin. Through her work there alongside a group experimenting with a cyclotron she was recruited in 1943, still only twenty-two years old, to work on the Manhattan Project at Los Alamos to produce the atom bomb.

She's very matter-of-fact about all this – to me – quite extraordinary story and unforthcoming about the sort of person she was then – maybe because it's a long time ago and maybe because it concerns herself as person she would rather not recall. Ruth Howes and Caroline Herzenberg in *Their Day in the Sun: Women of the Manhattan Project* provided more background. At Los Alamos, perhaps because of the physical separation behind barbed-wire secrecy, 'everyone worked hard, but they played just as hard'.[1] As Thomas MacMahon had related in his novel, there were dorm parties, square-dancing and poker games; on the sporty side there was golf, climbing,

Facing page: Joan Hinton in her days as a nuclear scientist.

hiking and skiing, at which Joan excelled. More sedately, she seems to have taken Putney's influence with her, playing violin in a quartet that included atomic scientists Edward Teller and Otto Frisch. She was clearly full of energy, focusing it on her work.

The location of Los Alamos, New Mexico, had been chosen to facilitate security and be far enough away from enemy bombing or centres of population in case of accidents. Robert Oppenheimer persuaded a team of scientists to join him and the US Army's, Leslie Groves, 'to produce a practical nuclear weapon in the form of a bomb in which energy is released by a fast neutron chain-reaction in one or more materials known to show fission'.[2] This followed on from Enrico Fermi's success at Chicago in building a self-sustaining reactor using uranium 235. The other known fissile material was the man-made element, plutonium. Both materials it turned out required radically different designs to create an atomic bomb. The former, dubbed 'Little Boy', was much simpler, consisting of a conventional explosive firing one mass of uranium into another; the other, 'Fat Man', consisted of a quantity of plutonium forced by a conventional explosive to implode and begin a critical reaction. Being more complicated, it was also known as the 'gadget'.

In early 1944 Joan Hinton joined a group under Fermi to test assemblies of enriched uranium and plutonium in reactors, one of which acquired its own nickname, 'the water-boiler', referring to the water that cooled it. She helped design the control rods. In Howes' book Joan states, 'I remember well when [the] first ball of plutonium arrived. Someone told me in a low voice, "The *pu* has come." It's in that little room. It feels warm to the touch. I went in and had a peek. I didn't dare actually touch it without anyone around to be sure I was doing right.'[3] The whole experimental process was fraught with danger and the unknown. 'Little Boy' was confidently expected to go off without a hitch but the plutonium bomb needed a test.

This was the first atomic explosion, 16 July 1945, the 'Trinity Test' that took place in the desert basin known as Jornada del Muerto, the 'Route of the Dead Man'. The actual date of the explosion was known to only a few, but the projected time and location was deduced. Joan set out on the back of a friend's motorbike, dodging military jeep patrols, to a small hill twenty-five miles from the detonation site. A little rain postponed the scheduled time at midnight but just before dawn the bomb was exploded. Joan described the explosion vividly to me, her slightly halting flow becoming more cogent:

First we felt the heat but then it was like we were in a sea of light, light all around. One person had calculated that the atmosphere would ignite, and then I thought that it had but I didn't feel anything. Then it was like as though the light had got sucked in to where the bomb had gone off and it turned into a horrible, horrible purple glow and then a cloud went up and up and up. Since the sun was just coming up as this cloud went up it hit the sun and it evaporated all the clouds in a perfect circle. It was glowing and just beautiful as we watched it. Then all of a sudden – WHAM, the noise came, quite a while before it arrived, the edge of it was so sharp – and then it rumbled and rumbled and rumbled. We'd been secret up until then and we thought there's no secrecy left now, the whole world can hear this.

Joan Hinton and her friend were not alone; the explosion revealed groups of others who had evaded security to witness the momentous event. Behind a strip of very dark glass that made 'the sun look like a little undeveloped potato', another witness, unlike her, had been officially invited and later gave his description:

At exactly the expected moment I saw a brilliant ball of fire which was far brighter than the sun. I saw it expand slowly and begin to rise, growing fainter as it rose. Later it developed into a huge mushroom cloud and soon reached a height of 40,000 feet. We next had to prepare to receive the blast-wave. When it came it was not very loud and sounded like the crack of a shell passing overhead rather than a high-explosive bomb. Rumbling followed and continued for some time.[4]

Quite remarkably, as we have recounted, Geoffrey Taylor, descendant of another of George Boole's daughters, was present. The Boolean diaspora had reconvened across the Atlantic at one of the defining moments of history. Taylor's expertise had led to the redesigning of the plutonium 'gadget' that Joan was also working on. She told me that she didn't know he might be there when she went to Los Alamos, but does recall meeting up: 'I remember climbing down the northern edge of the mesa with him and seeing a black bear in a tree and a flock of wild turkeys. I guess we must have talked about the gadget.'

Taylor's presence, in a dark but poetic way, made unintended sense of the 'Trinity' title taken by Oppenheimer from a poem by John Donne, relevant to the Boole/Hinton family. The third eminence, the ghost in

367

the machine, was Geoffrey Taylor's grandfather and Joan Hinton's great-grandfather, George Boole, the founder of the algebra used as the basis for the digital revolution and all computer systems. The world's first large-scale electronic digital computer, capable of being reprogrammed, ENIAC, was used for calculations regarding aspects of the plutonium device.

Three weeks after the test, on 6 August 1945, 'Little Boy', the uranium bomb, was dropped from the 'Flying Fortress' B29, 'Enola Gay', over Hiroshima. The 'yield' was equivalent to 18,000 tons of TNT. According to one source the other 'yield' was 66,000 people killed and over 69,000 injured. Three days later another B29, 'Bockscar', dropped the plutonium bomb 'Fat Man' over Nagasaki. It missed the exact target but nevertheless wreaked huge devastation.

Following my stay in Beijing in January 1983, and continuing my peace journey, I had flown on to Osaka in Japan aiming to talk to activists and visit Hiroshima.

28 January 1983
Remonstrate with Immigration officials who want an address when I'm not certain where I'm staying yet. Must be cooler with these type of people. Take the bus to the city centre marvelling at the motorways and traffic snaking along, the bright lights and well-dressed Japanese – such a contrast to Russia and China.

Phone a university professor whose number I've been given. He meets me and takes me back to their tiny flat crammed with books. Forget to take my shoes off as I enter – infidel! They make me a cup of

Hiroshima after the dropping of the atomic bomb.

Lipton's tea; a welcome change from the green or Georgian kind. In the morning we talk peace politics and I show the Prof. my Reitman document but he can't make out anything more than me. They ply me with gifts including a tissue-wrapped sweetmeat like a sugar-coated rissole which tastes awful. It's made of aduki beans but I have to munch away enthusiastically.

30 January 1983
It's an almost compulsory duty for me to attend the site of Hiroshima so I take the 'bullet train', the *Shinkansen*, from Osaka to Hiroshima, 200 miles in one hour twenty minutes. It feels like flying. Endless procession of noodle-sellers walk through. The countryside is only visible occasionally between tunnels – intensively cultivated rice paddies, barren at this time of year. Endless succession of small towns between small mountains, all very well kept, bright shining pantile rooves. Hiroshima is very modern and untidy – adverts on aerials everywhere. Everything is overwhelmingly consumerist but bright and colourful. Get stopped to sign for Hiroshima survivors, donate all my loose change.

Eventually at dusk I round a modern office block and am confronted with what remains of the Industrial Promotion building – the Dome from the Hiroshima bomb pictures. It's fenced off inside the Peace Park, girders twisted, skeletal roof. The rest of the park is a little unimaginative I feel, but nice coloured tassels made of paper birds are hanging from the children's memorial. At the rather ugly cenotaph Japanese are bowing their heads. I feel strangely distant from it all, the horror is hard to grasp. I walk on to the Peace Museum where the full awfulness of the A-blast hits home: a realistic model of the devastation, photos of burns victims, a watch that stopped at 8.15, pieces of glass that extruded themselves thirty years after their piercing of skin, pieces of fused metal, torn garments, photos of flash shadows of ladders and balustrades, whole scalps of hair that fell out after radiation. No mention of the reason for the bomb, only effects; but a veiled insinuation that the test apparatus found parachuted 15 miles away shows that the US was experimenting, at least in part. In another building see film about after-effects. Beginning to feel sick and clammy. The film is objective but grim, no factual memory, odd piano chords for soundtrack. Walk back to reality and yet another McDonald's.

There cannot be any way of adequately describing or recreating the scenes at either of the two cities after the bombs had exploded. They were at war but what happened without any discrimination came out of a clear blue sky and was over in two instants of Armageddon. The aftermath and its alienation, physical and mental, has been described by *hibakushas*, the survivors of the bombs, but Joan's own clinical description of one event during the experiments at Los Alamos somehow fits better. It was after the war that one US scientist had been accidentally irradiated with plutonium. 'He very gradually just disintegrated. All the cells in his body had been damaged way down inside. There was nothing the doctors could do. They packed him in ice. Finally his hair began to fall out and his mind began to be affected. It took Harry a month to die.'

Conventional historical wisdom maintains that the nuclear bombs, so expertly and urgently assembled, ended Japanese involvement in the Second World War. Their use thus spared American forces an invasion of mainland Japan that would have cost many lives. This is on the face of things true, but the issue is more complex: what were the real motives and were there other choices? Joan agreed in our conversation that the allied uncertainty as to whether the Nazis had developed the bomb was a key incentive to the weapon's development. Albert Einstein concurred in 1943 in a letter to Roosevelt. As well as research and development it was clear that the possibility of use was always envisaged; training of the pilots and the selection of targets had taken place in spring 1945. Hiroshima was one of them as it had been 'untouched'; the results would therefore be manifest. The surrounding hills would also focus the blast.

Against these plans a number of scientists were expressing their reservations. Joan was among those who agreed with a Chicago group that 'A demonstration of the new weapon might best be made before the eyes of representatives of all the united nations, on the desert or a barren island. America could say, "You see what sort of weapon we have but did not use."' Some politicians too felt that civilian targets were not appropriate, although after the saturation bombing of Dresden and Hamburg this principle had long been forgotten. The thousand-plane fire-bombing of Tokyo which preceded Hiroshima had brought a huge death toll. The question arose that if this massive destruction had not caused surrender, what would? To some the bomb seemed the answer.

For the Japanese there was a sticking point: the fate of the Emperor for whom so much had been sacrificed. Part of the terms of unconditional surrender that the Allies had made at Potsdam in May 1945 included the dismantling of the Peacock Throne. This gave little leeway to peace-seeking voices in Japan. Could it have been waived? The Allies knew from decrypted messages between Moscow and Tokyo that a distinct desire for peace existed and, according to Max Hastings in his account, *Nemesis*, had assessed that Japan was 'tottering'. Was unconditional surrender known to be a step too far that would justify using the A-bombs? Hastings bluntly argues that in any case after Pearl Harbor and the atrocities perpetrated by the Japanese why should the Allies have 'humoured the self-esteem of a barbarous enemy'?[5]

Equally important, Potsdam had re-confirmed that the USSR would enter the war in the East after three months, having been neutral until then. This they did in fact on 9 August. The dropping of the bomb, it has been maintained, was more of a hurried ultimate persuasion technique to bring rapid Japanese surrender in order to forestall post-war Soviet influence and land gains. It also, Truman believed, would impress Stalin as much as the Japanese to make settlements in Europe more favour-able to the West. Once again Hastings expresses the hawkish view that it was 'entirely reasonable that the US also wished to frustrate Soviet expansionism in the East and deliver a colossal shock ... to the leaders of the Soviet Union'.[6] It is only with hindsight he suggests that the atomic bombs represented something special. A view no doubt that would have assuaged war-mongering public opinion in the US at that time and justify the huge 2.2 billion dollars that had been spent on the Manhattan Project.

What could not have been clearly foreseen was that the events of August 1945 would set in train the competitive threatened terror of nuclear annihilation that became the Cold War for the next nearly fifty years. Behind the political posturing that the Peace Movement sought to deflate lay another insistent impetus described by Rhodes as already operating at Los Alamos: 'the technological imperative, the urge to improvement even if the objects to be improved are weapons of mass destruction ...'.[7] What Eisenhower later termed the 'military, industrial complex' was already up and running. After the dropping of the bombs, US Secretary of War Henry Stimson, despite earlier reservations, stated it represented 'the greatest achievement of the combined efforts of science, industry, labour and the military in all history'.[8] Edward Teller, the Austrian nuclear scientist, was already arguing for the next stage of research for the 'super', the hydrogen bomb.

The adage that mountaineers are driven to scale heights because 'they're there' applies to most people creatively driven. Ethel Voynich had written that an artist like Chopin 'must live under the compulsion of his art'. This applies to scientists as well. Joan had volunteered: 'We wanted to know whether it was possible or not.' A bomb that has never been exploded has never reached its summit. General Groves, who had been so successful in driving Manhattan forward, was adamant that the bombs should be used in combat. We still live under that same, seemingly dormant, threat that one day thousands of Hiroshimas will be unleashed. On 6 August 1945 the US announced its superpower status with a gigantic bang.

Joan learnt about the explosion of the Hiroshima bomb travelling on a bus to Santa Fe and was horrified. Somehow, like other scientists, she had managed to convince herself it would never be used. Her interest in nuclear physics did not dim but she became strongly opposed to further military development, joining a group at Los Alamos concerned to see science and technology used 'in the best interest of humanity'. One of the campaigns involved sending samples of the glassified sand they had named 'trinitite' from the Trinity Test site to mayors of some larger US cities. They were accompanied by the message, 'Do you want this to happen to your city?' Other Manhattan scientific colleagues questioned the whole issue too. Robert Oppenheimer, 'father of the bomb', who had originally accepted its use, was heard to utter the ominous words of the *Bhavagad Ghita* 'I have become Death, the Destroyer of worlds.' Suspected of communist sympathies, he was later investigated by a McCarthyite committee.

Joan joined Enrico Fermi and Edward Teller at Chicago University's Institute for Nuclear Studies in March 1946 to study for a doctorate. Her work on an accelerator was not without danger. On one occasion she accidentally received an electrical shock of 4,000 volts, luckily surviving with only a permanent scar to a finger. The moral scar of Hiroshima, however, was not going to heal over; she became further disenchanted with the military use of physics. The swelling paranoia of the Cold War was already apparent in Churchill's Iron Curtain speech at Fulton, Missouri in March 1946. Her sister Jean became vilified in the press for allegedly entertaining Russian diplomats at their Washington flat. 'All made up,' said Joan. 'I decided I wasn't going to spend my life making bigger and better bombs for the USA.' By the end of the decade the US military elite was in full control of nuclear weapons.

The call to China came from Erwin (Sid) Engst. He had grown up on an upstate New York dairy farm and been a room mate at Cornell with Joan's brother, Bill, studying agriculture. In 1946 he persuaded Sid to take up a post in Shanghai with UNRRA, the United Nations Relief and Rehabilitation Administration, and help start up a dairy programme. Disillusioned with the Nationalists (Kuomintang) under Chiang Kai-shek, Sid absconded to Yan'an in the northwest province of Shaanxi, then HQ of Mao's communist forces fighting for their own control of China. Under attack, he remained on the run during 1947, avoiding armed confrontation. Mao's tactic was not to engage with the enemy unless success was certain. Before Sid had left the US, a gradual friendship between Joan and him had solidified into an understanding that they would meet again in China after Joan had finished her doctorate, possibly with marriage in mind. Sid wrote to Joan urging her to come and join him. She needed no encouragement, slipping away secretly from the US, although with the knowledge of her nuclear physicist friends.

# PEKING JOAN

My stay in October 2005 at the dairy farm Joan Hinton managed just outside Beijing was not a Hilton experience, but entirely adequate. The constant running of water at the broken toilet cistern, flushed by pulling a piece of string, kept me awake on my creaking hospital-like bed, but as I was to hear, such details were trifling considerations compared to the hardships she had endured during her long stay in China. Every morning I followed her down to the bustling, humming dairy, built to her design, where she checked the milk yields. Joan showed me around both the cleverly arranged milking parlour and the ingenious layout for feeding and valeting the cows, a model that has been adopted widely in China. Although sprightly for an eighty-three-year-old, she tired quickly and was on medication for high blood pressure.

After our morning round and breakfast I'd set up my tape recorder for another session, to hear more about her remarkable life. Other people had interviewed her over the years; I was allowed the benefit because of the family connection. Joan was generously patient to my questioning, only too evidently based on the dimmest idea of the epoch she had lived through in China. A far more exhaustive re-telling of her and her husband, Sid's life was made by the American, Dao-yuan Chou, and was published in 2009 under the title *Silage Choppers and Snake Spirits*.[1] I have used it in places to supplement my own account. My meetings with her ended up as a BBC Radio 4 programme, *The Blonde Atomic Traitress*, broadcast in May 2007.

The luggage that accompanied her on her voyage from San Francisco to Shanghai in March 1948 consisted of a typewriter, violin, slide rule and sleeping bag. She had been recruited by Soong Chin-ling, the widow of Sun Yat-Sen, the former nationalist leader, who was operating a welfare fund for children. Joan was appalled at poverty on a scale she had never encountered before: people sleeping in the streets, houses with no roofs and a deep sense of hopelessness. Desperate to get into communist-held

Joan Hinton and her son, Fred, with Wang Jinhong at the Beijing dairy farm.

areas she joined up with another American, Sid Shapiro and his Chinese wife to reach Beijing, nationalist-held but communist-besieged. Here they holed up in my later haunt, the Beijing Hotel, before disappearing into the underground network of Mao supporters. After several forays attempting to leave the capital but having to turn back, she found herself actually arriving with the People's Liberation Army (PLA) as it entered in early 1949. 'I was going right along with the army,' she recounted, 'they were throwing confetti … and everybody was clapping and yelling.'[2] A little later Joan set out for Yan'an to meet Sid Engst through the ravaged countryside by truck, train and down the Yellow River. She stopped off to visit her brother Bill on the way, at the time teaching at a tractor school. Never having seen a fair-haired westerner before, she was everywhere an object of fascination.

Joan Hinton and her husband, Erwin 'Sid' Engst (1919–2003), outside a loess cave dwelling, Yan'an, 1948.

At Yan'an she was billeted in a loess cave to await Sid, who was working at an iron factory two days' walk away. Their reunion was not a Hollywood moment. 'Joan strode up to Sid and … gave him a big punch in the arm.'[3] Sid punched back. Marriage, although in the wings, was not a matter of romance; when asked eventually to give consent before witnesses she managed an 'I guess'. Joan's emotion was reserved for her wholesale endorsement of the revolution. 'I realised that I was part of something very very big, very very new … that the whole people of the world should know, but didn't. Suddenly I was in this family … of the human race. I just felt like "Oh boy, this is it."'[4]

For a honeymoon the new comrades-in-arms returned to take up work in the iron factory. Here everything they needed had to be fabricated: tools from US military parts, casts from wood and sand moulds. Her hands-on education at Putney School obviously came in useful. Joan showed me a photo of a windmill water pump she had produced from smelting scrap. At Los Alamos, she explained, you just got whatever you needed. Sid's important innovation was simpler but highly effective: a donkey cart that could actually be tipped.

After five months in 1949 living in caves they were asked to move 200 miles further north to Inner Mongolia to work on a state farm. The walk took a week, their carts loaded with cloth as a useful barter given the

Joan Hinton and Sid working with farm machinery.

uncertainty of any currency. Being so far away, unknown to them, a month later on 1 October 1949 the People's Republic of China was founded. Joan and Sid's task was to foster and breed a small herd of Holstein cows that had been rescued from the Japanese and brought over the mountains from Yan'an.

Their new domicile was within a stockaded fort named Chunchuan, built of twelve-feet-high adobe walls, circa 1900 during the Boxer Rebellion. Their house was a grain store, accompanied inevitably by a colony of rats that attempted to share their meagre basic diet of millet. The humans augmented it by rice, the ubiquitous napa cabbage and occasional mutton. Conditions were extremely primitive, without electricity in temperatures well below freezing at night. Heat came from a traditional clay *kang* sleeping platform absorbing heat from the sheep manure pies lit below. The Yan'an windmill they'd imported with them proved to be too effective in the windy plateau conditions: it exploded owing to Joan's design omission of a brake.

The pair set about learning to help improve the grassland beyond the fort. She described the scene in an article printed in *China Monthly Review*, in November 1950, signed not quite anonymously as 'J.H.E.'. In one paragraph she unintentionally parallels the vivid colours of a desert landscape that could have been not so long before the backdrop to the detonation in the Alamogordo desert:

> On all sides of the horizon are banks of sand hills. Sometimes they are almost white, sometimes deep grey. In the early morning they are a warm soft red. Just after the sun comes up they are particularly beautiful: the east face turns white as it dies while the west is still red with dew. Thus great sweeping patterns are formed as the sun keeps getting higher.[5]

Visiting a Mongolian household she good-humouredly details the ancient and laborious technique of churning milk with a pole. 'The first twenty strokes seem easy enough but by the hundredth the sweat is pouring down your face and by the thousandth your hands have become one big blister and your back feels like you've been lifting a camel.'[6] Joan clearly admired traditional ways, even the art in the robes of the congregation at the wooden Catholic Church within the compound. She noted its gaiety but felt that it should be possible to preserve yet detach it from its feudal origin. This essentially was her Maoist mission throughout her stay in China: to respect its culture but bring new technology and social relationships to enable it to lift itself from centuries of poverty and exploitation.

This theme recurred time and again in *China Monthly Review*. Articles extol the new society being built: China's first woman locomotive crew, a better life for China's miners, Kuo Tsan-ching army heroine, how nature is being tamed, the drive against spies and saboteurs – and more relevant to Joan, 'eliminating bandits in Hunan'. In the still unsettled times, outlaws attacked and looted wherever authority was shaky. Their fort was an obvious target for marauders. On the first threatened visit, pacifist Joan, whose principles were fast evaporating, asked for a gun to fight, but was told, 'you couldn't shoot a rabbit'. She was offered a baseball bat, before being eventually allocated a hand grenade.

When another attack was imminent Joan had a high fever thought maybe to be the plague. She had to be evacuated on a stretcher accompanied by the other occupants and a line of animals. By the time they reached a safer fort Joan was at death's door. Resignedly pondering her fate, she told herself, 'People have lived and died here for centuries, and that's the way life is.'[7] By good fortune some rescuing penicillin was found and the move made back to their own fort. It was untouched as the peasants had frightened off the invaders saying that the cows belonged to the PLA who must therefore be close at hand.

The drive to modernise peasant life was paramount, including improvements in medicine, veterinary treatment, plant breeding, making silage and the introduction of woollier sheep and the high-yielding black and white Holsteins. All such demonstrable gains helped in securing the peasants' trust to allow the challenging of other aspects of rooted traditional culture such as alcoholism and opium use. A movement was initiated against corrupt communist cadres; the long line of misdeeds of the local priest was revealed, resulting in his expulsion.

Joan had been living her exotic but hazardous life in China for two years before the rest of the world caught up with her. A routine subscription renewal from the Federation of American Scientists had found its way to Mongolia accompanying her sister Jean's mail. Joan's exact whereabouts had been known only to a few outside the family; those of her scientific colleagues who knew had continued to keep silent when questioned by the CIA. Her fulsome reply to the Federation enquiry firmly put her on the map when it re-appeared in the *People's China* magazine of 16 September 1951. In it she explained her motives for leaving the US after Hiroshima and her contempt for US support of the Kuomintang during the civil war. She ended up with an appeal to her fellow scientists: 'Use your strength wherever you can, to actively work for peace and against war. As long as there is war, science will never be free. Are we going to spend our lives in slavery for mad men who want to destroy the world?'

What the US government wanted to know was whether this particular scientist branded as 'Peking Joan' would pass on her intimate knowledge of the A-bomb to the Chinese. Paranoia broke out, omnipresent in the US at that time of the Korean War. On 10 March 1952, Owen Lattimore, a specialist on Asian matters and suspected communist, was interrogated by a McCarthyite committee about Joan and her family, including her mother, Carmelita. Joan's brother Bill, as we shall discover, was quizzed too in July 1954 by an official incredulous that a nuclear scientist could be working on a dairy farm in Mongolia. Joan's published letter putting the cat among the US hawks was nothing, however, compared to her personal appearance on the world stage a year later. It happened almost by accident.

Seven months pregnant in August 1952, she became seriously ill again and needed hospital attention. Bill was in Beijing, so she left the fort where, as one US journalist hack later revealed for *Fact Forum* News, 'this Mata Hari trudges in mud to the dairy-barn – nuclear cross-sections in her mind and quantum mechanics'.[8] Walking for a week back to Yan'an and on to the capital by train, she arrived at the same time in October 1952 as the Asian and Pacific Peace Conference was taking place. Discarding her padded peasant attire and borrowing a skirt, she took to the platform to address a Japanese audience. As a person who had 'touched with her own hands the very atom bomb that was used against Nagasaki she declared a deep sense of shame in this crime against humanity ... and the Japanese people ...'.[9] Her speech was well received among delegates – not so back in the US where it was announced that Mao himself had ordered the plane that had taken her there. 'Baloney,' Joan told me. After all the limelight, her baby was born in Beijing. She was urged to call the boy Kim after the Korean leader but opted for Heping, 'Peace', resonant with the conference, plus 'Fred', resonant with Sid's father.

I did ask her whether she would have helped with a Chinese bomb had she been asked. What interested her, she replied, was to make things useful to mankind, repeating a mantra of hers that it was 'millets and rifles', the stuff of everyday life, that had already beaten the Japanese occupation of China long before the bomb had been dropped. But, she added, the Chinese have always had the right to own and develop nuclear weapons as a matter of self-defence. Deterrence worked she implied: 'The only time the bomb was used was when no one else had it.' Despite sitting next to one of the very few who had actually created the weapons that drove the Cold War, I did not press her on her views on how to eliminate them. Her views, however, were pressed into service by the sensationalist press at the time.

Joan and Sid with the children at Caotan.

*Real* magazine in July 1953 posted an article, *The Atom Spy that Got Away*, by Rear Admiral M. Zacharias USN (ret'd) that included an artist's impression of a Lauren Bacall-like Joan in raincoat and neckerchief furtively writing notes behind the backs of military personnel.

*Fact Forum News* in March 1955, the vendetta still continuing, reprinted, under the heading *Atomic Power and the China Doll*, the whole Federation of American Scientists letter, prefaced by a warning to the innocent reader, 'the following document is of communist origin and was prepared to serve as communist propaganda'.[10] Just in case this primer failed, each page was stamped 'Communist' across it in heavy letters.

In May 1955 Joan, Sid and the now three children, Fred, Bill and Karen, accompanied by descendants of the Mongolian Holstein herd, moved southwards to another state farm, Caotan, beside the Wei river near the city of Si'an, capital of Shaanxi province. Here on land four miles by twenty, Sid assumed the role of vice-director of the dairy and Joan of technician. It was their role to oversee the mechanisation that would use marginal land for dairy production to help feed the nearby urban population.

Medical problems of TB and brucellosis were tackled and Soviet-designed milking machines introduced. Joan made important innovations in pasteurisation and refrigeration, but perhaps the most exemplary was the story of the silage chopper. Chopping corn and sweet potato vines to make silage for the cattle was a very labour-intensive process involving many men and a huge cutting knife before the material was transferred to a pit for fermentation. Sid, however, managed to secure an example of a Soviet mechanised chopper. It worked really well until a new four-foot blade had to be produced from the simplest of tools. A further problem was the excessive wear on the frame of the machine. Joan solved this by re-arranging its layout – the success of which she described in an awed religious manner, perhaps her life's motif: 'If you learn you can make it yourself, you think, "My God, I can not only make it but I can change it," and that's when you go into the world of freedom from the world of necessity.'[11]

That creative spirit met its match, however, stifled by its institutional opposite – bureaucracy. So successful was the revamped machine that the farm began to manufacture them, only to be told from above that it was not 'approved' despite a public competition between old and new models, which Joan's version won hands down. A decade later the Soviet model was still the only one recognised. As the system could become bogged down in the appropriately-coloured red tape, self-serving and kowtowing to party cadres, Mao in late 1956 tried to counter with the 'One Hundred Flowers' movement. One of many such campaigns of regeneration, it encouraged criticism politically and culturally. When it began to get out of hand, including anti-party agitation, a reversed 'anti-rightist' move was spawned. This lurch of extremes – some said only a ploy to winkle out opposition in order to suppress it – found its way to Caotan. An old vet with 'incriminating' one-time support for the Kuomintang was held responsible for fatalities in the herd of cows. Confined under guard he committed suicide, a small echo of the witch hunt that proceeded nationally.

Mao's seeming faith in grassroots peasantry and workers to motivate progress led in 1958 to the next grand policy to descend from Beijing. It was the period of the 'Great Leap Forward': his reckless bid to catch up with the West in only ten years by every self-reliant effort in agriculture and industry to increase production, modernise and innovate. Private ownership was gradually to be eliminated through stages of collectivisation. This was a huge process too and no doubt brutal, but perhaps less devastating than Stalin's method of merely extinguishing the *kulaks*. In China there was perhaps a stronger degree of commitment from the peasantry. This was just as well, as in most industrialisations it has been their wealth production ultimately that would finance it.

Mutual aid teams gave way to 'voluntary' co-operatives in the mid-1950s in which 150 or so family units worked together but retained private plots. To increase output the People's Communes were then introduced. These were much larger units of 10–20,000 in which families ate and slept together and co-owned most buildings, tools and private property. Members earned a wage and derived welfare benefits. Spurred on by the ideological fervour of Mao's appeal to the masses, communes exhausted themselves building roads, dams, terracing and industrial projects. In 1958 the infamous 'battle for steel' was launched in which any available metal was smelted in backyard furnaces – the end product mostly useless.

At Caotan, Joan's own ebullience is captured in an eleven-page letter dated October 1958 that she had sent to Geoffrey Taylor in Cambridge. She also sent it to Edward Teller, her nuclear scientist colleague from Los Alamos days, whom she had clearly not forgotten. Any real harmony from their music playing had by now dissipated as he had been implicated in spying accusations against their other colleague, Robert Oppenheimer. Teller had always been a 'Strangelove' advocate of building the even more destructive H-bomb, detonated in November 1952. 'Why,' asks Joan, 'are you fighting so hard to wipe out the socialist half of the world … to risk disaster to the whole human race, both East and West?' She entreats him that in order for society to realise the true freedom, the freedom from want …

Just as for the laws of physics, we must know the laws of society as well to harness the power of creative labour. Society can then develop continuously in an onward, upward spiral even if this is, of course, uneven. I wish you could just take a peek at China today and see with your own eyes ... the creative enthusiasm of the people ... the rate of construction is beyond description. The peasants of Shaanxi where we live built one half times more irrigation works than they had built in the 2,000 years before. The whole country is moving, 600 million people are on the march. Tell me what you are thinking ... tear what I have said apart.[12]

Joan related to me an episode in her own backyard of the conquering mood at the time when irrigation failed to cope with the bad weather. It reminded me unhelpfully of A. A. Milne's 'Piglet is entirely surrounded by water'. It rained and rained so much that the stream running through the farm breached its dykes, flooded and destroyed the bridge that took the milk production by mule and cart to Si'an. But in a mini-episode of self-reliance and fortitude, reminiscent of the model heroism that found its way into books, films and operas of the time, the milk got through. One intrepid youth managed to swim across the flood to attach a guiding rope to two posts either side and a raft was improvised from bed boards and oil drums. 'We didn't lose a single drop of milk,' said Joan proudly. In other episodes the outcome was not so successful as the Great Leap floundered and fell short.

Sid related in a transcribed tape entitled 'Remembrance of the strange pressures coming down in 1959', how intransigence at the Party top levels was passed down by ambitious visiting cadres. As he put it in unequivocal terms, it was 'when the shit really hit the fan'. One insisted that 100,000 ducks could be reared on the river, without suggesting what they could be fed upon. Another example was a directive to increase the variety of livestock on the farm. Without turkeys on hand, to introduce the species Joan cycled a day's ride to fetch a pair back in a rucksack with their beaks sticking out. An emissary was sent further afield to enquire about yak imports – a beast entirely unsuitable for their climate. An exhortation to multiply pig production beyond a realistic twice a year found Sid remonstrating to an official, 'You've got to wait till one litter comes out before you can start another.' Difficulties in producing more goats, in a foretaste of the Cultural Revolution to come, led to accusations of sabotage aimed at the farm's two foreigners.

Such stories were reproduced all over China as chaos sapped will and people became devious. Joan reports one incident exactly equivalent to my teenage shortsightedness selling the *Hayes News*, only rather more consequential. One frightened village piled valuable grain on a haystack to make it look as if they'd fulfilled their grain quotas and then had to pay a ruinous tax on the false value. The pattern of disaster was varied: one village simply starved, another had buried a life-saving cache of sweet potatoes. Joan's farm managed, just, on half a kilo of grain per month per person and a pint of their milk per day.

'Times were hard,' Joan told me in an understatement of epic proportion. Western sources state that millions died. There was famine, she admits, but this was due to drought as well as floods creating havoc. Alongside this the vindictive withdrawal of Soviet support had occurred. The ideological split between the USSR and China had long been on the cards; the two communist states had reached different stages of development. On the world stage, following Stalin's death and after Kruschev's 'secret' speech of 1956, the Soviet way was paved for material incentives, economic and 'peaceful co-existence' with the capitalist West. Mao's China was still set on a path of continuous struggle, its production still based on a peasant economy that endorsed collectivism and 'serving the people' – a sentiment Joan and Sid wholly subscribed to. Domestically, dissension was brewing between Mao and the incipient 'capitalist roaders', Liu Shaoqi and Deng Xiaoping, who favoured the individual economic unit and material incentive at the expense of class struggle. Joan adopted the rationalisation that they and others behind the scenes were trying to wreck the Great Leap by deliberately encouraging unrealisable goals.

On their own domestic front Joan and Sid found themselves split. When Sid accepted a farm proposal to re-house them in better style, Joan objected to it as elitist. The move happened anyway, leading to a further dispute over Sid's commanding of office space. They shared in the end but it was just one small example of elitism and its offshoot, sexism, that bugged her through her Chinese life. The kids, however, loved their new domicile and its porch allowing them to stage local-style operas with other farm kids.

It is incredibly difficult to get a handle on events of the magnitude of the Great Leap with little historical comparisons to guide one. British society since 1945 may have faced conflict – the miners' strike of 1984, the Cold War protests of the 1980s or the poll tax and race riots – but most of its citizens have never been threatened by the same level of overwhelming change. This was not the case, however, for my parents' generation. The Second World War brought death and destruction to city doorsteps as mournful telegrams arrived and bombs fell, but there was solidarity against a common enemy. Mottos were entreated, spies suspected, park railings melted and victory vegetables dug for the greater good as defined by a command economy and polity. People when exhorted and cajoled can pull together and liberate great energy. In Churchillian Britain, Mao's jingle about 'the boundless creative power of the masses' would not have been amiss. But this energy can dematerialise equally easily. The 'upward spiral' that Joan described to Teller had indeed become 'uneven' and calamitously reversed. And the weather, no doubt, didn't help either.

The tumultuous period of the Great Leap was not an unhappy one for everyone. In 2007 I stayed with Joan's daughter, Karen, living in Pau near the French Pyrenees. I took her a copy of *The Voynich Manuscript* as a present. What emerged from our long conversations was a childhood that would have put any soul to the test, yet as a child at Caotan in 1956 Karen mixed in with the other Chinese children, singing and playing in the fields, her rump frozen in the winter because of the traditional 'bottomless' trousers and at one point, as in my father's youth, without shoes. Despite no toys in a house, without running water, 'We had great fun.' She refused to learn English, but ubiquitous socialist politics crept in as criticism sessions within the immediate Hinton family: were they too nuclear? What about spanking for bad behaviour?

At the time, of course, she was too young to clearly remember major events but her later analysis echoed a general assumption that the Chinese need to 'save face' led to competitive, unreachable economic targets that confounded the central planners' ability to allocate resources and gain a clear picture of progress, or the lack of it. Karen remembered, however, tearfully, the emotional impact of collective camaraderie as she lay in her bed and saw the lights reflected on the wall from the tractor driven by a 'model worker' ploughing late into the night. 'People then wanted to make a difference,' she explained.

The overall failing of the Great Leap led to a period in 1962 of the 'Socialist Re-education Movement' when at mass meetings, cadres, assumed guilty until proved innocent, were asked to confess errors. In a glimpse of the Cultural Revolution to come, Karen recalled how the head of their farm was driven to drown himself in the river and how another, a fruit technician, was imprisoned for wrongly pruning peach trees, but was really a scapegoat due to his Taiwanese ancestry. Joan and Sid, however, as foreigners, were excluded from meetings. Normally only using bikes they borrowed the farm's car for a visit to Si'an to offset the slight. When Karen asked for this taste of privilege to be repeated in 'their' car, her fully collectivised parents were very shocked.

At another family meeting it was the children's turn to be tested. Perhaps because Joan and Sid were feeling excluded they wrote to the North Vietnamese government offering their services in a 're-education programme' aimed at compatriot captured US servicemen from the war then in full swing. With a faltering voice Karen added that the children were to be left behind, their parents trusting that 'the people will raise us'. When the Vietnamese peremptorily replied, asking of Joan and Sid, 'who are these two?' plans changed and the family moved reluctantly to Beijing in April 1966, the very beginning of the next momentous struggle to establish communism: the Cultural Revolution.

Despite talking at length with Joan and Karen about their lives during the 'Great Proletarian Cultural Revolution', I found it very difficult to build up either a theoretical or an empathic picture. Perhaps for Joan it was all too much to recall at her age – I was aware of how tired she became – but also, it seems, she has never been an overtly emotional person. Our dialogue was enlivened mostly by her endearing chuckle. I remained almost as much in the dark as before I met her. I was in good company: Anna Louise Strong, a lifelong communist and confidante of Mao, wrote to a friend, 'Can you tell me what is really happening in the Cultural Revolution?'

Now in Beijing the Hinton family were assigned a 'keeper' but refused to live under the same regime as many other foreigners in relative luxury at the Friendship Hotel five miles from the city centre. This had been built for the Russians but was now, in Joan's words, a 'Golden Ghetto'. Nevertheless they were housed in a hotel with other foreign correspondents and had to insist on not being chauffered to their offices, preferring bikes like other ordinary people. In early June the Hinton family heard a commotion from the top of the hotel roof. Below, a throng of people were marching, beating drums, carrying slogans and giant pictures of Mao protesting against his perceived enemies: the first opening shots in the Cultural Revolution.

The whole family were still disorientated by their move from relative rural tranquility. For Karen, in particular, her secure world as a peasant had been replaced by virtual house arrest in a strange city surrounded by alien objects. How did the people get inside the television, she wondered, why did the toilet run like a flood and the taps gush? She plotted to escape back to Si'an by following the railway tracks. Karen and her brother Billy were sent to an elitist boarding school nearby for foreign workers' children (their older brother Fred had been left back at the farm).

Here the regime was equally alien, one of strict rules and educational aspiration. She refused to bow to teachers and went on hunger strike; outside school Chinese kids threw rocks at her. As things began to hot up, the teachers, themselves challenging each other, were challenged in turn by their students. The school closed. 'The most beautiful moment of my life,' said Karen. But liberation did not last long. She was diagnosed as mentally ill, behaving aggressively to her parents and viewing them even as 'US imperialist spies'. She narrowly avoided being sent away. Her disaffection was eventually assuaged by the magical symbols given her by her doctor: a Mao lapel button and the *Little Red Book* of his quotations.

The family recuperated by living in a commune just outside Beijing, picking fruit and being treated as a normal working group. They returned on 16 August to an elite compound of embassies that at least housed Chinese and an area where the kids could play freely. They just missed witnessing another turning point: Mao's tour of the city in an open jeep that ended with his speech of endorsement to thousands of Red Guards in Tian'anmen Square.

If nothing else the Cultural Revolution was about 'struggle', this time nominally against the elites in power by those at the bottom, encouraged to turn the world upside down and 'bombard the headquarters'. It turned out to be a war of all against all. The Red Guards, Mao's sanctioned tide of youth, were to be the engine behind this assault, allowed until July 1968 a virtual free hand to criticise the hierarchies of power: parents, teachers, party cadres, anyone with 'suspect' backgrounds. The shifting list included traditional and western culture, anything symbolically bourgeois and 'old'. Temples and religious paraphernalia were fair targets and liable to be physically attacked. Only three institutions were to remain unassailable: the PLA, the Foreign Office and Radio China International.

Mao's wife, Jiang Qing, was one of the leading figures in deciding the parameters of criticism. Carma Hinton, Bill's daughter, who had left China in 1970 and become a filmmaker in the US, gave an example of changed sensibilities in her film *Morning Sun* (2003). Her great-great-aunt Ethel's story, *The Gadfly*, had now fallen out of favour. The heady revolutionary tale of sacrifice with its hint of foregone romantic love between heroic Arthur and noble Gemma had been as popular in China as in Soviet Russia. The Soviet film version lapsed into disfavour compared with the 1965 nationalistic rival spectacle of the song and dance epic, *The East Is Red.*

People with foreign backgrounds were high on the list of dubious characters. A deep mistrust of foreigners dated back to the Opium Wars of the mid-nineteenth century when China was forced to open up to western 'barbarians' and cede territorial rights. Foreigners had been allowed to take up various roles such as experts and advisors like the Russians until 1960. Foreign delegations were fostered to create favourable impressions. Sartre and Simone de Beauvoir visited in 1955, the latter publishing a glowing account that she later disavowed as being written only for the money. Any of the foreigners could, of course, actually be spies, especially at a time when there were fears of another civil war with the US-backed Kuomintang rump in Taiwan. Ideologues who chose to be resident in China must have seemed a strange bunch.

Joan and Sid had acquired the honourable status of being more than just 'international friends' and had escalated a rung to 'Friends of the Chinese People'. Their relative isolation in Mongolia and Si'an and their pedigree of being pre-1949 revolutionaries had provided a kind of immunity from the system's tagging. Joan was put to work for the Association for Cultural Relations with Foreign Countries, polishing the English of articles for consumption abroad. Sid polished foreign films at the Bureau of Movie Distribution. A life of being 'political typewriters', as she described it, did not suit their political sentiments, nor Joan's dyslexia.

Irritated at the way in which foreigners were being corralled away from events, in true fashion, the Hintons decided themselves to put politics in command. Over ten days with the help of Bill Hinton's ex-wife Bertha, and Carma, then sixteen and able in Chinese calligraphy, they set to compose a wall-poster or *dazibao* in that strange declamatory style. It asked,

Mao's *Little Red Book* held aloft.

'What monsters are behind the treatment of foreigners working in China?' It went on to list the five 'don't haves', including contact with the revolution, compared with the two 'haves' that included a rich seclusion. Such 'revisionist' thought, it stated, is designed to 'isolate foreign revolutionaries from their Chinese class brothers, break down their mutual class love and to undermine proletarian internationalism'. A list of 'requests' was ended with four 'long lives' including one to 'the Great Invincible Thought of Mao Zedong'.

The message of the poster found its way into a government office where it was eventually seen by Mao himself, who agreed with its contents. The verdict found its way back to Joan and others who wished to be in the thick of things. Her face lit up as she described all this to me. In Dao-yan Chou's account Joan queried, 'How did Mao know how they, and especially how I, felt? We were just such a tiny problem. He had to think about the whole world and all of China …'.[13] Twenty years earlier Oppenheimer had condemned atomic scientists including her for 'knowing sin' – here with the help of a long strip of paper she had experienced the sublime reverse: a starry-eyed communion with a godhead. George Boole would have regarded Joan's deification of Mao as idolatry. For her the poster and its success represented 'a genuine dictatorship of the proletariat, the first time in history … that the proletariat could rise up and write a *dazibao* and criticise the whole government'.[14] Passing by Tian'anmen Square she met a Tibetan and felt at one with him, as he'd been 'liberated' by Mao also.

Emboldened, Joan pitched in with criticism where she worked, objecting to the untruthful and propagandist nature of some of the material. Sid similarly objected to the 'Hollywood' slant of propaganda movies. Other struggles they talked about made less sense to me. Joan tried to explain: 'We accused whoever was in charge of us, not that we knew who they were.' This led unsurprisingly to accusations and counter-accusations. Factionalism in general became rife. Smaller groups allied with larger ones along ideological lines, or merely to settle grudges or further their power and influence. Despite the appeal for non-violence, vented passions led to victimisations and atrocities urged on by the rampant Red Guard army of youth.

Karen, only ten, joined a group of young teenagers whose first action was to acquire a rubber stamp with a moveable date as a token of legitimacy. Her group were too well-mannered to wreak havoc but a young neighbour was encouraged to cut off the ends of bourgeois pointed shoes and take scissors to overlong hair. 'It was really wild,' she enthused to me, 'you could do anything you wanted.' She and her brother found a more important niche when they joined an older Chinese college group with internationalist pretensions, glad to have real foreign members, especially those with an official rubber stamp. In late 1966 imbued with proletarian spirit she went to work in a light bulb factory. Carma joined a group retracing the Long March, later accompanied by Fred, now back from Si'an.

Joan and Sid joined a broad-based fighting group of foreigners known as the Bethune-Yan'an Rebel Regiment, centred on workplaces. Soon further factionalism emerged with strong-arm stuff. Sid wrote of the treatment of the Party Secretary at their workplace: 'About ten of us … drove out to Xueli's home … we whacked on the door and said "Get up, get up, you're under arrest."' They lined him up with other directors. 'We said formally that all power had been taken away from them and that they were to report … to do whatever work they were assigned to by the revolutionary masses.' He added, 'Those were the real days.'

Joan joined in with the giddy mayhem. Responding in July 1967 to anti-British riots in Hong Kong, she supported a crowd of Bethune-Yan'anites at the Beijing British Charge d'Affaires office demanding the acceptance of a petition. When thwarted, the mob broke windows and invaded, despite the PLA presence. Afterwards, given an attempt to set alight the Union Jack, Sid cut its rope fearing the flagpole would break with tugging. A month later on 22 August another mob, again watched passively by the PLA, burnt the building to the ground.

The two revolutionaries eventually found themselves in the opposing factions, 'Heaven' and 'Earth', that dominated Beijing. Unable and probably unwilling, to really explain to me how this came about she merely described it as 'artificial' and essentially the result of the 'capitalist roaders' who wanted to turn back the clock on socialism and discredit Mao. This was the other great struggle of the Cultural Revolution apparently: to eliminate the revisionist moves towards a more individualist materialist economy based on pragmatism, rather than a continuing collectivist ideology that reached beyond narrow 'bourgeois nationalism' towards Mao's 'socialist internationalism'.

Liu Shaoqi and his wife Wang Guangmei had become two such targets. Liu, as President, and second in command since 1959, had opposed the Great Leap and made the 'mistake' of steering recovery, as Lenin had in the Soviet early twenties, with private plots, free markets and autonomous enterprises. Wang had been part of the witch hunt of the Socialist

Education movement that followed and in 1966 had tried to repress dissent at Beijing's Qinghua University. Their opposition and 'bourgeois' lifestyle set them out. Both were publicly humiliated: Liu beaten, Wang made in April 1967 to stand before a crowd of thousands in a symbolic silk dress and a necklace of ping pong balls. Wang survived imprisonment for twelve years. Liu, expelled from the Party and denied medical treatment, died in 1969. Deng Xiaoping, also considered a capitalist roader, at least fared better, suffering from 1969 four years internal exile until recalled owing to Zhou Enlai's ill health.

As the Cultural Revolution spread nationally things began to get out of hand. Intense factionalism, for example, in the steel city of Wuhan on the Yangtse resulted in its division into two huge opposing forces of workers, students and Red Guards, one of which was backed by a rogue PLA general. Eventually intervention in July 1967 by Zhou Enlai and the PLA itself restored order but not before many had died.

To add to the ideological foment of the time there were jockeyings for power centred on Mao's succession. Lin Biao, a veteran military leader, had encouraged rebellion among the Red Guards and even within the PLA. He was rewarded for his loyalty by being appointed Mao's successor in 1969. Accused of an assassination plot, however, in 1971 he fled and was killed when his plane crashed over Mongolia. Weaving in and around the continuing clashes, the 'Gang of Four', including Jiang Qing, Mao's wife, by controlling the media and propaganda, zealously upheld Mao's cult of personality – aiming, it was later claimed, to take power on his death.

The well-documented violence of the dictatorship of the proletariat that shook China does seem to me to be not one in which the party as destined class vanguard holds tight power from above, but one, far more risky, where, within limits, it allows a welling up and anarchic rampaging, its sole anchor point the deified charisma of Mao. The only tiny wisp of events that I can think of in my own life took place in that same contagious era. Students were again a driving force, having no material stake or institutional power base of their own to defend and therefore able to discredit whatever was deemed as opposition.

At Leeds University in 1968 I joined the occupation of the Brotherton main building amid euphoric solidarity and good intentions. Senate representation for the student body and for workers such as porters and cleaners seemed like a genuine demand. A giddy sense of empowerment was in the air. Symbols bolstered solidarity – Marcuse was our bible, Guevara our sweatshirt emblem. When a faction of local trades unionists visited the university carrying their roped banner aloft it felt as if times really were changing. They didn't very much.

The somewhat more tumultuous days in China led to major conflict as Mao tried to reign in the worst excesses of revolutionary zeal. The Red Guards were dispersed into the countryside alongside intellectuals to 'learn from the people'. In January 1968 in another *volte face* foreigners were expressly forbidden to be involved in Chinese affairs. Their vulnerable anomic status in the power struggles left them open to scapegoating, withdrawal of privileges, house arrest and imprisonment. Joan and Sid took to DIY and repairing things to pass the time. The Bethune-Yan'an Rebels decided to close down but three of its leaders were confined. In October 1967 Marni Rosner's father-in-law, David Crook, was picked out for victimisation and jailed, spending five years in solitary confinement.

When, in another major change of direction, Mao decided to turn away from international isolation and open table tennis links with the US in April 1971, the Hintons, long-standing and useful allies, played their part. Joan's brother Bill arrived back in China from the US in May after an absence of eighteen years. The whole family met with Premier Zhou in a series of meetings. At one Joan raised the problem of women's identity in China, an issue that had bugged her for years. The rapprochement period led to the invitation to Joan's mother, Carmelita, and a group of students to come over. Bill most wanted to see Long Bow again, the village of *Fanshen*; the others were encouraged to encounter the model village of Dazhai in the yellow loess hills of Shanxi province.

The 'Dazhai way' was held out by Mao as the key to generate increased agricultural production integrated with small-scale industry, through self-reliance, collective action and common ownership under 'correct' leadership. Thousands of peasants and workers had been bussed into the showpiece. The hillsides of the drought-prone area had been terraced and irrigated by viaducts, dams and reservoirs, the soil and crop yields improved by composted matter. Dazhai seemed to be the Great Leap Forward made manifest. The posters showing the neat housing surrounded by orchards, beehives and small workshops must have filtered back to us in the West, proving that utopia was possible. Joan's encomium to me was ecstatic.

The Hintons pitched in, working during their three-month stay. Karen joined the 'Iron Girls' team, Billy and Fred the 'Iron Boys'. Joan, however, began to feel gender-sidelined once again and when she overheard Wang Yuwen, the police chief, declaring the limits of what women could do, she grabbed his ankles and wrestled him to the ground. I find this a most endearing image to temper her otherwise unbounded triumphalism. Years later the Dazhai way was replaced by individual, market-driven policies. The village was downgraded as yet another Mao folly subsidised by the state with artificial yields, failed irrigation and soil erosion.

Whatever the rights and wrongs of the divergent policies, the Hinton family continued to be most at ease in an agricultural setting. Joan and Sid returned to their lives in peasant agriculture by moving in 1972 out of the city to the nearby Red Star Commune that Howard Hinton had visited. The pair resumed their own private hierarchy: Joan as 'technician' under Sid as Vice-Chairman of the Revolutionary Committee. Both, however, found the drive to technically innovate frustrated by bureaucracy and power politics at the expense of grassroots involvement. A glaring example was finding the old Soviet silage cutter from Caotan days still insisted upon. Western machinery was now in vogue lauded, by Deng Xiaoping who in 1975 was back in a position of power.

A high-tech milking machine called a rotolacter, a kind of bovine merry-go-round, was brought in despite Sid's justified warnings of likely breakdown. Joan took on a challenging project to build a combine harvester from photos of a US model. Using her attention to detail, stubbornness and mechanical wits she tried to get it finished by harvest time. Her workforce suddenly disappeared, however, siphoned off to another project personally endorsed by 'Gang of Four' supremo, Jiang Qing. Joan's combine was eventually made and despite teething problems became a great success, enough for it to be displayed at an exhibition about self-reliance. It was blocked by Jiang Qing as being ineligible if designed by a foreigner. Joan's other important designs for modern milking parlours, pasteurisation and feedstuff production were eventually adopted in many places, but not until she became a member of the Academy of Agricultural Machinery in 1978 was she given full recognition.

It was only through the intercession of Zhou Enlai that Joan's son, Billy, was allowed in 1972 to join other young people, 'learning from the peasants' at a tea plantation in Anhui province south of the Yangtse. Karen went too; feeling that her essentially peasant background entitled her to feel that she had something to offer as well. Despite the real improvements in rural lives since the revolution, old habits still shocked her: night-soil pots rinsed in the same water as dishes and the endemic patriarchy. Although this too had changed on paper, she baulked at the minutiae of everyday inequality: the village bath used by men first, the objection to swimwear or hanging out underclothes. Karen contracted malaria and was very ill while Billy fared even worse with heart problems.

In among all their struggles there was some countervailing easing of pressure. In March 1973 Sid, Joan, Fred, Karen and the Crooks were invited to meet Zhou Enlai again at the Great Hall of the People. The apology Zhou made for their treatment was marred for Joan by his quoting Sid's name in Chinese and hers in English. Apart from her objection to institutionalised sexism, we can perhaps discern some latent 'bourgeois' ego at work here.

Around that time, Fred, working at a plywood factory, decided to visit the USA. After overcoming passport difficulties he left with Sid in 1974, flying to Philadelphia to meet his cousin, Carma, now living with her father Bill on the Pennsylvania farm. En route at San Francisco airport he found himself lost, unable to speak much more English than 'Long Live Chairman Mao'. He stayed on the farm for a year during which, with a slightly improved grasp of the language, he accompanied Sid on a speaking tour of thirty-five US cities. Owing to the recent Nixon/Mao engagement, audiences were keen to hear about China. Sid was not reciprocally impressed with American corporate agriculture.

1976 proved to be a momentous year for the Hintons and their adopted country. In the first month Zhou Enlai died after a long battle with cancer. He had been their ally and nationally a well-respected moderating force during the turmoil. It prompted a large demonstration in Tian'anmen Square that was put down violently and later blamed on the 'Gang of Four'. Natural events followed that seemed to the battered population to represent portents. In March a four-ton meteorite fell in north-eastern China; in late July a huge earthquake at Tangshan killed a quarter of a million people. Six weeks later on 9 September the 'Great Helmsman', Mao Zedong, died aged eighty-two. In early October the 'Gang of Four', including Mao's widow Jiang Qing, scapegoats for the Cultural Revolution, were arrested, later tried and given death sentences, reprieved to imprisonment.

A decade-long catastrophic upheaval came to an end with the ultimate victory of the 'capitalist roaders' under Deng Xiaoping. China emerged into the present hybrid of a market economy commanded by the Communist Party. As with the Great Leap Forward, Mao had hugely set back China trying to maintain his own power and the vision of an anti-bourgeois/elitist society forged by the hammer of class struggle. The dictatorship of the proletariat that Joan lauded had descended into a violent factionalism that had cost countless lives. For her all this was the result of the destructive machinations from the past that had distorted Mao's true path. Deng Xiaoping's famous dictum of not caring about whether a cat was white or black but whether it caught mice, highlighted the headlong rush into the restoration of private ownership, the downplaying of collectivisation and the drive to industrialise regardless of the effect on the agricultural peasantry. Ideology was transmuted into the quest for wealth and associated corruption.

Joan, like Sid, had no problems returning to the US briefly in 1977 with her son Billy to give an extensive lecture tour to Chinese-American Friendship societies. The next in line to confront airport staff at San Francisco, ill and emaciated, Billy was almost taken as a vagrant by security while Joan, absent from the US for twenty-nine years, was trying to relearn how to make a phone call. Like Sid's tour earlier they were well received, although Joan was shocked by the racism of the South. In Virginia she visited an American version of a commune at Twin Oaks founded in 1967 with 100 members living on 100 acres producing hammocks and tofu. One wonders what she made of a social system in which clothes were collectivised but no structured ideology enforced discipline. She also made a pilgrimage back to Los Alamos where she viewed the 'water boiler' again that she had helped to construct and lectured on her exploits since then, re-making contacts with the scientific coterie of thirty years before. Ever pragmatic, she returned to China with a consignment of Holstein bull semen in a tank of liquid hydrogen.

Karen returned to Beijing in 1976 to study biology at university, enabled, despite her uneven education, by a programme giving preference to the children of workers, peasants and the army. In 1980 she left to study biochemistry in the US where both her brothers had preceded her. (Billy later remained in the US and became an engineer; Fred returned to China and teaches economics in Beijing.) In 1983 Joan and Sid moved to their final destination, again just outside Beijing, to set up an

Karen leaving the commune to attend Beijing University.

395

experimental dairy. It was here that I stayed with her and was shown around in 2005 – I could have done so probably in 1983 given any awareness of our Boolean connection.

Both mother and daughter harboured deep reservations about capitalist society, its waste, inequality, and for the latter, eventually engaged in postgraduate research into plant pathogens, the overarching influence of big corporations. Karen became involved in a campaign investigating US covert activity in Central America. It was also a period of questioning her own life and its upheavals. She understandably went through a period questioning her parents' lives when politics was on an equal footing with family life. Her deep affection for and solidarity with the Chinese people is adumbrated by its political past. 'Why did the Communist Party talk so beautifully yet perform so poorly?' she asked. Although on return there she lauds the new freedoms, she is angry that it has been at the expense of the welfare state that Mao created. Underlying everything, inevitably, are her personal divided loyalties and insecurity. She profoundly recognises the adage, 'to be a citizen of everywhere and yet nowhere'. Karen suggested optimistically that the wider family should all one day meet up at Lincoln cathedral in England to pay their respects to George Boole and the stained glass window dedicated to him. (This took place in August 2015 when her cousin, Marni Rosner and her family visited to celebrate the 200th anniversary of his birth.)

Joan too expressed respect for their venerable ancestry. In a letter to Geoffrey Taylor in October 1964 she writes, 'I feel a very special attachment for the five Boole sisters. The fact that … they had such fine qualities and such fine brains makes me secretly proud – it is proof in practice of what woman can be.'[15] Karen also obviously has a deep admiration for her parents and their commitment. Sid died in Beijing after a long illness in 2002. His ashes were spread under three fir trees around the Beijing farm and at Yan'an. Karen enthused to me over Sid's lovable and humorous nature, ever-practical, polite and understanding, even if he had endorsed spanking as a recipe for her badness. Joan complemented him with her physicist's more theoretical view of the world, but also possessing the nuts-and-bolts ability to implement what might exist on paper. The immediate material world of clothes, personal possessions and lifestyle meant little to Joan. She didn't really cook, and as one critic said, rather severely, making a meal for her was as uncalled-for as 'making dumplings for pigs'.

Joan continued to make political statements to the last. One of them was to be seen on a T-shirt in 2006 bearing the message, 'F—K Bush' in Chinese. In a number of interviews she stated her opposition to the new China. She remained entirely scathing about the 'capitalist roaders'. Mao for her remained unimpeachably correct. In 2008 she recognised one of the major events of her life by visiting Hiroshima. Could Joan's ideological tunnel vision, one wants to ask, be another example of the Boole/Hinton taint: the search for some inviolate ultimate truth?

On the last day I visited Joan I took her a bunch of lilies as some sort of token of thanks and admiration. Characteristically she enquired, 'What shall I do with them?' She wasn't asking about which vase to put them in (if there had been one) but really, why would anyone want to make some kind of individual emotional connection? Life seemed to accord for her with the Chinese notion of respecting what you do. The flowers might have meant more with a card that phrased her own assessment, 'I have taken part in two of the greatest events of the twentieth century – the development of the atom bomb and the Chinese revolution. Who could ask for anything more?' Joan died on the 8 June, 2010. I felt very privileged to have met her.

# FANSHEN

Bill Hinton aged eighteen and restless, left Carmelita's Putney School in 1937 and set off to work his away round the world. Having hitchhiked across the States he embarked on a freighter to Japan, scrubbing its decks to earn his passage. In Tokyo he worked as a journalist, then on through Korea, north-east China, across Russia on the Trans-Siberian Railway (going the opposite direction to me), through to Europe and back home. En route he had worked at whatever came his way: dishwasher, brick-cleaner and skivvy.

After two years at Harvard University he transferred to Cornell to study agronomy and dairy husbandry, graduating in 1941. When the Second World War started, Bill had, like his sister Joan, adopted a pacifist stance but later tried to enlist. He was turned down on medical grounds and recruited instead into the American Office of War Information. Roused by reading Edgar Snow's *Red Star Over China* and its account of Mao's Long March in 1936 to establish a base in Yan'an, he returned in 1945 to China and its turmoil. At the unsuccessful Chongqing peace talks in August 1945 between the Communists under Mao and the Nationalist Kuomintang under Chiang Kai-shek, he met Mao and Zhou Enlai on several occasions. Until Japan's defeat a truce had held between them but now civil war broke out.

In 1947 the United Nations Relief and Rehabilitation Administration donated some tractors and sent them with volunteers. Bill ended up in Kuomintang-occupied territory and was appalled by the general corruption and cruelty of the regime. He left to cross lines and continued to work in the communist-held and liberated area of Hebei province. When the UN programme finished he asked to stay on and ended up teaching English at Changzhi University in south-east Shanxi. This in turn led to volunteering work with a university team to act as an observer of land reforms in a nearby village, Changchuang, which he called 'Long Bow'. The eight months in 1948 he spent working and living with the peasants provided

1,000 pages of detailed notes on the stages of often bitter and violent struggle against tradition and landlord power. This 'turning over' and calling to account of the years of the abuse of the peasantry gave the title of his book, *Fanshen*. Like the Chinese revolution it did not have an easy birth.

In a talk on his eightieth birthday at Columbia University in April 1999, he described some of the horrors of quitting China in 1948 with his precious material:

> We had three hours to get out of the university with the notes on my back … and the planes came to bomb us. When they came we ran out into the fields and threw our bedrolls in the middle and lay face down because they said it looked like a manure pile from the air. I was looking up and you could see the pilots' faces sometimes.

Finally leaving in 1952, he once again took the Trans-Siberian across Russia, carrying a 'heavy piece of paper' stamped with exit visas in lieu of his expired passport. Via Prague, where he applied for a new passport, he ended up in England only to be interrogated for nearly a day by members of British Intelligence. Hinton was a marked man. Knowing that he was likely to be searched and that his papers could be used to help make forgeries to send agents into China, he took the document to the toilet and ate it. 'It took a lot of chewing,'[1] he remarked. Yet this episode was a lot more palatable than those that awaited him on his return to the US.

In 1953 at the height of the McCarthy period his notes were confiscated by the notorious Senate Committee on Internal Security. The rumours concerning his sister Joan in China cannot have made his position at all secure. Hinton's civil liberties were seriously infringed: his passport was taken, movements recorded, phone tapped and his character defamed. He, however, seems to have been undaunted and took on the US Constitution itself. Keeping one step ahead of subpoenas he traversed the entire country giving many talks on his Chinese experiences, until eventually he was forced to appear before the infamous Committee.

Unlike the Hollywood Ten in 1947 who refused to give evidence about their beliefs, citing the First Amendment, Hinton used the Fifth, enshrining the right not to have to testify against oneself. It produced surreal Marx Brothers conversations. Having said 'no' to allegations about his Communist Party affiliations and uttered 'Same answer' to a further eight enquiries, he was asked, 'Have you ever engaged in espionage while

William 'Bill' Hinton (1919–2004), appearing before the Senate Committee in 1955.

a member of the Communist Party?' Hinton protested that this was a very serious charge to which the Committee replied, 'We only seek information about the internal security of this country. Can you answer or not?' Hinton answered that of course he had not. The next question lobbed over the net was, 'Have you ever engaged in research for members of the Communist Party?' Hinton volleys back, 'Same answer.' Reply: 'The same answer as what. The last answer?' Hinton: 'I decline to answer.'[2] He eventually won his notes back after a lengthy and expensive lawsuit. In 1966, having been turned down by most major publishing houses except the Marxist *Monthly Review*, *Fanshen* was launched, a book he felt that most people could understand and appreciate.

Its subsequent popularity stems in part from that quality of first-hand storytelling that compels much more than an anthropological study. The often shockingly violent dramas that unfolded in Long Bow in 1948 during the land reforms grip the imagination. Characters appear almost as participants in a soap opera. The story narrative, within which Hinton was not afraid to address the reader, made it an ideal subject for the play of the same name written by David Hare, first performed in London in 1975.

Long Bow village consisted of about 1,000 people living off 1,000 acres of the hilly, un-irrigated, winter-frozen, summer-boiling land in south-eastern Shanxi province. Its staple crop of maize was garnered when times were propitious with mostly wooden implements and animal power whose manure supplemented the fertility basically dependent on human waste. The conversational greeting, 'Have you eaten?' summed up the nexus of poverty. The land and the social wealth and power that it provided was divided up extremely inequally between rack-renting landlords and grades of peasantry descending to the landless hired labourer. Traditional clan affiliations bolstered the class-based authority of the village elders, who dominated local administration and collected taxes.

Long Bow was but one example of a vast feudalism ossified in the resigned adage that expressed the hopelessness of change, 'Can the sun rise in the west?' For the majority 'village idiocy' perpetuated the never-ending grind of daily survival. When Hinton arrived in Long Bow he writes, 'I felt very much like a visitor at a gallery being led to a hall of living exhibits.'[3] He set about amassing first-hand recollections from the villagers about the major changes of recent times.

Shen Fa-liang, indentured to a rich peasant for seven years to pay off his father's debts, told him his story:

> The worst days of my life were when I was a child. I often had nothing to eat. In the winter I had no padded clothes … one suit had to last for many years. It was patched over and over again. It wore so thin that it was no better than a summer jacket. When we didn't have any millet, we drank hot water. What was the happiest day of my life? I haven't passed any happy days.[4]

During recurrent famine things became even worse. Another peasant recounted what happened as part of the 'chronic social tragedy':

> We ate leaves and the remnants from vinegar-making. People were fighting each other over the leaves on the trees. We were so weak we couldn't walk. My little sister starved to death. My brother's wife couldn't bear the hunger and ran away and never came back. My cousin was forced to become a landlord's concubine.[5]

Hou, found guilty of stealing a few ears of corn, was strung to a tree and flogged fatally. Another, objecting to the rape of his wife, was hung by his hair until his scalp separated from the skull. 'Violence was endemic at all levels of human relationships. Husbands beat their wives, mothers-in-law beat their daughters-in-law, peasants beat their children, landlords beat their tenants and the militia beat anyone who got in their way.'[6]

Against this hellish background stalked civil and international war. The Japanese invaded Long Bow in the summer of 1938 as part of an effort to control the nearby railway corridor. From here they attempted a typical 'kill all, burn all, loot all' campaign to pacify outlying areas under guerrilla control. Their presence was nullified by the Japanese surrender following Hiroshima and Nagasaki in August 1945. With American support, Nationalist forces under Chiang Kai-shek attempted to maintain Japanese control of much of northern China rather than see it handed over to Mao's communists. Nevertheless, Long Bow was liberated from the Japanese in August 1945 after eight years of oppression, ushering in a period of uncertainty as nationalist and communist forces vied for control. Behind the scenes it was the communist cadres who were instigating change and *fanshen*.

After the settling of accounts with collaborators, the first of a series of social adjudications began against the village landlords. Hinton writes, 'Story followed story. Many wept as they remembered the sale of children, the death of family members, the loss of property.'[7] Buried bitterness welled up into beatings and attacks on the accused. The old social order was savagely brought to book: the head of the Catholic Church was beaten to death, clan assets were seized. Soon the time for redistribution rather than retribution arrived. Hinton's unashamedly poetic prose sets the scene.

> March came in cold and clear. The sun moved in brilliant splendour across a cloudless sky but cast so little heat upon the earth that it did not even begin to melt the light mantle of snow that had fallen in the night. The glistening snow miraculously transformed the dusty, crumbling, adobe village and turned it into a fairyland of black and white, as pure and clean as the day the world was born.[8]

The new world order could be seen physically arranged in a yard where implements, furniture and clothing confiscated from the gentry were set out. Each poor family was allowed to choose a single item. More fundamentally, land and housing were equalised and livestock shared in an Orwellian variant, 'four legs good, one leg better than nothing'. The elation of unexpected economic justice spread to personal relationships: peasants began to call each other 'comrade', a Women's Association was formed to fight for the long-overdue rights of 'one half of China'.

As with the Bolshevik Revolution of 1917, control of the land by the peasantry, 'land to the tiller', was fundamental but was implemented, Hinton maintains, pragmatically so as not to harm 'middle peasants' and open up serious social rifts. These shrewd principles were not necessarily shared by the peasantry, who engaged on a round of further expropriation of the 'feudal tails' revealed both by ancestral affiliation and physically hidden wealth and treasure. At the end of this period with no more easily gained 'fruits of struggle' revolutionary momentum slackened, worsened by the incipient trait of party cadres using their position to feather their own nests and issue orders peremptorily.

With the dramatic style that enlivens *Fanshen*, Hinton asks the most pertinent question of any revolutionary vanguard that Kropotkin, Charlotte Wilson and George Orwell would have echoed: 'Could they abolish petty advantages won through the lever of leadership, lead all the poor to stand up, and unite the whole population around that vast programme of private mutual and public production which alone could lift Long Bow out of the miasma of the past?' Or would the egalitarian dynamic lead to a new porcine tyranny?[9]

Despite the advances of the previous period when Hinton's team from the university arrived, the poorest peasants were still enduring suppurating head sores, malarial fevers, slow deaths from TB and venereal disease. Rubbing shoulders with the peasantry literally meant catching their parasites. Unable to face bursting the lice that infected his padded jacket, using chopsticks, Hinton picked them out of the lining 'as if I were picking delicacies off a banquet-table'.[10] He soon learnt a more robust attitude to the myriad aspects of rural privation. Endless meetings continued to take stock of social inequalities, revealing 'a succession of tragedies, incidents full of pathos, greed, rollicking humour, cruelty and kindness …'.[11] These were lightened for Hinton by his affinity with local children in rough play – something their parents would never indulge in.

In April 1948 the purge of Long Bow's communist cadres was begun by an elected committee of peasants. This was a dangerous step for the former, not only because of the possible findings but because such a process publicly exposed their existence to any Kuomintang forces that might yet return to the village. Man-hsi had become known as the 'king of the devils' for his petty thefts and bullying ways including rape and beatings-up. Ch'un-hsi revealed his selfish appropriation of land and how he had grafted money and beaten people. Such opportunism paled, however, as ex-bandit, ex-Catholic, Wang Yu-lai, like Orwell's Napoleon, utilised witch hunt accusations of liaison with the Kuomintang to discredit and isolate any opposition to his power. Adopting honesty and repatriation rather than vengeance as its guide, the peasants decided he must pay back an equivalent to his corrupt rake-off after the harvest. The Communist Party itself should decide on his likely suspension as a member. Man-hsi's failure to convince the interrogators of the truth of his confessions led to his exclusion from any active role in the village. Of Long Bow's twenty-six party members, twenty-two were cleared.

More importantly, the process itself created a democratic solidarity in the village that touched Hinton himself. 'The power of the revolution to inspire and remould people had stirred me.'[12] He sees the possibility of 'objectivity' both for one's own behaviour and others' assessment; of not going with the crowd nor standing aside in arrogance. The 'objectivity' of class position needed to finalise *fanshen* was rather more complicated, depending on varied factors of labour and land ownership. The motive was to distinguish rich peasants from middle and lower in order to make further change. After endless hearings a master list was posted on the main street leading to fierce disputation as to its findings.

A new round of criticism of cadres brought Man-hsi to the hustings once more. Devastated by the previous verdict, he managed to convince a general meeting of peasants that he was now a sincerely reformed character. 'Man-hsi's face lit up with joy'. "Do your best in future," they enjoined.'[13] Yu-lai, his son, Wen-to and two other henchmen, however, who had declared their brutality towards women, were only saved from harsh punishment by the policy of 'saving the patient' as well as 'curing the disease'. They were repatriated in a school for penitent cadres. All of this local self-criticism was being watched by the Party. The dismantling of feudalism had often been over-harsh, alienating landlords and middle peasants who might have then turned to Nationalist forces. This was the 'leftist deviation' of pursuing equality too hard, when times weren't ripe.

The diagnosis was soon put to the test when freak rain and a flash flood interfered with human efforts as it had done in so much of China's past. A kind of lethargy pervaded the whole population. Some with traditional fatalism blamed the storm on the God-threatening changes of recent years. Despite a growing sense of collectivity, 'leftism' had also produced a feeling of each-man-for-himself that sapped the energy of party cadres and diminished mutual aid in the fields. The antidote to all these down-turnings proved to be the setting-up of an elected Village People's Congress for the masses from which delegates could be sent to the highest levels, reversing the hierarchy of previous ages. The position of the Party in relation to this new democratic form was to be that of a leading role acting by example and persuasion. In July 1948 it was the Village Congress that met to finalise the class status of every family in Long Bow.

In an episode of *Fanshen* that makes palatable the heavy-going reading of party directives and theories, Hinton relates the story of Yu Pu-ho, a rich widow. Knowing that her classification would entail loss of some of her riches, she craftily hid and displaced them. A village group decided to raid her house to pre-empt this. 'Obviously the visit came as no real surprise. The widow's round, usually smooth face was furrowed with anxiety, but she pretended a hearty welcome as she talked without pause in a tone sweet with assumed geniality. We filed silently past her into the cavernous dark dwelling.'[14] The deputation found the cupboard bare; her rich, fine clothes, for example, had been re-sown to fit her daughter. Exasperated by her trickery, Man-hsi proposed to beat her and throw her into the village pond. The committee reluctantly rejected such violence; renewed enquiries managed to make up a list of her dispersed grain and goods to be re-distributed to poor peasants.

After months of copious note-taking and involved activity Hinton left Long Bow in August 1948. Despite the turmoil, he declares:

Feudalism had indeed been uprooted, and nothing could be the same in south-east Shanxi. The democratic reforms, the consolidation of the Party and the establishment of new organs of political power now guaranteed the new egalitarian base of rural society and set the stage for a great advance.[15]

Land reforms, Hinton asserts, had only removed the feudal barriers to production; illiteracy, the absence of medical care and the primitive methods of cultivation had yet to be tackled. These were tasks Mao set for progressing after his power was consolidated.

Hinton's book *Fanshen* captured a grassroots mood in the West in the 1960s and '70s and as a result became very influential. No one had any illusions that conditions in a largely peasant China bore much resemblance to our western societies, but the belief that personal and social lives were capable of transformation was seductive. By the end of the 1960s an educated and well-fed generation had grown up disillusioned with orthodox politics and its capacity to motivate and organise. The events in Paris 1968, the anti-Vietnam demonstrations worldwide, the beginnings of feminism all attested to a belief in movements rather than parties. The only parties around were celebrated during the 'summer of love'. The anarchistic idea that wholesale changes in values had to underlie legislation had surfaced. Equal pay and other rights for women, although crucial, had, for example, to be a part of a far more fundamental move from patriarchy, changing the very basic attitudes of men. The self-criticism sessions at Long Bow were not a world away, in principle, from the stern examinations of domestic life and work by women contesting the sexism embedded in the details of the everyday. If 'comrade' became the new currency in Shanxi, 'sister' became part of the mother tongue in Europe.

The key words 'alternative', 'libertarian' and 'counterculture' were appended to many aspects of life at that time. Schools were to be 'free', as were attempts to create a free press. Law centres were set up to help people fight for their rights. Complementary medicine declared itself opposed to the drug companies and advocated self-health. A small 'cultural revolution' had occurred that didn't involve the masses or a political vanguard. In China the capital-lettered Cultural Revolution seemed at the time to be motivated by the same urges of democratic criticism and a back-to-the-people impulse, even if it was on a vast scale and lacked individual expression. It wasn't until much later that the amount of damage done in China was revealed.

The one aspect that particularly interested me and others, for which *Fanshen* was an inspiration, was the question of land. Some of us wanted to be peasants. We were modern *narodniks*. Compared to the days of *Fanshen*, when Mao was preparing for land reform to finance the great leap forward, we were preparing the great leap backward into what many saw as rural torpor. The innocent adage 'small is beautiful' extolled the virtues of communities like Long Bow in which face-to-face interaction would enable control of life in all its aspects. Mao had been envisaging something similar, we surmised, with the launch of his communes in 1958. Pictures of the Dazhai model looked idyllic. They seemed to represent the Blake, Owen and Morris utopian visions that had often been sentimentalised. For the British public it was diluted into the innocent froth of the TV sit-com, *The Good Life*.

During the middle of his troubled McCarthyite period Hinton took his family to a farm bought by his mother, Carmelita, in Bucks County, Pennsylvania. Here he put his university training to good use growing corn and soya beans. On his frequent visits back to Long Bow he helped design and demonstrate irrigation practices and small-scale tools such as ploughs

韓丁在張莊水泥廠掄錘粉碎石料
Han Ding breaks stone with a sledge –
hammer in Long Bow Cement Plant.

"他在中國百姓間!"韓丁與農民情同手足,以心溝通
"He's among the Chinese folks." Han Ding talks
warmly with Long Bow villagers.

韓丁駕駛拖拉機在張莊田間作業
Han Ding drives a tractor, plow-
ing in the fields in Long Bow.

韓丁向張莊農民傳授農業耕作技術
Han Ding teaches Long Bow's peasants some farming methods.

Bill Hinton, known as 'Han Ding', working at Long Bow village.

and corn-hulling machines that could be bought cheaply or made locally to increase production. In the West it was known as 'intermediate technology' and espoused in monthly journals, in which were found details about the building of woodstoves from milk churns, energy from windmills and home-made methane as well as the power of comfrey and the soy bean. Bill Hinton would have applauded much of the agenda.

My parents were unstinting in their lack of approval for my discarding of a university education. As a child my father had experienced deprivation in the slums of Sunderland, subsisting with six siblings on the meagre earnings of a lame, tailoring father. We listened uncomprehendingly to his tales of searching for sea coal on an icy North Sea beach. He

Bill Hinton with cadres at Long Bow.

savoured the punning of 'iffit pie': 'If it goes round you'll get some.' At the back of our Hayes semi he kept chickens and grew sprouts, beans and leeks as another reminder of times when the cupboard might be bare. For him to grow them as some kind of political act was complete madness. The struggles of the factory shopfloor and a commitment to Labour Party socialism were his priorities.

From Bill Hinton's progressive background, a sense of social commitment was inevitable. In a speech at a Harvard reunion in 1991 he declared, 'What has meant the most to me is the world outlook of Marxism.' This would not have found favour with back-to-the-landers. Hinton's Marxism was unorthodox, however. He accepted that revolution was possible in a non-industrialised society like China. Furthermore, his personal vision of 'the good life' embraced the continuation of agricultural values and the dignity of labour. His hands-on practicality distinguished him from many a Marxist, too often from their armchairs unable to distinguish a spanner from a dibber. His individuality and by all accounts likeable, rugged personality I find attractive.

Perhaps of all the Hinton family, although not living in China permanently, Bill Hinton was the most actively engaged in writing about the country for a worldwide readership. When he returned to Long Bow in 1972 with Carma Hinton for the first time after a long gap, he was barely able to recognise where he had stayed. His impressions were conveyed in an equally monumental sequel *Shenfan* – connoting the idea of the deep-digging of ideology since the Great Leap Forward and after.[16] His writing style is similar, allowing actors to speak for themselves. The most important change was the industrialisation of the landscape and the changed cosmopolitan atmosphere brought by nearby immigrant workers. Long Bow inhabitants had benefited by being able to earn sideline income as contract workers and dabble in speculation and minor commodity dealing.

Hinton found himself, as we did in the 1970s, thinking of Blake's 'pastures green'. There had been compensating improvements: a co-operative shop, flourishing school, two doctors and a pharmacy, domestic electricity and piped water from a reservoir built during the Great Leap. Streets had been drained and cindered, the roadside privies once open to manure donations from passers-by now relocated privately. The once-patchworked land had been collectivised and worked by six teams. Although yields had doubled in the 1950s, they had since stagnated owing to the limits of the poor-quality soil.

Long Bow had not been immune from the grand political changes of the Cultural Revolution. The first fatality in Shanxi province had occurred nearby when a Red Guard was shot while requisitioning from a grain store. Traditional clan loyalties in the village, divided north/south, had been transmuted into allegiances to the 'Stormy Petrel' and 'Shankan Ridge' factions. In 1967 Brigade offices were invaded to obtain the seals that conveyed the power to authorise actions. Wang Jinhong, then the Communist Party vice-secretary, was deposed but reinstated. The rebels, slurred by the customary accusation of being 'landlords, rich peasants, counter-revolutionaries and bad elements', were arrested, beaten and driven out, not to return until 1969.

This did not end factionalism. A work team from Changzhi investigating the events deposed Wang Jinhong and reinstated the rebel leader. Wang had now to exile himself briefly but returned. Like Joan and Sid he bided his time with DIY and repairing bikes. His broad practical skills, learnt from working all over China as a construction worker, were supplemented by his leadership qualities. After a self-criticism session in 1973 he returned to office to tackle some of the village's problems.

It was found that the soil could be conditioned by applying large amounts of ash from the power plant nearby. New irrigation projects were launched and in a burst of energy they ventured into small-scale industry polishing saw blades and making saw handles. Needing to grind phosphate fertiliser they built a mill, leading in turn to cement production. Mechanisation was needed on the land too. Bill suggested the kind of basic tools found on his own American farm: an auger to lift grain for storage, designs for drying it, a rotating crop spray. Through pushing ahead and using their own initiative without waiting for bureaucratic approval, by 1979 there had been major increases in productivity.

Hinton remained very outspoken against the Deng Xiaoping market reforms of the 1990s and the dismantling of collective land into 'noodle-strips' too small for mechanisation. In Long Bow large fields were broken into smaller plots. Flying from Beijing to Shanghai he commented, 'I looked down in growing disbelief and wept.' While accepting the short-term gains of increased rural production, he remained deeply critical of the long-term outcome of the worldwide neo-liberal agenda. The domination of the cash nexus over collective spirit would, he claimed, lead to growing social and economic inequality, corruption, speculation, cultural regression and environmental depredation. In particular the countryside would be degraded, leading to a mass movement of the peasantry into the city. The Communist Party is led, he declared by 'newly-constituted bureaucratic capitalists busy carving the economy into gigantic family fiefs' – a reversion to the nationalist days of Chiang Kai-shek. In *The Great Reversal*, another of his books, Hinton quotes a rhyme he encountered in rural Shanxi: 'In the 1950s we helped one another, In the 1960s we denounced one another, In the 1970s we doubted one another, In the 1980s we swindled one another.'

Bill Hinton kept up his connections to Long Bow, visiting almost every year until he died in 2004, aged eighty-five. In 1995 he had moved to Mongolia with his third wife. His enthusiasm for hands-on agriculture persisted in his advocacy of farming without actual tilling. His daughter, Carma, also continued the connection, making five films about the village. Wang Jinhong visited Bill in the US in 1987 and in the year before Bill died presented him with a lavish appreciation, *An Old Friend of Chinese*

*People*, prefaced by David and Isabel Crook. Its pages show clearly how he was loved for his humour and warm-hearted sociability as much as for his political and agricultural knowledge.

While staying at his sister Joan's, I was urged to visit Long Bow, the site of *Fanshen*. The village is almost twinned as a Hinton family seat with Putney School in Vermont. Diminutive She Jenping was accordingly allocated to take me there. We met at Beijing's South Station. As I have consistently found on my travels, a nation's railways and metros are a reliable indicator of its recent history; this station was no exception. When I departed from it in 2005 it consisted of a ramshackle brick building and concourse strewn with peasants lying on and carrying huge packs, the barely-roofed platforms just beyond. The new South Station, opened in 2008, is the second largest in Asia. British-designed, it consists of a vast ellipse of solar-panelled glass, housing twenty-four platforms from which run two routes of 200-mph high-speed trains. It features underground car parking, retail stores and accompanying fast food outlets. In the words of modern design-speak, its 'unifying contemporary form makes possible a fully integrated multi-modal transportation hub acting as a catalyst for new developments to surrounding areas'.

We were to travel, however, the 400 miles southwards to Changzhi in Shanxi province on a seventeen-hour hard-sleeper, open-compartment train with bunks stacked three high, in the top two of which one could not sit upright. Despite the noise of the swaying and rattling, the piped music and the clouds of cigarette smoke, I slept fairly well. I'm ashamed to say that as an honoured visitor, saddled with claustrophobia, I claimed the lowest bunk.

We were met at Changzhi by Bill Hinton's old friend Wang Jinhong. His slightly hunched back was apparently the result of bearing over-heavy

Long Bow village scenes, 2005.

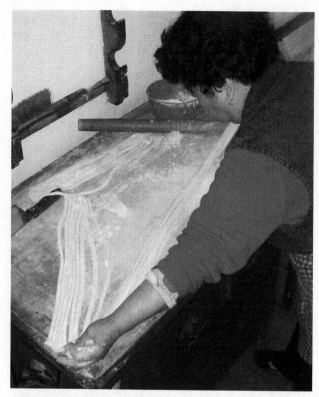

Left: Making noodles at Wang Jinhong's house.

Below: Old agricultural implements kept by Wang Jinhong.

loads carried by pole when a youngster. The booming voice and ever-smiling visage demonstrated that those times were now over. We were taxied to his fine house with its large courtyard and spacious rooms. Things for him at least had moved on since the 1970s. Later he showed me his collection of old wooden agricultural implements – shoulder ploughs and carrying vessels – all of which would have broken the will of any back-to-the-soil westerner. We breakfasted on yellow bean soup and dumplings dipped in thick vinegar washed down by green tea. I took to a washing-down a little later with colourless, mighty-strong sorghum wine. As Mao was an inspiration to Joan, Bill Hinton was to Wang; all over the house were hung trophy newspaper clippings of him, a poster for David Hare's production of *Fanshen* and for good measure a three-foot-high statue of Mao.

Having caught up on some sleep, She Jenping and I wandered the village. It was completely different from the folksy childlike sketch-map of it Hinton had drawn forty years earlier in *Fanshen*. Long Bow had been engulfed by the once fortress, now industrial, city of Changzhi, losing 40 per cent of its land in the process. It seemed almost housed and gated as a theme park, like a Chinatown in a western city. Much had obviously changed once again. The adobe one-storeyed houses were now built of brick and privately owned, the stream in the middle of the long, broad main street culverted, and the pond concreted over into a meeting square, still emblazoned with a portrait of Mao. Some houses now sported TV aerials, solar roofs and glazed-tile walls but very few had cars outside them.

Wang escorted me round the large primary school, where the dutiful children beamed and chanted, 'Good afternoon guest.' I tried in vain to teach them how to say, 'Hello foreign imperialist.' One essential port of call was to the museum that celebrates Bill Hinton – his name transliterated as Han Ding. Photos of him taken at various times are everywhere: tall, gangly, buck-toothed, his mop-head of hair blanching as he aged. Many show the contribution he made to the community in the shape of farming equipment that he introduced (and helped pay for), such as seed planters and manure spreaders. Wang made an enthusiastic running commentary – our city interpreter was also impressed: 'Hinton know everything,' he declared.

The background commentary once outside again was the clanking of the nearby marshalling yard where trains load the local shale-coal and limestone to be sent to distant power stations. The agricultural connection is still there, however: the fields around growing maize, the cobs drying in dusty courtyards and on pavements alongside ubiquitous piles of household rubbish. Street life was Lowry-esque: small knots of people, chugging three-wheeled motorbikes bearing piled cabbage, a bicycle adapted to mend boots, a rag-and-bone woman on another. Traditional culture had returned since the Cultural Revolution; next to a shrine to the Earth-god a group of men were setting off firecrackers. In the evening a travelling street-theatre group loudly amplified a folk-based play. The village's famous stilt walkers, once banned, have returned. We went to visit a fortuneteller who had unpropitiously gone out. Times change. For all I know Long Bow may have moved on dramatically since then – the village may now itself be a 'fully-integrated modal hub'.

What can an outsider like myself make of Joan and Bill Hinton's lives and the other westerners who gave up their own histories to take part in a foreign culture? In 2005, at the time I was in Beijing, the debate was opened up by the publication of Jung Chang's Mao: The Untold Story,[17] a sequel to the personal account of her family's sufferings in Wild Swans. Despite a mixed reception by western academics, it looked set to become part of a received history. Its unrelenting tenor left me feeling that something was missing. Was China under Mao's leadership nothing but an unmitigated saga of virtual genocide at his whim? Was he directly responsible for 70 million deaths, 38 million of them in the Great Leap Forward of 1958, three million in the Cultural Revolution 1966–76? Did, as Chang alleges, the peasantry see no improvement in their lot during his lifetime? How relevant to Chinese history are his personality traits? Might there not be something more to be said, à la Monty Python: 'What did the Romans ever do for us?' Through visiting Joan Hinton I was able to meet three other broadly pro-Mao, veteran westerners, long based in China, to gather their responses.

The anonymity asked for by one of them was well suited to the lofty apartment in which she lived – yet another of the tower blocks that look out on Beijing's vast sprawl. Both she and her husband, alongside the Hintons, had attended the 1971 meeting with Zhou Enlai that had 'rehabilitated' them. 'Anne' felt that although China needed Mao in a society that had always enshrined the 'top man decides principle', he never emancipated himself from traditional feudal thinking and could be 'cold-hearted, callous, vindictive and a mass of contradictions'. His maxims such as 'the masses are never wrong' or the 'more the merrier' attitude to population growth were grossly inflexible. She stopped a long way short, however, of Chang's list of Mao's crimes and misdemeanours.

Although the Great Leap was pushed too far, she suggested many communes worked well, and highly successful policies such as the rural 'barefoot doctors' were developed. This involved programmes of family planning and immunisation by basically-trained commune health workers. It contributed to a decline in mortality and avoidable diseases. Was that mentioned in Chang's book on Mao, she wondered? The Cultural Revolution was a 'blunder'; people she knew had been beaten or even committed suicide. Nevertheless, she thought, there had been a degree of valuable democratisation as workers took places on factory boards; lessons in participation at her college had been learnt.

Incongruently perhaps for a 'Maoist', 'spry' is a word that comes to mind to describe Sidney Shapiro, then in his mid-nineties. We talked at his charming courtyard house close to Beihei Lake just off the tourist rickshaw route. (We met him previously in our narrative when he accompanied Joan to Beijing in 1948.) For years he worked as a translator for the Foreign Language Press. His autobiography, was published under the title, *I Chose China*.

His take on Mao broadly echoed Mao's own estimation of being, '30 per cent bad 70 per cent good', rather than Chang's more probable, '95 per cent wicked'. His greatest admiration centres on the 1940s when Mao 'had it right on the nose' in abolishing feudalism and colonialism in a way that Chiang Kai-shek could never have done. According to him the communists at that time were welcomed with open arms as land reform created possibilities undreamt of before. The Great Leap Forward is regarded by Shapiro with less enthusiasm: 'The need to pull up socks productively, despite some very good ideas and real advances, relied too heavily on a naïve spirit of patriotism.' The blame, said Shapiro, lies squarely with the Communist Party and Mao. The Cultural Revolution he found 'as exciting as a great jamboree'. Many people got 'bashed', but as to how many, 'it could have been 300, 3,000, 30 million. Pick a figure out of the air. People followed Mao blindly and his perhaps understandable semi-divine status, which was a dangerous path.' Deng Xiaoping's reforms that headed China towards a socialist market he found more realistic, but at the expense of 'a now evaporating spirit of help thy neighbour. No country has succeeded to serve both God and Mammon,' he warned.

How valid is it, one wonders, to create a league table of monstrosity, headed by Mao with Hitler and Stalin trailing on toll-average? Must he not inevitably have been constrained by historical contexts and forces? Winston Churchill, the closest to being a godhead in British society, has been accused of direct responsibility for the deaths of three million in the Bengal famine of 1943 by denying the shipping that would have carried relief grain there. Do we add him to the league tables? Does catastrophic loss of life by human agency only register when one man can be held responsible? Those who perished in the imperialist conflagration of the First World War are no less victims because the political systems that conflicted were more democratic than dictatorships.

Marni Rosner's mother-in-law, Isabel Crook, would no doubt agree. She and David Crook had stayed on in China from 1940 and seen it all. They had undertaken two field studies in rural Shanxi in 1947 and 1959. Their conclusion, like that of Howard Hinton, was that the commune movement had in many respects been a success and mitigated the effects of the great famine of 1960. 'There are more tragedies in the capitalist than the socialist world when you come down to it,' she commented to me. As I write in 2013 one reads of 1,000 workers dead under a collapsed clothes factory in Bangladesh.

Since Jung Chang's book, two other books by Frank Dikotter of the University of Hong Kong have been published updating the debate.[18] *Mao's Great Famine* (2010) adds another 7 million to Jung Chang and puts the death count at 45 million, three million of which were due to political repression. In *The Tragedy of Liberation* (2013) Dikotter tackles the earlier period with which Bill Hinton was acquainted. It was not a prelude to a golden age, he alleges, but the beginning of a period when the peasantry were beaten into submissive slavery to Mao's policies. *Fanshen* was a period of excessive violence, he claims, during which cadres enforced killing quotas to intimidate the peasantry. Uprisings were ruthlessly quashed and mass public executions held as private property was expropriated and the economy ruined.

Were all the people I met in China just dupes unable to get beyond the notion that omelettes need broken eggs, that revolution is no picnic? Their situation is not unlike that of the *Proletariat* party members in Poland that Wilfrid Voynich allied himself to. Honest motives are not enough when history condescends to render you obsolete. All of the participants in China's affairs from the West that I met had lived their lives in good faith to an ideal that seemed right at the time. It is never easy to admit that decades of dedication might have been wasted. None of them at least can be accused of not getting their hands dirty – in Joan and Sid's case they were in it up to their elbows. All of them laboured hard to aid China's continuing struggle for a better life for its people. I can only attest to their integrity and admire their unwavering sacrifice and commitment to their ideals.

# THE VERMONTER

14 March 2011

There are few more unattractive major railway stations in the world than New Street, Birmingham. Penn Street, New York is one. It handles twice as many trains as Grand Central, not far away, but unlike its heavenly-blue, starred cavernous roof, Penn's graceless, diminutive, circular concourse reminded me of a betting shop. The only hint of a railway is the fluttering indicator board, beneath which huddle groups of punters waiting for the off. To a Briton the information carries a certain romantic aura, with names like the Acela Express and the Empire Service. American popular song once celebrated the mystique of the age of the train: Chattanooga, Tulsa and St Louis. Do US visitors quiver at King's Cross station in London viewing destinations such as Stevenage, Peterborough and Doncaster? At Penn Street all routes lead down escalators to Hades where the grimy alleyways alongside the silver carriages are so narrow that you can only shuffle off to Buffalo sideways.

I was leaving 'All aboard' on *The Vermonter* bound for Brattleboro, Vermont, 200 miles away to visit Marni Rosner, daughter of Jean (Rosner) Hinton, at Carmelita's famous Putney School. Part of the mission was to look at archives but I also carried a prized Boole/Hinton relic – the charming oil painting of Ethel Boole as 'Treacle Baby'. Somehow its ownership had become confused. At one time it had belonged to Wilfrid Voynich but no one was sure who he had passed it on to. Putney School's art gallery seemed to be the best public space to house it.

As if to compensate for the subterranean first few miles of track leaving Penn Street, the railway soon rises majestically and curves gracefully high over the East River on massive spans – only to descend to an unending trackside litter and suburban sprawl all the way to Newhaven. Here I could have popped off and paid my respects to the *Voynich Manuscript* again, housed at Yale University. A huge, throbbing diesel engine replaced the silent electric one – somehow a more pleasing form of propulsion running on its own steam as it were. We set off at a steady 55 mph through empty snow-covered scrubland and endless silver birch. It reminded me of the Trans-Siberian except for the stubby sticks of last year's maize, arrayed like tiny anti-tank defences.

It is hard to imagine that New England was once the agricultural cornucopia of the US until Carmelita's Midwest prairies took over. The railways too have been eclipsed by road; once-magisterial stations are ghostly remnants of the past. At Springfield the train is clumsily now forced to reverse direction; the folksy conductor warned us over the intercom not to get out, 'or the crocodiles will get you'.

At Brattleboro I did get off to meet Marni, who I'd last seen in Beijing, her normal residence. We drove on to Putney, a few miles away, through the small town and upwards past acres of apple trees cultivated by the school and into the campus; it really is a small village of its own. Near a magnificent and cathedral-like red barn that houses the dairy herd, a small contingent of inmates were tapping maples for sugar. One of the party was Kola, Karen Hinton's son, who I'd last met at her home in Pau, southern France.

The Hinton connection with Putney is still strong. Nearby stands the wooden, single-storey house built by Carmelita, once housing pupils' bunks and still owned by the family. From its floor-to-ceiling windows there is a stunning view of a sequence of valleys culminating in the peak of Mount Monadnock.

I'm taken on a tour: down through woodland to the old wooden theatre, hued, like the barn, in that faded vermilion that is so distinctively attractive against a monochrome background. Near to the original old farmhouse, now an administrative block, is the old pottery and the KDU hung with flags. Here pupils take their turns on a rota of cooking and serving. Two new buildings have arrived since Carmelita, a gymnasium where fitness is encouraged to accompany the more sedentary activity of the new Arts block nearby. It is alive with creative ideas: a skeleton sits playing a piano, a drum kit awaits pounding, artwork is on display, murals encapsulate the mythic history of Putney. On one wall the legend 'To make school a less self-centred venture' chimes with the co-operative ethos of the place. A poster extols, 'math isn't just a four-letter word' (in the UK it isn't just five). Outside, beyond a group of iron sculptures are the fields where pupils in the summer will be growing much of the food consumed by them. Cows will roam among an array of solar panels and later the hay will be collected for winter feed for the herd. The scene will perhaps look like an upmarket version of Dazhai.

The KDU, kitchen dining room, at Putney School.

The cow barn at Putney.

The author and Ethel Weinberger meet again at Putney School after thirty years.

On the intervening weekend of my stay Marni drove me up to Woodstock in New York to find Camelot, the house on the Byrdcliffe estate that the Hintons and the Booles had often inhabited. The famed town was not particularly impressive at this time of the year. Where the main road turns ninety degrees opposite a pretty, steepled church lies the euphemistic village green. Here a couple of guitarists perched valiantly trying to make a tune, their fingers not working in the cold. In the summer Woodstock would be thronged with burgered youngsters hanging out to soak up the shrine-vibes that linger still, and buying the hippy trinkets on offer.

Byrdcliffe, not far away, was also somewhat disappointing. Whitehead's house, 'White Pines', was empty and uncared for. We couldn't find Marie Little's shack, 'The Looms', where Ted Hinton had often lazed beside her fire. Questing for Camelot, we set off by car up a soggy winding dirt-track road and stopped at one of the scattered houses to ask directions. Apparently, we were actually on Camelot Road and told to continue ascending through the woods. The overgrown road, however, soon petered out and became impassable. I was forced to continue walking solo while Marni drove back. I hadn't gone far when the word 'bears' poked its nose into my unwilderness-trained English mind. It had been mentioned that they might be quite peckish at this season. On encountering one, I wondered, should I run, stand and stare or climb a tree? I was flummoxed, but then Piglet (again) did a brave thing … and sauntered gingerly forwards. Mission bravely accomplished, at the end of the track lay Camelot, not shining with fabled towers but a tidy gleaming white clapboard house where many Booles and Hintons had sojourned. Sadly no one was in to allow us to look around.

Later we learnt that the house I had enquired at was where Bob Dylan had lived from 1965 to 1969 and composed *The Basement Tapes*. There was a time when I was not a fan of his. On first hearing 'Blowing in the Wind' back in the sixties on an LP it had been sung by Pete Seeger. I was disappointed with Dylan's own bear-growl version, not realising it was his composition. Very fortuitously that same evening after my escapade, 'where the grizzle wasn't', Pete Seeger, that stalwart redwood legend, was making a rare appearance with Peggy Seeger to benefit the Byrdcliffe Trust. The small hall in Woodstock was packed with appreciative backwoodsfolk. Seeger at ninety-two still stood erect, singing and talking to you straight as if you were an old friend. The last time I had seen him was in LA on my arrival in the US in 1983. I wrote in my diary how the concert, with its hundreds of radicals, seemed like a revivalist meeting. I was overcome, as in the song, and spent much of the time looking up at the ceiling.

Back at Putney there was one other revivalist meeting to be had. I managed to contact Ethel Weinberger, still living not far away in Hartland. She and her husband drove to the school for lunch. They had hardly changed; like me they were just greying round the edges, still living in their eco-home. They brought with them a collection of cuttings from local newspapers dated 1983. Unlike our Shropshire peace group's slightly over-ambitious idea for a tripling, theirs had sent a specially-made box containing items descriptive of the town, to be delivered somewhere in the USSR. The *Eagle Times*, 10 April 1983 reported that 'more than 200 citizens had put together projects to show Soviet people what small-town America is really all about. Women stitched together quilts, spun wool from native sheep and children wrote about themselves.' Other contents were packets of seeds, maple syrup, placemats, photos and poetry. The Soviet cultural attaché in Washington was both 'startled and surprised' by the offering. The box was passed on eventually. Somewhere in Russia it is now either gathering dust or on show as a curiosity.

One Hartland first-grader had written for the box, 'Peace is me and a boy from a different country. We can get along and play on the monkey bars.' How delighted Carmelita Hinton would have been to feel that Ted's imagination had created a healing mechanism able to generate a simple joyful pleasure worldwide.

Back in London in April 1983 I had finally delivered Mark Reitman's smuggled document on the mathematics of peace to a peace academic. It had travelled around the world with me in just over eighty days. It too probably lies unregarded in an archive somewhere, another relic of the Cold War. Both the peace box and the document, wherever they may reside, are testament to the continuing struggles of ordinary people to try and gain a small purchase on their lives, given the overwhelming indifference of those who wield power.

And the Moscow Trust Group? In 1999 I met again some of its members now living in the US to make a BBC Radio 4 programme with Nigel Acheson in his *Document* series. Sergei Batovrin had segued into the New York art world, Yuri and Olga Medvedkov were teaching at Ohio State University. They showed me proudly their outdoor spa bath. Valodya Brodsky emigrated to Israel. Mark Reitman in Boston was by then rather frail from his diabetes. Many of the others who launched the group in 1982 had been exiled and resettled in the US.

# AFTERWORD

Over a dozen years have elapsed since I began my pursuit of the Booles and Hintons following that first remark regarding the *Voynich Manuscript* at my Aunt Doreen's funeral in 2000. It has indeed been a long journey, many times longer than the three months I spent in 1983 on the global tour that so serendipitously traced their lives. My ancestral connection to the two dynasties exists right enough, but it was not the claim of some form of ownership that kept me going – I found their lives intrinsically fascinating. Regarding my own Boole side, the surname ensuing from George's brother, William, has in fact all but disappeared. His other brother, Charles, keeps it going. At reunions in London, Cork and Lincoln, I met Kevin Boole and his family who had come all the way from Australia.

The Hinton link rests primarily with George's eldest daughter Mary Ellen's marriage to Charles Howard Hinton. Another earlier link, buried under hearsay, we have established in the liaison between Mary Everest and James Hinton. Despite the Atlantic separation, what is unusual and gratifying is that the Booles and Hintons kept very much in contact and paid mutual homage. Their residence in very different cultures, nevertheless produced outlooks remarkably of a piece. It was highly fortunate as far as I was concerned as biographer that their shared perspectives I shared too. I can't imagine anything worse than having to write about people you have no empathy with.

To me what is most attractive about them all is their rare ability to combine intellect with doing, personified perhaps by Carmelita Hinton who set out explicitly to educate young people on this basis at her Putney School. Ideas motivated them, often of wide implication on a grand scale: George Boole himself, for example, and his wife, Mary Everest; James Hinton and his son Charles Howard Hinton. The latter's offshoots were drawn to ideas enacted on a grand scale in Mao's China. As free-thinkers they all found it difficult to work within the set bounds of institutions. Even Geoffrey Taylor at Cambridge, an establishment body *par excellence*, worked necessarily free from restriction; maverick Howard Hinton at Bristol similarly. Both, as Taylor described it, had the Boole/Hinton fire within 'frightfully strong'. Too strong sometimes, as it led them to 'a want of judgement and balance'. Their intellects were mostly in the service of radical ideas in the guise of an anarchistic humanism that, even when paying lip service to a deity or an ideology, was grounded in the faith that individuals could promote social change. Whatever may have 'tainted' their blood also carried an optimistic and sanguine utopian current. They believed that reason, science and art could be put at the disposal of the collective good.

One other strength, which I found wholly resonant, is that these – to me – admirable characteristics were found in and fully enacted by the women of the clans. All five Boole daughters, most especially Ethel perhaps, exhibited talents and qualities that derived from their mother Mary Everest. It would be difficult to imagine any more dynamic women than Carmelita, Jean and Joan on the Hinton side.

If one were to institute a small museum dedicated to the Booles and Hintons to demonstrate the 'doing' side of their natures it would not be difficult to furnish it with artefacts. A Boolean computer representing George would probably take pride of place, these days no bigger than the ophthalmoscope with which James Hinton pioneered aural surgery. Margaret Boole's exemplary drawings of the insides of diseased ears would sit beside Alice's amazing four-dimensional geometric models. They could possibly be loaned from the museums in which they are already on show. Lucy's research papers into croton oil would sit nicely alongside the scripts of Ethel's novels and musical scores. Their mother Mary Everest's curved-stitching cards would demonstrate her practical application of geometry, while a pride of place label for functionally elegant design would be put next to Geoffrey Taylor's CQR anchor. One of Wilfrid Voynich's rare books, foraged on his European safaris, might be loaned from the British Library.

Charles Howard Hinton's cubes would look impressively incongruous stacked next to his baseball gun, although his son Ted's jungle gym would present daunting problems owing to its sprawling size. The Mexican Hintons branch from his other son, George, would be able to present a sample of his 'floating cement' building blocks and examples of mined geological specimens, or illustrations of the plethora of plant species his offspring have discovered. Howard Everest Hinton's exhibit could be one of his jarred specimens (or maybe a dried-out tortoise). Ted's children,

Jean, Bill and Joan, would require a special wing of the museum if it housed agricultural machinery such as the model milking parlour that I saw at Joan's Beijing farm. The ingenious tools of brother Bill's agricultural endeavours could be loaned and shipped from the Long Bow Museum in China.

I would hope to have some say in the collection, not on account of this present tome but to insist that the 'treacle-baby' painting of Ethel Voynich that I rounded up and took to Putney School could be released for display. Carmelita's own presence would no doubt be manifested by any testimonial of the hundreds of pupils who have passed through her school. The only exhibit that could never find its way to the exhibition would be that most singular and valuable document the *Voynich Manuscript*. A facsimile would be more likely, alongside a visitor book for browsers to inscribe their own version of its elusive meaning.

Eminent biographer, Michael Holroyd has written, 'There is a paradox about research: the more you do, the more you appear to give yourself to do.' I have found this to be only too true. It is difficult to know when to stop and abandon any new likely discoveries. The possibility that they might exist draws one on addictively. In the case of the *Voynich Manuscript* there are thousands of people all over the world fascinated by its continuing mystery. I've done the best I can to add to our knowledge by trying to uncover more about its discoverer, Wilfrid Voynich. For all I know, by the time this book reaches a public the tantalising curiosity may have been 'solved'.

New theories do continue to make headlines. One recent item struck a fanciful chord. In January 2014, A. Tucker, a US botanist, and R. Talbert published an analysis of the manuscript's plants, suggesting that they show distinct resemblances to those found in Mexico. Furthermore, the *Voynich Manuscript* appears similar to a sixteenth-century codex from Mexico City that uses loan words from native languages prior to the Spanish conquest. They match the manuscript's writing, it has been claimed. I don't actually know whether the Mexican Hintons ever saw any copies of their Aunt Ethel's husband's famous document. It is intriguing to think that in all their literally ground-breaking discoveries of unknown plants and voyages into ancient territories they might have formulated a similar idea decades ago.

After Ethel Voynich's triumphant resurrection as a novelist in 1955 and the accolades she received in the communist world, it is unsurprising that her devoted companion, Anne Nill, felt that her musical talent would also eventually be recognised. As noted in Chapter 16, Raymond Leppard was not unkind to the example of her music he had looked at in 1967. I decided to put Ethel's case to the test again and sent a leading English composer, John Joubert, some pages of her 'royal white elephant', the oratorio *Babylon*. In the 1920s it had received some endorsement. His verdict was sadly that 'She hasn't the overall technical grasp to deal with such an ambitious subject. The general impression is one of gaucheness and amateurism.' He notes that in her orchestration she is 'unclear what instrumental forces she is writing for'. Both she and Anne would have been more than a little disappointed.

Ethel's fame, however, as a novelist does live on. *The Gadfly* refuses to lie down and pretend to be mutton. Shostakovich's beautiful love theme from his *Gadfly Suite* was used as the signature tune for the TV series about Sidney Reilly, *Ace of Spies*. In 1983 *Ovod* transformed into a Soviet rock musical; in 2005 the novel was re-filmed in a joint venture between Ukraine and China. More surprisingly, adherents to the social network *Facebook* have taken Ethel and her novel to its heart. She has 3,000 'likes', mostly young women, from around the world.

James Hinton, for my purposes the co-founder of the Boole/Hinton dynasties, is likely to remain largely an eccentric ghost. His phantom presence, however, is still to be found spectrally on London's East End streets. In 1987 both James and his son, Charles Howard Hinton, were resurrected in Ian Sinclair's novel, *White Chappell, Scarlet Tracings* and Alan Moore's graphic novel, *From Hell*. Both Hintons are vicariously linked in the books to the Ripper murders of 1888 – a 'sacrifice of women' on the grisliest scale. James, as we know, found a great cause in prostitution and had connections with the East End of London. He also had known Sir William Gull, Queen Victoria's surgeon, also mired in the saga. We learn from Sinclair of the interdependency of events, that they 'all co-exist in the stupendous whole of eternity', a notion implied in one of Charles Howard's own novels. The very idea of the fourth dimension, it seems, allows any narrative a vast space for fanciful notions.

In a more down-to-earth fashion Sinclair's use of psycho-geography embellishes the conjuring of myth from landscape. The historical agent is the *flâneur* tramping London's streets, his nose to the ground. The geometric pattern at least resonates well with a central conceit of the Boole/Hintons: the storyteller progresses in a spiral fashion, 'stool-snuffing, circling back on itself'. It may well be, less fancifully, that some of the characters in this volume and the paths they trod will come to be sniffed over once again in the future. For my part, in 1983 I did circle back on myself, globally passing within a whiff of the Booles and Hintons. My concern with them now, however, is following a definitely linear trajectory – and has (probably) arrived at its end.

# NOTES

## Chapter 1

1. George Orwell. *Coming Up for Air*, (London, Victor Gollancz, 1939).

## Chapter 2

1. Gerry Kennedy and Rob Churchill, *The Voynich Manuscript* (London: Orion, 2004).

2. H. P. Kraus, *A Rare Books Saga: the autobiography of H. P. Kraus* (New York: Putnam, 1978), p. 12.

3. William Romaine Newbold, *The Voynich Roger Bacon Manuscript: transactions of the College of Physicians*, 3rd series, vol. 43 (1921), p. 461.

4. Mary D'Imperio, *The Voynich Manuscript: an Elegant Enigma*, (National Security Agency, 1978), p. 11.

5. All the folios can be viewed on Beinecke's website.

6. R. S. Brumbaugh, *The World's Most Mysterious Manuscript* (Carbondale: Southern Illinois University Press, 1978), p. 136.

## Chapter 3

1. Robert Harley, 'George Boole, FRS', *British Quarterly Review*, vol. 44 (July 1866); *Collected Logical Works of George Boole* (La Salle: Open Court Publishing Company, 1916), p. 427.

2. Daniel Cohen, *Equations from God: pure mathematics and Victorian faith* (Baltimore: Johns Hopkins University Press, 2007), p. 78.

3. E. P. Thompson, *The Making of the English Working Class* (Harmondsworth: Penguin, 1968), p. 895.

4. Ibid., p. 23.

5. Ibid., p. 783.

6. Ibid., p. 816.

7. Ibid., p. 812.

8. Ibid., p. 44.

9. Harley, 'George Boole, FRS'.

10. Thomas Cooper, *The Life of Thomas Cooper* (Leicester: Leicester University Press, 1971), p. 362.

11. Ibid., p. 117.

12. Cohen, *Equations from God*, p. 77.

13. Desmond MacHale, *The Life and Work of George Boole* (Cork University Press, 2014), p. 22.

14. Mary Everest Boole, *Collected Works*, vol. 1, ed. E. M. Cobham (London: C.W. Daniel, 1931), p. 92.

15. Sir Francis Hill, *Victorian England* (Cambridge: Cambridge University Press, 1974), p. 147.

16. George Boole, *An Address on the Genius and Discoveries of Sir Isacc Newton*, Feb 5, 1835, p. 10.

17. Charles Clarke, *Sixty Years in Upper Canada* (Toronto: William Briggs, 1908), p. 17.

18. Ibid.

19. MacHale, *George Boole*, p. 29.

20. Harley, 'George Boole, FRS'; *Collected Logical Works of George Boole*, p. 458.

21. Kenneth C. Dewar, *Charles Clarke: pen and ink warrior* (Montreal and Kingston: McGill-Queen's University Press, 2002), p. 51.

22. Harley, *George Boole FRS*, p. 440.

23. George Boole, *The Right Use of Leisure* (London: J. Nisbet. 1867), p. 14

24. Ibid. p. 9.

25. Mary Everest Boole, *Collected Works Vol. 1*, p. 81.

26. Cynthia Jolly, 'Rescuing the Fallen', *Lincolnshire Life* (November 2005).

27. MacHale, *George Boole*, p. 60.

28. Ibid., p. 77.

29. G. C. Smith, *The Boole–de Morgan Correspondence, 1842–64* (Oxford: Clarendon Press, 1982), p. 62.

30. Boole, *Collected Works*, p. 40.

31. University College, Cork, *George Boole Papers*, IE/BP/1/221/7.

32. Ibid., BP/1/221/15.

33. Unpublished letter, courtesy John Rollett.

34. MacHale, *George Boole*, p. 73.

35. Cohen, *Equations from God*, p. 81.

36. George Boole, *The Laws of Thought* (New York: Dover Publications, 1958), p. 1.

37. University College, Cork, *George Boole Papers*, BP/1/232.

38. Ibid., BP/1/99.

39. Boole, *Collected Works*, p. 18.

40. Ibid., p. 20.

41. MacHale, *George Boole*, p. 180.

42. Boole, *Collected Works*, p. 22.

43. Ibid., p. 23.

44. Ibid., p. 17.

45. Ibid., p. 32.

46. Ibid., p. 30.

47. Ibid., p. 36.

48. F. D. Maurice, *The Claims of the Bible and of Science: correspondence between a layman and the Reverend F. D. Maurice* (London: Macmillan, 1863), p. 56.

49. Ibid., p. 142.

50. Jeremy Morris, *F.D. Maurice and the Crisis of the Christian Church* (Oxford: Oxford University Press, 2005).

51. J. F. Porter and W. J. Wolf, *Toward the Recovery of Unity: the thought of F. D. Maurice* (Greenwich, CT: Seabury Press, 1964).

52. Courtesy John Rollett.

## Chapter 4

1. Mary Everest Boole, *Collected Works*, vol. 4, ed. E. M. Cobham (London: C.W. Daniel, 1931), p. 1510.

2. Mary Everest Boole, *Collected Works*, vol. 3, ed. E. M. Cobham (London: C.W. Daniel, 1931), p. 1072.

3. Mary Everest Boole, *Collected Works*, vol. 1, ed. E. M. Cobham (London: C.W. Daniel, 1931), p. 1033.

4. Ibid., p. 82.

5. T. R. Everest, *A Popular View of Homeopathy* (Allentown: Academical Book Store, 1834), p. 13.

6. Boole, *Collected Works*, vol. 4, p. 74.

7. Ibid., p. 1515.

8. Ibid., p. 1517.

9. Ibid., p. 1524.

10. Ibid., p. 1521.

11. Ibid., p. 1519.

12. Ibid., p. 1519.

13. Boole, *Collected Works*, vol. 1, p. 1540.

14. Ibid., p. 236.

15. Trinity College, Cambridge, *Papers of Sir Geoffrey Taylor*, A131(14).

16. Boole, *Collected Works*, vol. 1, p. 77.

17. Ibid., p. 78.

18. F.W. Daniels, *A Teacher of Brain Liberation* (London: C.W. Daniel, 1923), p. 16.

19. Maurice's conviction of a divine order, interestingly, emerges from his interpretation of Job hearing God speaking to him out of a whirlwind, convincing him of His majesty.

20. Boole, *Collected Works*, vol. 1, p. 51.

21. Boole, *Collected Works*, vol. 2, p. 478.

22. Boole, *Collected Works*, vol. 4, p. 1307.

23. Boole, *Collected Works*, vol. 2, p. 717.

24. Ibid., p. 513.

25. Boole, *Collected Works*, vol. 1, p. 306.

26. Boole, *Collected Works*, vol. 2, p. 403.

27. Ibid., p. 647.

28. Ibid., p. 402.

29. Boole, *Collected Works*, vol. 3, p. 1093.

30. Boole, *Collected Works*, vol. 1, p. 332.

31. Ibid., p. 106.

32. Ibid., p. 98.

33. Ibid., p. 170.

34. Ibid., p. 173.

35. Ibid., p. 223.

36. Elaine Kaye, *The History of Queen's College, London* (London: Chatto & Windus, 1972), p. 102.

37. G. K. Batchelor, *The Life and Legacy of G. I. Taylor* (Cambridge: Cambridge University Press), p. 18.

38. Boole, *Collected Works*, vol. 4, p. 1362.

39. Boole, *Collected Works*, vol. 3, p. 1107.

40. Boole, *Collected Works*, vol. 4, p. 1362.

41. Ibid., p. 1365.

42. Ibid., p. 1363.

43. Ibid., p. 1364.

44. Ibid., p. 1366.

45. Boole, *Collected Works*, vol. 3, p. 1004.

46. Boole, *Collected Works*, vol. 2, p. 728.

47. Edith Somervell, *A Rhythmic Approach to Mathematics* (National Council of Teachers of Mathematics, 1906), p. 11.

48. Boole, *Collected Works*, vol. 3, p. 943.

49. Boole, *Collected Works*, vol. 2, p. 537.

50. Ibid., p. 531.

51. E. M. Cobham, *Mary Everest Boole: a memoir* (London: C.W. Daniels, 1931), p. 45.

52. Another character fits better: Altiora MacVitie, who inhabits a similar house where all sorts of interesting people congregate, 'from the obscurely efficient to the well-instructed famous and the rudderless rich'.

53. Elaine Showalter, *A Literature of Their Own* (Princeton: Princeton University Press 1977), p. 248.

54. Dorothy Richardson, *Revolving Lights* (London: J.A. Dent, 1923), p. 371.

55. Helen Douglas and her sister Lady Low, wife of Sir Hugh Low, diplomat, and the Marchioness of Ailsa, who found 'poise and refreshment' at number 16. Others included Jewish and Indian friends, adherents of Grieg and Ibsen. Sir Arthur Somervell, composer, and Julia Wedgwood, novelist and critic, were also part of the circle.

56. Mrs Henry Cust, *Echoes of a Larger Life* (London: Jonathan Cape, 1921), p. 105.

57. Ibid., p. 91.

58. Ibid., p. 155.

59. Ibid., p. 96.

60. Ibid., p. 246.

61. Boole, *Collected Works*, vol. 1, p. 101.

62. Boole, *Collected Works*, vol. 3, p. 969.

63. Boole, *Collected Works*, vol. 1, p. 225.

64. Boole, *Collected Works*, vol. 3, p. 1156.

65. Boole, *Collected Works*, vol. 4, p. 1435.

66. Ibid., p. 1426.

67. Ibid., p. 1424.

68. Ibid., p. 1441.

69. Ibid., p. 1429.

70. Ibid., p. 1430.

71. Ibid., p. 1229.

72. Ibid., p. 1441.

73. Ibid., p. 1431.

74. Ibid., p. 1432.

75. Some of her declarations would sit well in an anarchist anthology. Desmond MacHale collects samples of what he describes as her 'insight, common sense, perception, educational innovation, long-winded banality, incoherent confusion between philosophy and maths and … complete nonsense' (p. 290). (We haven't even touched on Atlantis, black magicians and angels.) He asks the reader to judge his selection for themselves. One of the mix (p. 291) includes her visualisation of three symbols of authority: the slave-driver's whip, disguised as sceptre, mace, truncheon and cane, signifying obedience: the shepherd's crook, modified into a bishop's crozier that persuades a corralled flock not to question, but accept their destination at the slaughter house: the conductor's baton, however, brings people together for a performance of a common cultural heritage leaving them otherwise free to play their own tunes or combine with others. She asks in typical style, 'Friends, under which symbol will you serve? And by which will you prefer to rule?' This fits with a libertarian tradition that Peter Kropotkin would have recognised.

76. Cust, *Echoes of a Larger Life*, p. 156.

77. Boole, *Collected Works*, vol. 4, p. 1277.

78. Unpublished letters. Courtesy John Rollett.

79. Cobham, *Mary Everest Boole: a memoir*, p. 106.

80. Ibid., p. 115.

81. See G.K. Valente, 'Giving Wings to Logic: Mary Everest Boole's propagation and fulfilment of a legacy', *British Journal for the History of Science*, vol. 43, no. 1, pp. 49–74; also Luis M. Laita, 'Boolean Algebra and its Extra-logical Sources', *History and Philosophy of Logic*, no. 1 (1980), pp. 37–60.

82. Boole, *Collected Works*, vol. 4, p. 1558.

## Chapter 5

1. Cobham, *Mary Everest Boole: a memoir*, p. 122.

2. Boole, *Collected Works*, vol. 3, p. 905.

3. Cust, *Echoes of a Larger Life*, p. 155.

4. Anne Fremantle, *The Three-cornered Heart* (London: Collins, 1871), p. 59.

## Chapter 6

1. Ellice Hopkins (ed.), *Life and Letters of James Hinton* (London: Kegan Paul, 1878), p. 47.

2. Ibid., p. 60.

3. Ibid., p. 72.

4. Ibid., p. 105.

5. Ibid., p. 105.

6. Ibid., p. 165. His thoughts and enquiries did indeed produce a vast body of work, an overflowing torrent of observations and insights described by one admirer as 'beams of light glancing here and there'. Mary Everest Boole, hardly reticent herself, described him as a 'thought-artist'. His four volumes of *Selections from Manuscripts* based on his early musings were printed privately between 1870 and 1874; they run to 2,360 pages. There are another six volumes of unpublished hand-written notes in the Havelock Ellis collection in the British Library amounting to another 2,000. Under sections such as 'Sermons', 'Science', 'Women' and 'Life' they wander repetitiously but rarely give a glimpse of the actual man.

    Fortunately for us after his death he was favoured by four acolytes who plumbed the brimful cistern to present their own interpretations. Caroline Haddon edited a selection under the title *Philosophy and Religion* in 1881 and his wife Margaret another entitled *The Law-breaker* in 1884. Edith Lees/Ellis (wife of Havelock) who took up Hinton's cause much later, regarded him less as a philosopher and constructor of systems and more as a seer. As such, in her work *Three Modern Seers* (1910) she ranked him alongside Nietzsche and Edward Carpenter. She followed with another appreciation, *James Hinton: a sketch*, published in 1918 after her death, using many unreferenced extracts from his *Selections*.

7. British Library, *Havelock Ellis Papers*, f17.

8. Caroline Haddon (ed.), *Selections from the Manuscripts of the Late James Hinton*, vol. 1 (London: Kegan Paul, Trench & Co., 1881), p. 59.

9. Caroline Haddon (ed.), *Selections from the Manuscripts of the Late James Hinton*, vol. 3 (London: Kegan Paul, Trench & Co., 1881), p. 321.

10. James Hinton, *Life in Nature* (London: Smith, Elder & Co., 1875), p. 87.

11. Ibid., p. 95.

12. Caroline Haddon, *The Larger Life: studies in Hinton's ethics* (London: Kegan Paul, 1886), p. 210.

13. Hopkins (ed.), *Life and Letters of James Hinton*, p. 139.

14. Ibid., p. 172.

15. At Guy's he enjoyed the company of Dr William Gull, later to attend as surgeon to Queen Victoria, and another aural surgeon, Joseph Toynbee, father of the economic historian Arnold. Joseph had been a friend of Hinton's and medical mentor for many years. Both of the Toynbees were much influenced by James' ideas.

The connection was not to last long: Toynbee was found dead at his surgery in Savile Row in 1866, a bottle of chloroform and a phial of prussic acid nearby, which he had disastrously inhaled trying to cure tinnitus. James took over the practice and worked diligently and expertly in his field, in 1868 studying in Germany and practising in Vienna. Hinton made 500 dissections during his career and had been the first to successfully conduct a mastoidectomy. In 1874 his *Atlas of the Membrane Tympanum* was considered definitive.

16. Hopkins (ed.), *Life and Letters of James Hinton*, p. 220.
17. Haddon (ed.), *Selections from the Manuscripts of the Late James Hinton*, vol. 3, p. 486.
18. Haddon, *The Larger Life*, p. 139.
19. Hopkins (ed.), *Life and Letters of James Hinton*, p. 221.
20. R. H. Hutton, 'The Metaphysical Society: a reminiscence', *The Nineteenth Century* (1885).
21. Hopkins (ed.), *Life and Letters of James Hinton*, p. 277.
22. Mrs Havelock Ellis, *James Hinton: a sketch* (London: Stanley Paul, 1918), p. 93.
23. Ibid., p. 107.
24. British Library Add MS 70530 f37.
25. Ibid., 70535 f21.
26. Ibid., 70532 f107.
27. Ibid., 70532 f3.
28. Ellis, *James Hinton: a sketch*, p. 101.
29. British Library Add MS 70530 f10.
30. Ellis, *James Hinton: a sketch*, p. 291.
31. Ibid., p. 142.
32. Ibid., p. 283.
33. Ibid., p. 180.
34. Ibid., p. 238.
35. Hopkins (ed.), *Life and Letters of James Hinton*, p. 369.
36. Mrs Havelock Ellis, *Three Modern Seers* (London: Stanley Paul, 1910), p. 46.
37. British Library, Add MS 70524 ff5–7.
38. This reputedly scurrilous and damning volume is presumed to have disappeared. I found it in the not so surprising location of the Havelock Ellis Papers 1874–1951 at the University of California, Los Angeles.

**Chapter 7**

1. Anne Fremantle, 'Return of the Gadfly', *Commonweal*, vol. 74 (1961), pp. 167–71.
2. Ibid., p. 169.
3. Evgenia Taratuta, *The Fate of a Writer and the Fate of a Book* (Moscow: [publisher], 1964), p. 11.

4.   Ibid., p. 15.

5.   Sergei Kravchinsky, *Underground Russia* (New York: Scribner, 1883), p. 161.

6.   Ibid., p. 284.

7.   Ibid., p. 38.

8.   Ibid., p. 39.

9.   Taratuta, *The Fate of a Writer and the Fate of a Book*, p. 16.

10.  Ibid.

11.  Kravchinsky, *Underground Russia*, p. 139.

12.  Fiona MacCarthy, *William Morris* (London: Faber, 1994), p. 469.

13.  Julia Briggs, *A Woman of Passion* (New York: Hutchinson, 2000), p. 63.

14.  Ibid., p. 141.

15.  S. Hinely, 'Charlotte Wilson: anarchist, Fabian and feminist', Stanford University PhD dissertation, 1987, p. 197.

16.  John Quail, *The Slow-burning Fuse* (London: Granada, 1978), p. 58.

17.  Taratuta, *The Fate of a Writer and the Fate of a Book*, p. 26.

## Chapter 8

1.   Harvey Kushner, *The Future of Terrorism* (London: Sage Publications, 1998), p. 185.

2.   Soloman Volkov, *St Petersburg: a cultural history* (London: Sinclair-Stevenson, 1996), p. 48.

3.   Ibid., p. 52.

4.   Taratuta, *The Fate of a Writer and the Fate of a Book*, p. 6.

5.   Anne Fremantle, 'The Russian Best-seller', *History Today*, vol. 25, no. 9 (September 1975), p. 634.

6.   Ibid.

7.   Taratuta, *The Fate of a Writer and the Fate of a Book*, p. 56.

8.   Ibid., p. 34.

9.   Ethel Voynich, *Olive Latham* (London: J.B. Lippincott, 1904), p. 55.

10.  Ibid., p. 56.

11.  Richard Pipes, *The Degaev Affair* (New Haven: Yale University Press, 2005).

12.  Taratuta, *The Fate of a Writer and the Fate of a Book*, p. 46.

13.  Her husband Vasily's career, however, was teased out by no less a personage than V. I. Lenin. Writing in 1910 in an article, *The Career of a Russian Terrorist*, on the occasion of Karaulov's death he recalls Vasily's past career (Lenin, *Collected Works*, vol. 17 (Moscow: Progress Publishers, 1974), pp. 46–48). Lenin correctly casts doubt on why he had received a relatively light sentence at his trial in Kiev in 1884 compared to his colleagues. Lenin quotes with disdain a comment of Karaulov's in 1905 decrying 'the notorious slogan of the dictatorship of the proletariat'. To Lenin this was a red rag to a bull; he condemns him as 'a renegade and despicable counter-revolutionary'. In 1906 Karaulov was elected to the third Duma parliament as a democrat. Lenin had no time for them any more than the *narodniks* with their erroneous utopian

programme masquerading as socialism. He respected at least their selfless struggle against tsarism. A decade after Lenin's article the Bolshevik proletarian dictatorship began a campaign of state terrorism that would last over half a century.

14. Taratuta, *The Fate of a Writer and the Fate of a Book*, p. 52.

15. Ibid., p. 57.

## Chapter 9

1.  Rosa M. Barrett, *Ellice Hopkins: a memoir* (London: Wells Gardiner Dalton, 1908), p. 152.

2.  Havelock Ellis, *My Life* (New York: Houghton Mifflin, 1939), p. 131.

3.  Ibid., p. 183.

4.  Yaffa Claire Draznin (ed.), *My Other Self* (New York: Peter Laing, 1992), p. 183.

5.  Ibid., p. 54.

6.  British Library Add MS 70524 f41.

7.  Draznin (ed.), *My Other Self*, p. 93.

8.  British Library Add MS 70524 f41.

9.  Draznin (ed.), *My Other Self*, p. 72.

10. British Library Add MS 70524 f47.

11. Draznin (ed.), *My Other Self*, p. 368.

12. Caroline Haddon, *The Larger Life* (London: Kegan Paul, Trench & Co., 1886), p. 165.

13. Ibid., p. 61.

14. Caroline Haddon, *Where Does Your Interest Come From?* (London: John Heywood, 1886), p. 5.

15. Ibid., p. 9.

16. Caroline Haddon, *The Future of Marriage: an eirenikon*, p. 22.

17. Ibid., p. 24.

18. *Pall Mall Gazette*, 6 July 1885.

19. University College, London, Karl Pearson Papers, 10/1.

20. University College, London, Hacker Papers, Box 4.

21. Ibid.

22. Ibid.

23. Ibid.

24. Ibid.

25. Ibid.

26. Karl Pearson Papers, 10/28, 16 October 1886.

27. Karl Pearson Papers, Letter to Miss Sharpe, undated.

28. Havelock Ellis, *Studies in the Psychology of Sex: vol. 6 – Sex in relation to Society* (Philadelphia: J. A. Davis, 1910), p. 16.

29. Draznin (ed.), *My Other Self*, p. 512.

## Chapter 10

1. Taratuta, *The Fate of a Writer and the Fate of a Book*, p. 58.

2. Ibid., p. 67.

3. Ibid., p. 71.

4. Ibid., p. 91.

5. Ford Madox Ford, *Ancient Lights* (London: Chapman & Hall, 1911), p. 121.

6. Isabel Meredith, *A Girl Among the Anarchists* (London: Duckworth Press, 1902), p. 22.

7. Ford, *Ancient Lights*, p. 121.

8. Ford Madox Ford, *Return to Yesterday* (Manchester: Carcanet, 1999), p. 61.

9. Meredith, *A Girl Among the Anarchists*, p. 188.

10. Meredith, *A Girl Among the Anarchists*, p. 23.

11. Taratuta, *The Fate of a Writer and the Fate of a Book*, p. 277.

12. Anne Fremantle, 'The Russian Best-seller', *History Today*, vol. 25, no. 9 (September 1975), p. 633.

13. Taratuta, *The Fate of a Writer and the Fate of a Book*, p. 80.

14. Ibid., p. 83.

15. Ibid., p. 90.

16. Hoover Institution, Volkhovsky Papers, File 4, 'Voinich'.

17. Taratuta, *The Fate of a Writer and the Fate of a Book*, p. 92.

18. Annie Besant, *An Autobiography* (Adjar, Madras: Theological Publishing House, 1939), p. 425.

19. E. Douglas Fawcett, *Hartmann the Anarchist* (London: Tangent Books, 2007).

20. Sergei Stepniak (Sergei Kravchinsky), *The Career of a Nihilist* (New York: Harper, 1889), p. x.

21. Ibid., p. ix.

22. Ibid., p. 253.

23. Ibid., p. 320.

24. *The Arizona Republican*, 21 December 1902, p. 11 (via Dana Scott).

25. Taratuta, *The Fate of a Writer and the Fate of a Book*, p. 290.

26. Ibid., p. 44.

27. Ibid., 287.

28. Ibid., p. 104. Rachkovsky, although diligent, was not immune to mistakes. In a report of April 1894 he stated that Voynich had married Volkhovsky's daughter. Another agent, Alexander Evalenko, was more successful. Sent to New York in 1891 to work with Goldenberg and the US Russian Free Press Fund he gained his and Wilfrid's confidence by donating large sums of money and ordering quantities of books to be sent to Russia which he promptly destroyed. An expensive but successful ploy; the US operation closed down soon after.

29. Barry Johnson, *The Diary of Olive Garnett*, vol. 2 (London: Bartlett, 1993), p. 2.

30. Ibid., p. 1.

31. Barry Johnson, *Olive and Stepniak* (London: Bartlett, 1993), p. 27.

32. Ibid., p. 28.

33. Taratuta, *The Fate of a Writer and the Fate of a Book*, p. 296.

34. Ibid., p. 139.

35. Nothing is known about Stein, but Taratuta refers to 'a recent arrival from Russia … a person of energy and initiative … a scientist, business man and revolutionary' (p. 115). Might he too have been an *Okhrana* plant? To raise money he had become involved in some venture involving soap. A small clutch of letters in the Wellcome Medical Library, London bears out the fact that Voynich was becoming interested in business as much as politics. In August 1895 a Mrs Van der Weyde had written to Henry Wellcome, the wealthy pharmacist and medicine manufacturer, to endorse a 'Russian gentleman Mr Voynich whom you met at our house', who wishes in turn to contact him. Voynich indeed wrote to Wellcome proposing that, 'I and a friend of mine … are anxious to obtain work in connection with a chemical or drug manufacturing.' He mentions that the friend 'has been assistant to Prof. Frezenius'. This must refer to Carl Fresenius the analytical chemist. Was the friend the shadowy Stein? More than this we don't know, but it is clear that Wilfrid, not yet five years in London, is moving purposefully among its higher echelons.

36. Taratuta, *The Fate of a Writer and the Fate of a Book*, p. 117.

37. Evgenia Taratuta, *Our Friend Ethel Lilian Voynich*, translated by Séamus Ó Coigligh (Cory City Library, 2008), p. 29.

38. Ibid., p. 31.

39. Ibid., p. 28.

## Chapter 11

1. Ellice Hoskins, *James Hinton: life and letters* (London: Kegan Paul, 1878), p. 251.

2. Edwin Abbott, *Flatland: a romance of many dimensions* (Oxford: Oxford World Classics, 2006).

3. Martin Gardner, *Mathematical Carnival* (New York: Vintage Books, 2007), p. 52.

4. Charles Howard Hinton, *A New Era of Thought* (London: Swan Sonnenschein, 1888), p. 92.

5. Ibid., p. 79.

6. Ibid., p. 78.

7. Ibid., p. 86.

8. Ibid., p. 75.

9. Ibid., p. 89.

10. Mark Blacklock, *The Fairyland of Geometry*, https://higherspace.wordpress.com

11. Charles Howard Hinton, *Many Dimensions* (New York: Dover Publications, 1980), p. 69.

12. Charles Howard Hinton, *An Episode in Flatland* (London: Swan Sonnenschein, 1907), p. 76.

13. Ibid., p. 156.

14. Ibid., p. 73.

15. Ibid., p. 139.

16. Hinton, *Many Dimensions*, p. 76.

17. Charles Howard Hinton, *Stella* (London: Swan Sonnenschein, 1895), p. 32.

18. Ibid., p. 105.

19. Charles Howard Hinton, *An Unfinished Communication* (London: Swan Sonnenschein, 1895), p. 112.

20. Ibid., p. 114.

21. Ibid., p. 97.

22. Ibid., p. 99.

23. Ibid., p. 104.

24. Ibid., p. 105.

25. Hinton, *Stella*, p. 31.

26. University College, Cork, George Boole Collection. The following seventeen references are from BP/1/339.

27. *New York Times*, 8 December 1895.

28. Bertrand Russell, *Mind*, vol. 13, pp. 573–4.

29. Hinton, *Many Dimensions*, p. 32.

30. University of California, Berkeley, Gelett Burgess Papers (September 1906). Thanks to Mark Blacklock.

31. Ibid., 18 September 1906.

32. Ibid., 22 August 1906.

33. Ibid., *Travels of an Idea*. p. 6.

34. Ibid., p. 14.

35. Ibid., 108.

36. Ibid., p. 124.

37. Letter to George Hinton, 7 April 1907. Courtesy Patricia Hinton Davison.

38. *Washington Post*, 1 May 1907.

39. Bruce Clarke, *Energy Forms* (Ann Arbor: University of Michigan Press, 2001), p. 177.

40. David Toomey, *The New Time Travellers* (New York: Norton, 2007), p. 177.

41. Peter Ouspensky, *Tertium Organum* (London: Arkana, 1981), p. 258.

42. Ibid., p. 166.

43. Ibid., p. 257.

44. Hinton, *A New Era of Thought*, p. 85.

45. Ouspensky, *Tertium Organum*, p. 281.

## Chapter 12

1.   Lucjan Blit, *The Origins of Polish Socialism* (Cambridge: Cambridge University Press, 1971), p. 121.
2.   Archive kindly provided by Marni Rosner.
3.   Ethel Voynich, Letter to William Bishop, 20 October 1931.
4.   Taratuta, *The Fate of a Writer and the Fate of a Book*, p. 82.
5.   James Westfall Thompson, 'Progress of Medieval Studies in the USA and Canada', *Bulletin*, vol. 8 (1931), pp. 90–2.
6.   Millicent Sowerby, *Rare People and Rare Books* (London: Constable, 1967), p. 9.
7.   Taratuta, *The Fate of a Writer and the Fate of a Book*, p. 83.
8.   Ethel Voynich (ed.), *Chopin's Letters* (New York: Dover Publications, 1988), p. vi.
9.   Lydia Loiko, *From Land and Freedom to the All-Union Communist Party* (Moscow: 1928), p. 86.
10.  Ibid., p. 96.

## Chapter 13

1.   Quoted in Fremantle, 'The Russian Best-seller', p. 629.
2.   Taratuta, *The Fate of a Writer and the Fate of a Book*, p. 64.
3.   Barbara Garlick, *Atheism and the Problem of Textual Production*, p. 5.
4.   Ethel Voynich, *The Gadfly* (London: Granada Publishing, 1973), p. 66.
5.   Ibid., p. 75.
6.   Ibid., p. 223.
7.   Ibid., p. 225.
8.   Ibid., p. 229.
9.   Ibid., p. 231.
10.  Ibid., p. 232.
11.  Ibid., p. 250.
12.  Ibid., p. 253.
13.  Ibid., p. 142.
14.  Ibid.
15.  Ibid., p. 145.
16.  Ibid.
17.  Ibid., p. 146.
18.  Ibid., p. 38.
19.  Ibid., p. 153.
20.  Ibid., p. 249.
21.  Fremantle, 'Return of the Gadfly', p. 169.
22.  Joseph Conrad, *Letters*, vol. 1, p. 395. Conrad complained contrarily of Voynich being 'hollow-eyed' as a woman-writer, providing the Gadfly's kid-glove rejections as motives for the plot's over-melodrama. His seeming sexism doesn't fit well with Ethel's almost a-gendered robustness, avoiding a love

story and dishing up a ripping yarn. Some of her themes – South America, revolution, silver mines and a charismatic Byronic hero – appeared in his own later novel, *Nostromo* (1904).

23. *The Times*, 25 March 1930.

24. Arnold Hunt, *Out of Print and Into Profit*, ed. E. Giles Mandelbrote (London: The British Library/Oak Knoll Press, 2006), p. 248.

25. Frances Larson, *An Infinity of Things* (Oxford: Oxford University Press, 2009), p. 66. The pair actually traded and haggled with each other regularly. In July 1914 Voynich offered him an 'extraordinarily rare' Polish herbal of 1595, the first to use native names for plants. In September Voynich sent him a list of some fifty books, many of them incunabula; Thompson bought twenty at £65-11-0 less 15 per cent discount. Like Voynich too he made detailed catalogues of his acquisitions for Wellcome.

26. A. J. Bowman, *A Critical Edition of the Private Diaries of Robert Proctor* (New York: Mellen Press, 2010), p. 62.

27. Ibid., p. 132. On 20 July Proctor writes, 'Voynich turned up on a flying visit home, he returns to Italy at once.' His next noted arrival was on 1 October: 'Voynich came in high spirits; great accounts of his capture of the Italian book trade.' This may have been Wilfrid's bluster but clearly he had been away on safari there for several months – at least the third year in a row building up expertise and contacts. This didn't necessarily bring him good returns all the time; on 8 February 1901 Proctor notes of Voynich's third catalogue that 'the books are dull and exorbitantly dear'. A year later in October the book-hunter seems depressed and complaining of not getting paid, but 'doesn't like asking for money'. Another Voynich collection is being viewed in April 1903. In Proctor's last entry, 28 August 1903, he writes, 'I went to bed early being weary. What shall I be when I open this book again three weeks hence?' Sadly for him there was to be no more taking up of his pen: Proctor died shortly after in an accident climbing in the Tyrol.

28. Giuseppe Orioli, *Adventures of a Bookseller* (privately printed, London, 1937), p. 94.

29. Sowerby, *Rare People and Rare Books*, p. 32.

30. Ethel championed the fiction of Waclaw Seroshevsky and helped his friend Bronislaw Pilsudski in 1910. Both of the Poles in Siberian exile, surrounded by unknown cultures, had become, like Voynich, experts in ethnography. Bronislaw, brother of Joseph Pilsudski, who we recently encountered, had been sentenced to exile on Sakhalin Island in far Siberia for his part in an abortive plot on the Tsar in 1887. While there he had made a study of the indigenous Ainu people. In 1910 he brought sample cylinders of their music to London. The Voyniches obtained performances at the Royal College of Music. Notations of them were made, possibly by Ethel, as at that time she was renewing her interest in music.

31. Ethel Voynich, *Jack Raymond* (London: Heinemann, 1901), p. 47.

32. Ibid., p. 59.

33. Ibid., p. 106.

34. Ibid., p. 189.

35. Ibid., p. 233.

36. Ibid., p. 272.

37. Another rather obscure and wild reference to the novel comes from Lytton Strachey no less, author later of *Eminent Victorians*. In a letter to Leonard Woolf in February 1907 he complains of a 'dim embryo' (Cecil Francis Taylor), who he hopes will not be elected to the elite, secret Apostles Club of Cambridge University. Strachey erroneously writes that Taylor is the grandson of George Boole and also the nephew 'of that dreadful woman Mrs Voynich who wrote "Jack Raymond" and other ghastly works'. Rare praise indeed. Paul Levy (ed.), *The Letters of Lytton Strachey* (London: Penguin, 2006), p. 119.

## Chapter 14

1. Marlene Rayner-Canham and Geoffrey Rayner-Canham, *Chemistry Was Their Life* (London: Imperial College Press, 2008), p. 12.

2. Alice's son Leonard Stott (1892–1963) became one of the pioneers of Papworth village in Cambridgeshire treating tuberculosis. In 1921 he became Resident Medical Officer there, echoing his cousin Julian's career in medicine. Its revolutionary principle (continuing the Boole/Hinton utopian mutualist strain and hands-on putting theory into practice) revolved around the idea that families should not be separated but live together within a settlement where rigorous supervision of health and diet could be monitored and co-ordinated. The prophecy that children there would inevitably be infected was disproved. The unit even later offered employment in the production of military and other public vehicles. The principle of occupational therapy continued when the population became focused on the problems of the disabled. Like his cousin Geoffrey, Leonard had an original and inventive mind, producing a system of navigation based on trigonometry and an artificial pneumothorax apparatus.

3. H. S. M. Coxeter, *Regular Polytopes* (New York: Dover Publications, 1973), pp. 258–9.

4. Letter to Geoffrey Taylor, 1911.

5. Unpublished memoir of Carmelita Hinton.

6. I. and B. Hargittai, *Candid Science V: conversations with famous scientists* (London: Imperial College Press, 2005), p. 12.

7. H. MacNaughton Jones, *Atlas of Diseases of the Membranum Tympanum* (London: Churchill, 1878).

8. Ethel Voynich, *An Interrupted Friendship* (London: Granada Publishing, 1974), p. 234.

9. Ibid., p. 239.

10. Pamela Blevins, *Ivor Gurney and Marion Scott* (Woodbridge: Boydell Press, 2008), p. 45.

11. Edmund Blunden, *The Poems of Ivor Gurney* (New York: Hutchinson, 1954), p. 12.

12. Michael Hurd, *The Ordeal of Ivor Gurney* (London: Faber & Faber, 1978), p. 34.

13. R. K. R. Thornton, *Ivor Gurney: collected letters* (Ashington: MidNAG Press, 1991), p. 155.

14. Ibid., p. 180.

15. Ibid., p. 8.

16. Ibid., p. 9.

17. Ibid., p. 58.

18. Ibid., p. 57.

19. Ibid., p. 365.

20. Ibid., p. 470.

21. Ibid., p. 469.

22. Ibid., p. 470.

23. Taratuta, *The Fate of a Writer and the Fate of a Book*, p. 299.

24. Ibid., p. 300.

## Chapter 15

1. Sowerby, *Rare People and Rare Books*, p. 12.

2. Ibid., p. 11.

3. Ibid., p. 12.

4. One figure of later high eminence Voynich apparently met at some point was Achille Ratti, the future Pope from 1922–39, Pius XI. The connection lay in Ratti's role as librarian and scholar first at the Ambrosian Library in Milan (1888–1911) and later at the Vatican Library until his elevation in 1922. His renown as a paleographer and open personality would have drawn Voynich to him when he was in Italy. Ratti also made visits to London in 1906 and again to Oxford in June 1914 to attend the Royal Society celebration of the seventh centenary of Roger Bacon's birth. At the Vatican, Ratti had discovered unpublished manuscripts of Bacon; their interests clearly overlapped. Did, one wonders, Wilfrid consult with Ratti over his own Bacon document?

5. Hunt, *Out of Print and Into Profit*, p. 252.

6. Colin MacKinnon, an author of a novel inspired by the Voynich Manuscript, unearthed the fascinating and illuminating record.
    Voynich had re-met resident British author William Booth in July that year. Voynich reminded him of the Bacon cipher they had discussed some years before, mentioning that the War Department and a Professor Manly at Chicago University were working on it. At a worrying time concerning German spies, Booth was suspicious and took the matter up with the authorities in two letters.
    In one he states that Voynich is 'Austrian by birth, pro-German in outlook and a pretty slick article'. In the other he stated that when he knew him after his arrival in the US he was 'distinctly a boaster, a very good salesman … he told fascinating, mysterious stories of his adventures in search of hidden or lost libraries'. To compound this Voynich had been already misreported to officialdom in 1916 as possessing the cipher of the American War Department following a dinner with Walter Lichtenstein, a book-buying senior librarian at North West University. Lichtenstein, from German stock, suspected of pro-German sympathies himself, must have taken a dislike to Wilfrid.
    Agent Fry on case 33354 reported that Voynich had 'put it over most of the wealthy Chicagoans and has not an honest hair in his head'. Another

enquiring at the opulent Waldorf Astoria where he was staying in New York counter-rumoured that he was a valued guest and a gentleman. His bank was questioned – revealing his financial association with his UK friends the Levetus family – and his office searched. In December 1917 Voynich replied in an official interview that he merely thought that the Bacon cipher might have been of use to the government. Manly, now working for US military intelligence, valiantly tried to put things straight mentioning that Voynich had actually supported movements to secure Polish liberation. He suggested that Lichtenstein may have been acting out of book-dealing malice towards Voynich. The Bacon cipher he asserted was bought by Voynich circa 1907 for $5 million from a royal European family. A. W. Pollard at the British Museum, clearly an ally of Wilfrid from earlier days, backed up his *bona fides*.

Things went quiet for a while until in July 1918 some neighbourhood snoops from the vigilante American Protection League reported to the Bureau. They had interviewed book-dealers who variously averred a lack of trust in Voynich; one suspected him of tampering with the truth. This resurrected enquiries when Wilfrid was applying for a visa to return to the US in late 1919. Fortunately a very respectable friend from Chicago calmed troubled waters by attesting that both he and Ethel had been organising large sums of money for European relief agencies. One last splutter from the APL snoops in October 1919 mentioned that Voynich 'has all the appearances of untold wealth; wearing one of the most expensive fur coats'. A week later case 33354 was closed. Back in the UK, however, where despite awareness that the subject is 'strongly anti-Bolshevik', an official letter of September 1920 declared that the police were keeping him under surveillance. This prompted a letter to J. Edgar Hoover no less suggesting that it should continue on American shores also.

Wilfrid was of course not an enemy to either side of the Atlantic, but his own worst enemy. His egoistic flamboyance had got him into deep trouble. His self-importance and loose talk in the wrong company echoed all the way back to his London days of the early 1890s working with Russian exiles. Exiled from their trust in 1895 he very nearly had himself exiled from his latest adopted country. Underneath that, however, his loyalty to his compatriot Poles shines through.

7. Collection of letters between Edward Levetus ('Pote') and Wilfrid and Ethel Voynich, courtesy Margaret Till.

8. A. G. Little (ed.), *Roger Bacon: essays contributed by various writers on the occasion of the commemoration of the seventh century of his birth* (Oxford: Clarendon Press, 1914), p. 375.

9. Levetus Collection.

10. Ibid.

11. Ibid.

12. William Romaine Newbold, *The Cipher of Roger Bacon* (Philadelphia: University of Pennsylvania Press, 1928), p. xiii.

13. Levetus Collection.

14. Ibid.

15. Ibid.

16. Ibid.

17. Ibid.

18. Courtesy Marni Rosner.

19. Wilfrid writes to her that he had received Erla's announcement of her marriage years before to a Pole and had tried to call on her in Lemberg (Lviv), then in Austria. Interestingly, Wilfrid mentions to her, 'I am in England during the War,' indicating that his imminent trip to the US was not necessarily pre-planned. Erla returned to the US and worked for the YWCA, based in New York as a social worker.

20. Rosner collection.

21. Ibid.

22. Ibid.

23. Ibid.

24. Ibid.

25. Ibid.

26. Levetus Collection.

27. Rosner Collection.

28. Sowerby, *Rare People and Rare Books*, p. 33.

## Chapter 16

1. Asmolean Museum, Pissarro Collection.

2. Ethel Voynich, *Put Off Thy Shoes* (New York: Heinemann, 1946), p. v.

3. Ibid., p. 42.

4. Ibid., p. 48.

5. Ibid., p. 64.

6. Ibid., p. 80.

7. Ibid., p. 119.

8. Ibid., p. 99.

9. Ibid., p. 81.

10. Ibid., p. 145.

11. Ibid., p. 147.

12. Ibid., p. 197.

13. Ibid., p. 199.

14. Ibid., p. 346.

15. Ibid., p. 385.

16. Ibid., p. 389.

17. Ashmolean Museum, Pissarro Collection.

18. According to Maggie Armstrong in *The Irish Times*, 30 July 2010, *The Gadfly* gave solace to prisoners during the Irish Civil War in 1922.

19. Although in the West it was re-published in 1927 and 1938, the book aroused little interest and Ethel Voynich was not mentioned in the *Encyclopaedia*

*Britannica.* She was, however, in *Who's Who.* In Russia the trail of confusion went right back to 1898. By a strange twist it involved two activists involved in St Petersburg *Narodnya Volya* circles. Ivan Popov and Peter Yakubovich wrote of Lily Voynich, author, living among the Karaulovs in 1882, five years before she actually arrived. Other reference works claimed that she was a Pole named after the town 'Voynich' or born in Russia and had emigrated to England. She was assumed to be now dead.

20. Taratuta, *The Fate of a Writer and the Fate of a Book*, p. 215.

21. Ibid., p. 223.

22. Ibid., p. 253.

23. Ibid., p. 255.

24. Ibid., p. 265.

25. Trinity College, Cambridge, Geoffrey Taylor Papers, A132.

26. Kindly supplied by Patricia Hinton Davison.

27. Letter to Geoffrey Taylor, 26 August 1960.

28. Bruce Lockhart, *Ace of Spies* (London: Hodder, 1967), p. 23.

29. Taratuta, *The Fate of a Writer and the Fate of a Book*, p. 135.

30. It also made possible sense of a puzzling letter of early 1978 from Ethel's daughter Winifred to Edward Levetus. At that time Fremantle was asking everyone for sources of information. Winifred suggested that he should be wary of giving away too much to her and mentioned that on her last visit to New York Ethel had said that 'all she had told me about the Russian affair was to die with her'. Winifred added that 'the horrible articles ... on ELV are fit for the fire and I shall be sorry if it is all revived'. Was 'the Russian affair' with Georgian-born Sidney Reilly?

31. (The following may be of interest only to Voynicheros.)
    According to Andrew Cook there could even have been some involvement as lovers or acquaintances after Reilly's settling in London in 1896 and his departure for China in June 1899. There is some substantial although debatable evidence. He asserts that Reilly's task, according to the papers of William Melville, Head of Special Branch, was to infiltrate the Society of Friends of Russian Freedom; he may therefore have met both Wilfrid and Ethel Voynich. Cook suggests that either Wilfrid or Reilly (and maybe together?), sharing a professional relationship as chemists, might have also indulged their mutual interest in medieval books and art to produce forgeries. Cook is circumspect but it was this implication, abetted by their both being members of the British Museum Library where information on inks, etc could be obtained, that partly sowed my own suspicion that the *Voynich Manuscript* might have been one of Wilfrid's concoctions. In neither case, however, are there any records of which books they may have consulted.
    Did Reilly's infiltration inevitably mean a connection at all? Cook maintains that after 1896 Ethel was 'an active courier for the Free Russia cause and travelled abroad frequently'. This is probably incorrect. As we have noted, according to Theodore Rothstein, in 1894/5, alongside the financial crash of the Voynich/Stein venture and the complete failure of the Voyniches rival book-smuggling organisation, Voynich had catastrophically fallen out with the

Russian émigré colony. This is congruent with the way the couple's future developed. Ethel got on with publishing her novel, travelling abroad sometimes with Wilfrid; he established his antiquarian book enterprise, producing his first list in 1898.

But Cook does have one important piece of documentary evidence that no one else had come across. Dimitry Belanovsky in Moscow in his researches for Cook in the State Archives discovered an Okhrana list, compiled by agent Rataev, of nearly 150 emigrés 'worth paying attention to' (not specifically 'members of the SFRF', as Cook implies) drawn up between 1898 and 1903. Among them are numbers 117, Sigismund Rosenblum, 119, Rothstein, and 25, Voynich. The inclusion of Rosenblum indeed implies that he was spying for Melville; details of number 25, however, reveal a more startling connection. Number 117 is apparently 'a close friend of Voynich's, especially his wife's. He accompanies her everywhere, even on her trips to the continent.' The details on Voynich himself declare that he 'formerly took an active part in the revolutionary movement, but is now more inclined to literary work, also on revolutionary issues. Holds an annual international revolutionary library. His wife is a novelist.' The use of 'formerly' suggests that Rothstein was right in maintaining a complete split: the 1898 document date does fit with the couple's later lives as book-dealer and writer. There seems to be no reason after then, or from 1896, why Reilly would have 'accompanied her everywhere'. If true, it remains a mystery.

It's worth noting that Okhrana agents were not known for their accuracy, always justifying their usefulness with self-importance and alarm. Hardly any of the birth dates of the suspects on the 1898 list, such well-known characters as Kropotkin, Volkhovsky and Chaikovsky, are correct. In 1894, for example, as noted, Wilfrid was supposed to have married fellow SFRF member Volkhovsky's daughter, and in April 1895 when Ethel went to Italy a clerical mistake led arch-agent Rachkovsky to identify her in a report as a dangerous terrorist named Kupreyanov.

If the Okhrana were keeping tabs on Ethel, Rataev could, of course, have been simply backdating exaggerated rumours of Ethel and Reilly's Florence tryst. With a retrospective time frame it might also similarly suggest that Wilfrid could have had dealings with Reilly before he finally came to London in 1895. Might Reilly be the 'friend' with whom Voynich approached Henry Wellcome interested in chemical and drug manufacture leading to the commercial deal that crashed in 1895? Cook discovered that under his Rosenblum name Reilly did set up a patent medicine firm in 1896 – was fellow chemist Voynich an accomplice? Or perhaps Reilly was the mysterious Russian chemist, S. Stein, according to Taratuta 'a recent arrival … a person of energy and initiative', and partner in a soap enterprise who joined Voynich as an entrepreneur at that time. The Rataev file on Reilly records his interest in a soap factory near Bristol. Cook suggests that the company could have been a front for 'smuggling money and … other commodities'. Maybe an association between Voynich and Reilly explains how Wilfrid acquired the funds to finance his otherwise inadequately explained rise in the book trade?

## Chapter 17

1. Ethel Voynich, *Olive Latham*, p. 38.
2. Ibid., p. 39.
3. Ibid., p. 52.
4. Ibid., p. 53.
5. Ibid., p. 55.
6. Ibid., p. 177.
7. Ibid., p. 182.
8. Ibid., p. 187.
9. Ibid., p. 214.
10. Ibid., p. 226.
11. Ibid., p. 240.
12. Ibid., p. 261.
13. Ibid., p. 161.

## Chapter 18

1. Batchelor, *The Life and Legacy of G. I. Taylor*, p. ix.
2. G. I. Taylor, *Proceedings of the Royal Irish Academy*, vol. 57 (1945–6), p. 73.
3. Trinity College, Cambridge, Papers and Correspondence of Sir Geoffrey Taylor, A135/5.
4. R. V. Southwell, 'G.I. Taylor: an autobiographical note', in Southwell, *Surveys in Mechanics* (Cambridge: Cambridge University Press, 1956), p. 2.
5. Batchelor, *The Life and Legacy of G.I. Taylor*, p. 2.
6. Ibid., p. 60.
7. Ibid., p. 65.
8. Ibid., p. 67.
9. Ibid., p. 94.
10. Ibid., p. 116.
11. Ibid., p. 175.
12. Ibid., p. 189.
13. Graham Farmelow, *Churchill's Bomb* (London: Faber & Faber, 2013), p. 283.
14. G.I. Taylor, 'Trying Out the Bomb', *The Listener*, 16 August 1945.
15. Batchelor, *The Life and Legacy of G. I. Taylor*, p. 78.
16. Ibid., p. 20.
17. Ibid., p. 191.
18. Thomas McMahon, *The Principles of American Nuclear Chemistry: a novel* (Chicago: University of Chicago Press, 1970), p. 81.
19. Batchelor, *The Life and Legacy of G. I. Taylor*, p. 98.
20. Trinity College, Cambridge, Papers and Correspondence of Sir Geoffrey Taylor, A123/1.

21. Ibid.

22. Batchelor, *The Life and Legacy of G. I. Taylor*, p. 254.

**Chapter 19**

1. Bristol University, Howard Everest Hinton Papers, DM 1718.
2. Ibid.
3. Graham Greene, *The Power and the Glory* (London: Penguin, 1940), p. 18.
4. From correspondence courtesy of Patricia Hinton Davison.
5. Bristol University, Howard Everest Hinton Papers.
6. Ibid.
7. Ibid.
8. Ibid.
9. Pierre Trudeau, and Jacques Hebert, *Two Innocents in Red China* (Oxford: Oxford University Press, 1968), p. 81.
10. Ibid., p. 124.
11. Ibid., p. 140.
12. Ibid., p. 152.
13. Bristol University, Howard Everest Hinton Papers.

**Chapter 20**

1. Courtesy Marni Rosner.
2. Anita M. Smith, *Woodstock: history and hearsay* (New York: Woodstock Arts, 2006), p. 90.
3. Nancy E. Green (ed.), *Byrdcliffe: an American arts and crafts colony* (New York: Cornell, 2004), p. 236.
4. Edward Yeomans, 'Shackled Youth', *Atlantic Monthly Press* (1931), p. 2.
5. Susan M. Lloyd, *The Putney School* (New Haven: Yale University Press, 1987), p. 9.
6. Ibid., p. 12.
7. Ibid., p. 13.
8. Rosner Collection.
9. In another of those daft coincidences that emerged writing this book, one of my mother's cousins, Winifred Wallace, and her husband Vivian ran a private boys' school in Putney, London, until 1986. They lived *à la* Mary Everest with their children on the top floor.

**Chapter 21**

1. Defying the video cameras hidden in the huge lamp standards, I was making some sound recordings asking people if they'd heard of Ethel Voynich and *The Gadfly*. This was not completely stupid as her celebrated novel was as famous here as it was in the Soviet Union. Wandering around the vast space, I was pestered by souvenir sellers: 'Would you like to buy some stamps?' asked one.

It was a cheap, heaven-sent moment. 'No thanks,' I replied. 'I'm told philately will get you nowhere.' She grinned uncomprehendingly.

2. R. Howes and C. Herzenberg, *Their Days in the Sun* (Philadelphia: Temple University Press, 1999), p. 55.

3. Ibid., p. 53.

4. Batchelor, *The Life and Legacy of G. I. Taylor*, p. 209.

5. Max Hastings, *Nemesis: the battle for Japan, 1944–45* (London: Harper Perennial, 2008), p. 503.

6. Ibid., p. 516.

7. R. Rhodes, *The Making of the Atomic Bomb* (New York: Simon & Schuster, 2012), p. 516.

8. Jeff Hughes, *The Manhattan Project* (Cambridge: Icon Books, 2002), p. 92.

## Chapter 22

1. Chou Dao-yuan, *Silage Choppers and Snake Spirits* (Quezon: Ibon Books, 2009).

2. Ibid., p. 123.

3. Ibid., p. 131.

4. Ibid., p. 129.

5. *China Monthly Review* (November 1950), p. 69.

6. Ibid., p. 70.

7. Dao-yuan, *Silage Choppers and Snake Spirits*, p. 162.

8. *Fact Forum News* (March 1955), p. 41.

9. Dao-yuan, *Silage Choppers and Snake Spirits*, p. 184.

10. *Fact Forum News* (March 1955), p. 47.

11. Dao-yuan, *Silage Choppers and Snake Spirits*, p. 211.

12. Trinity College, Cambridge, Papers and Correspondence of Sir Geoffrey Taylor, A127 3(18).

13. Dao-yuan, *Silage Choppers and Snake Spirits*, p. 349.

14. Ibid., p. 348.

15. Trinity College, Cambridge, Papers and Correspondence of Sir Geoffrey Taylor, A127 4(1).

## Chapter 23

1. Talk at Columbia University, April 1999.

2. Hearings, 17 July 1954.

3. William Hinton, *Fanshen* (London: Penguin, 1972), p. 296.

4. Ibid., p. 45.

5. Ibid., p. 49.

6. Ibid., p. 59.

7. Ibid., p. 154.

8.  Ibid., p. 172.
9.  Ibid., p. 282.
10. Ibid., p. 340.
11. Ibid., p. 359.
12. Ibid., p. 432.
13. Ibid., p. 536.
14. Ibid., p. 654.
15. Ibid., p. 712.
16. William Hinton, *Shenfan* (London: Random House, 1984).
17. Jung Chang and Jon Halliday, *Mao: the untold story* (London: Jonathan Cape, 2005).
18. Frank Dikotter, *Mao's Great Famine* (Bloomsbury Publishing, 2010) and *Tragedy of Liberation* (Bloomsbury Publishing, 2013).

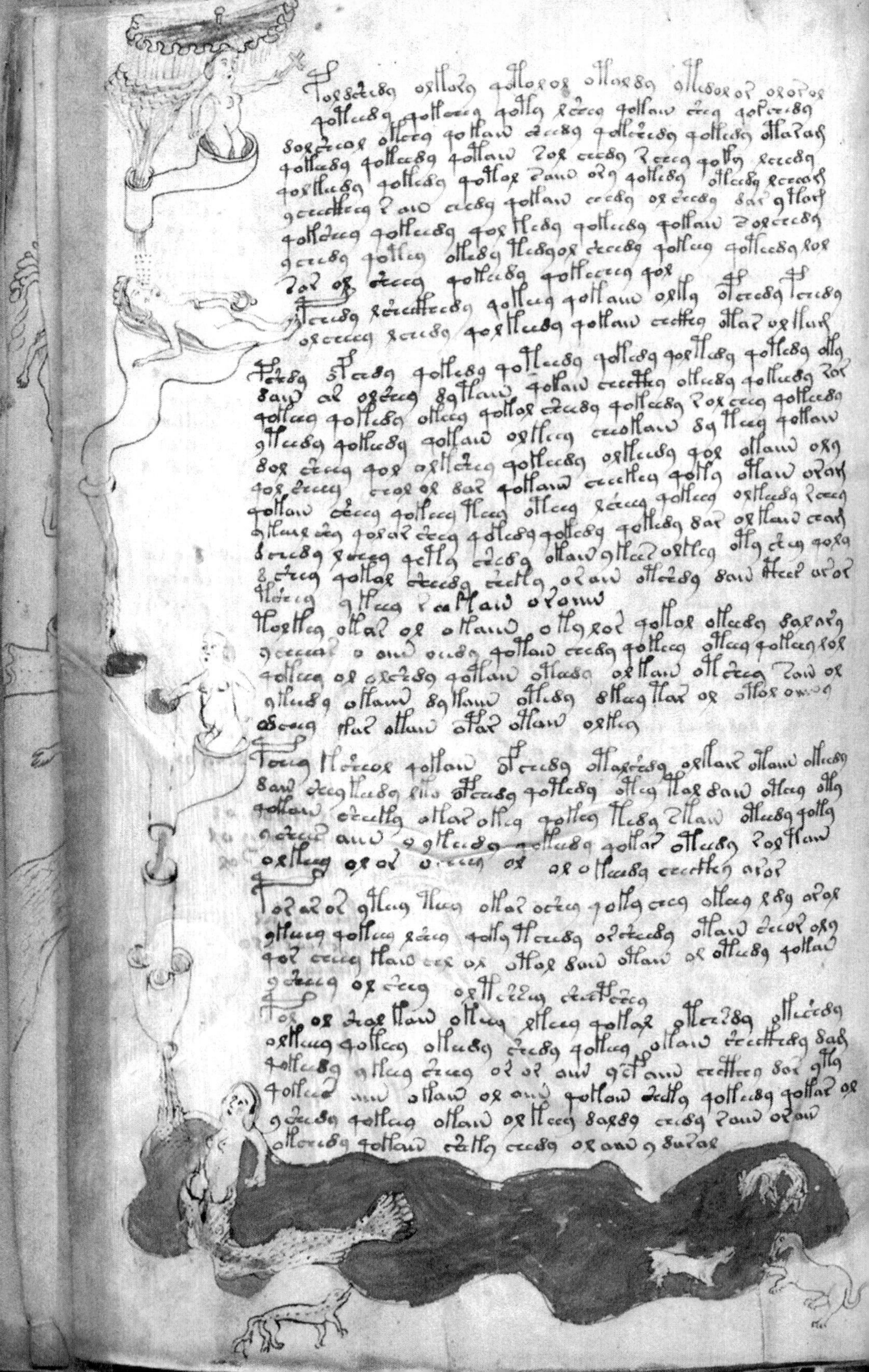

# ILLUSTRATION CREDITS

Images courtesy of University of Bristol Library, Special Collections on the following pages: p. ii (DM1718/A105), p. 14, (both DM1718/A103) p. 60, (DM1718/A105), p. 76 (DM1718/A105) and (DM1718/A106). p. 84, (DM1718/A103), p. 129 (DM1718/A103), p. 150, (DM1718/A105), p. 182, (DM1718/A93), p. 227, (DM1718/A106), p. 230, (DM1718/A93), p. 234, (DM1718/A101), p. 273, (DM1718/103), p. 274, (DM1718/A103), p. 308, (DM1718/A103), p. 315, (DM1718/A95), p. 320, (DM1718/A92), p. 324 (DM1718/A103).

Images courtesy of University College Cork Library on the following pages: xi (BP/1/351), 232 (BP/1/347).

Images courtesy of the author on the following pages; 1, 5, 33, 34, 35, 49, 59, 87, 119, 126, 133, 137, 185, 201, 206, 243, 280, 288, 289, 290, 291, 294, 296, 297, 298, 338, 353, 359, 361, 363, 411, 412, 419.

Images from, Mrs Havelock Ellis, *James Hinton: A Sketch*, (London: Stanley Paul and Co. 1918) on the following pages: 89 (frontispiece), 91 (p. 96) 145, (p.224).

Images from Claude Bragdon, *A Primer of Higher Space*. (Rochester, New York: Manas Press, 1913) on the following pages: 174 (plate 3), 175 (plate 5).

Images courtesy of Marni Rosner and Hinton family on the following pages: 238, 252, 303, 337, 339, 342, 346, 351, 365, 375, 376, 377, 380, 395, 400, 407, 408.

Images courtesy of Moorfields Eye Hospital, London on the following pages: 235, 236.

Images courtesy of Library of Congress, *Look* magazine, July 8 1958, Vol 22, no. 14 on the following pages: 267 (p, 68), 269 (p, 70).

Images courtesy Beinecke Rare Book and Manuscript Library, Yale University, on the following pages: 16, 17, 19, 452.

All other images courtesy of copyright holders as detailed below:

P. 7 courtesy Trinity College Library, University of Cambridge, G.I. Taylor Collection A112/47.

P. 29 https://commons.wikimedia.org/wiki/File:Portrait_of_George_Boole.png. *Illustrated London News*, 21 January 1865. CC-PD-Mark.

P. 32 John Britton, 'Picturesque Antiquities of the English Cities' ( London: M.A. Nattali, 1836), plate 16.

P. 55 https://en.wikipedia.org/wiki/Frederick_Denison_Maurice. Photograph from 1865 original edited in Photoshop CS2. 18 July 2010, Acabashi. CC-BY-3.0.

P. 58 courtesy Kevin Boole.

P. 59 courtesy Wellcome Library, Creative Commons, attribution 4.0 International Licence. Cover, Thomas Roupell Everest, *A Popular View of Homeopathy* (Philadelphia: J.G.Wesselhoft 1835).

P. 63 https://en.wikipedia.org/wiki/George_Everest. Uploader, Hephaestos, from English Wikipedia. CC-PD-Mark. .

P. 100 https:commons.wikimedia.org/wiki/File:Portrait_of_Havelock-Ellis_(18591939)_Psychologist_and_Biologist_(2575987702). Uploaded by Magnus Manske. Smithsonian Institution Libraries of History and Technology.

P. 107 https://commons.wikimedia.org/wiki/Category:Mikhail Bakunin#/Media/File:Bakunin_Nadar.jpg. New York Public library. MSSCol3040. CC-PD-Mark.

P. 108 courtesy National Portrait Gallery, *Sergius Stepniak* x15590.

P. 109 https://commons.wikimedia.org/wiki/File:Kropotkin2.jpg. Koroesu. CC-PD-Mark.

---

Facing page: Folio 79v, 'nymphs bathing', from the *Voynich Manuscript*.

P. 111 https://en.wikipedia.org/wiki/Alexander_II_of_Russia#/media/File:Attentat_mortal_Alexander_II_(1881).jpg.The Assassination of Alexander II of Russia. Uploader Jagarn at sv.wikipedia. Zeitung Bd 76 (1881) 5. 262 (Leipzig, Marz/April 1881). Gustav Broling, CC-PD-Mark.

P. 113 courtesy Nigel at: www.hampsteadheath.net.

P. 116 https://commons.wikimedia.org/wiki/File:Farewell_Europe!. Alexander Sochaczewski (1843–1923). http://tradytor.pl/node/90. CC-PD-Mark.

P. 117 Michael J. Schaack, *Anarchy and Anarchists*, (Chicago: F.J. Schulte, 1889), p. 142.

P. 122 courtesy Ann Pettit.

P. 131 https://en.wikipedia.org/wiki/Nevsky_Prospect. The view on Znamenskya church on Nevsky Prospekt in St Petersburg (Russia) nineteeenth-century photochrome print (1896). Reproduction number: LC-DIG-ppmsc-03886 from Library of Congress Prints and photographs division.

P. 132 left-hand image: https://commons.wikimedia.org/wiki/File:Vsevolod_Mikhailovich_Garshin_1885.jpg. Russian writer Vsevolod Garshin. Памяти В. М. Гаршина. Художественно-литературный сборник. — СПб:Типография и фототипия В. И. Штейн, 1889.

Right-hand image: https://commons.wikimedia.org/wiki/Category:Mikhail_Yevgrafovich_Saltykov Shchedrin#/media/File:Nikolaj_Alexandrowitsch_Jaroschenko_Mikhail_Yevgrafovich_Saltykov-Shchedrin.jpg. Mikhail Shchedrin 1886 Nikolai Yaroshenko (1886-1898). CC-PD-Mark.

P. 134 https://commons.wikimedia.org/wiki/File:Maria_tsebrikova.jpg. http://dlib.rsl.ru/viewer/01003713486. Author unknown, pre 1903.

P. 142 https://commons.wikimedia.org/wiki/Olive_Schreiner#/media/File:Olive_Schreiner00.jpg. Photo of Olive Schreiner (1855–1920). South African author, pacifist, political activist. *A Standard Encyclopedia of South Africa*. 1973. CC-PD-Mark.

P. 147 https://commons.wikimedia.org/wiki/Category:Karl_Pearson#/media/File: Karl_Pearson,_Biometrician,_at_his_desk,_1910_Wellcome_M0011713.jpg. Wellcome images M0011713. CC-BY-4.0.

P. 155 https://commons.wikimedia.org/wiki/File:William_Morris_age_53.jpg. William Morris, portrait of, aged 53. http://books.google.com/books?id=0ZQOAAAAIAAJ Google Books edition of J.W. Mackail, *The Life of William Morris*, London, Longmans, Green and Co. 1899. CC-PD-Mark.

P. 157 Donald J. Senese, *Stepniak Kravchinsky: The London Years* (Newtonville, MA USA: Oriental Research Partners), 1987. Plate 1.

P. 160 https://commons.wikimedia.org/wiki/File:Matchgirl_strikers.PNG. Photo of match girls participating in a strike against Bryant&May, London 1888.http://www.dailymail.co.uk/tvshowbiz/article- 500624. CC-PD-Mark.

P. 172 https://commons.wikimedia.org/wiki/File:Flatland_cover.jpg. Cover of *Flatland*. Transferred to Commons by Maksim. CC-PD-Mark.

P. 180 Charles Howard Hinton, *The Fourth Dimension*. (London: Swann and Sonnenschein, 1904). p.124.

P. 214 https://commons.wikimedia.org/wiki/File:Joseph_Mazzini-portrait.jpg Giuseppe Mazzini. Jules Claretie, *Histoire de la Révolution de 1870–71 illustrée*, Paris: Librairie Polo, 1874, p.161. 1873 Bocourt (dessin) and Tourfaut (graveur). CC-PD-Mark.

P. 217 left-hand image: https://commons.wikimedia.org/wiki/File:Brisighella,_Panorama_-_panoramio.jpg. Brisighella, http://panoramio.com/photo/4335220. CC-BY-SA-3.0. right-hand image, https://upload.wikimedia.org/wikipedia/commons Сцена из спектакля «Овод» Ленинградского государственного театра им. Ленинского комсомола. В роли Ривареса

Ленинградского государственного театра им. Ленинского комсомола. В роли Ривареса Михаил Боярский, в роли Монтанелли Роман Громадский, режиссёр-постановщик Егоров Геннадий Gennady Egarov. CC-BY-SA-3.0.

P. 247 *The Tatler* no.152, 25 May 1904.

P. 251 *Philadelphia Enquirer*, 10 January 1921.

P. 271 https://en.wikipedia.org/wiki/File:Sidney_Reilly.jpg,http://bookhaven.stanford.edu/tag/Sidney-reilly.

P. 284 https://commonswikimedia.org/wiki/File:Ilya_RepinUnexpected_visitors.jpg. 'The Unexpected Stranger by IlyaRepin'. http://.allpaintings.org/v/Realism/Ilya+repin+-+ Unexpected+Visitors.jpghtml?g2_imageViewsindex=1. Tretyakov Gallery, Moscow. CC-PD-Mark.

P. 310 https://commons.wikimedia.org/wiki/Filetrinity_College_Cambridge_1690.jpg. Bird's eye view of Trinity College, Cambridge, with Great Gate and Great Court in the foreground, Nevile's Court and Wren Library in the background. Trinity College Cambridge 1690, David Loggan, *Cantabrigia Illustrata*, Cambridge 1690, Plate XXIX (cropped). PD-Art.

P. 328 courtesy Patricia Davison.

P. 368 https://www.google.co.uk/search?q=hiroshima+wiki+commons&espv=2&biw=1280 &bih=939&tbm=isch&imgil=TXempUBRhI4uDM%253A%253B_QThqD4W7LzPBM%253Bhttp s%25253A%25252F%25252Fcommons.wikimedia.org%25252Fwiki%25252FFile%25253A AtomicEffectsHiroshima.jpg&source=iu&pf=m&fir=TXempUBRhI4uDM%253A%252C_QThqD 4W7LzPBM%252C_&usg=__baeWCZVIzpGFgTMBp1SGL7WT7IA%3D&ved=0ahUKEwjNgtSb 9_zLAhXJcRQKHXkeAzQQyjcIQQ&ei=tIgGV83YIcnjUfm8jKAD#imgrc=TXempUBRhI4uDM%3A. media/File:AtomicEffects-Hiroshima.jpg.  Effects of the atomic bomb on Hiroshima. View from the top of the Red Cross Hospital looking northwest. Frame buildings recently erected.1945. Ibiblio.org and centerfor thepublicdomain.org. US Government. Post-work:User:W.wolny. PD US Military.

P. 388 https://commons.wikimedia.org/wiki/File:Red_Guards.jpg. Image shows three Young Chinese Red Guards from the Cultural Revolution. 26 December 1971. Scan of cover of non-copyright elementary school textbook from Guangxi 1971. Villa Giulia.

# INDEX

Note: illustrations are indicated by page numbers in **bold**.